AF899970

UV-Visible Photocatalysis for Clean Energy Production and Pollution Remediation

UV-Visible Photocatalysis for Clean Energy Production and Pollution Remediation

Materials, Reaction Mechanisms, and Applications

Edited by Xinchen Wang, Masakazu Anpo, and Xianzhi Fu

WILEY-VCH

Editors

Prof. Xinchen Wang
Fuzhou University
State Key Lab. Photocatalysis on
Energy and Environment
No.2, Xueyuan Road
350116 Fuzhou
P. R. China

Prof. Masakazu Anpo
Fuzhou University
State Key Lab. of Photocatalysis on
Energy and Environment
No.2, Xueyuan Road
350116 Fuzhou
P. R. China

Prof. Xianzhi Fu
Fuzhou University
State Key Lab. of Photocatalysis and
Environment
350116 Fuzhou
P. R. China

Cover Image: Scheme by Prof. Masakazu Anpo

All books published by **WILEY-VCH** are carefully produced. Nevertheless, authors, editors, and publisher do not warrant the information contained in these books, including this book, to be free of errors. Readers are advised to keep in mind that statements, data, illustrations, procedural details or other items may inadvertently be inaccurate.

Library of Congress Card No.: applied for

British Library Cataloguing-in-Publication Data
A catalogue record for this book is available from the British Library.

Bibliographic information published by the Deutsche Nationalbibliothek
The Deutsche Nationalbibliothek lists this publication in the Deutsche Nationalbibliografie; detailed bibliographic data are available on the Internet at <http://dnb.d-nb.de>.

© 2023 WILEY-VCH GmbH, Boschstraße 12, 69469 Weinheim, Germany

All rights reserved (including those of translation into other languages). No part of this book may be reproduced in any form – by photoprinting, microfilm, or any other means – nor transmitted or translated into a machine language without written permission from the publishers. Registered names, trademarks, etc. used in this book, even when not specifically marked as such, are not to be considered unprotected by law.

Print ISBN: 978-3-527-35050-6
ePDF ISBN: 978-3-527-83797-7
ePub ISBN: 978-3-527-83798-4
oBook ISBN: 978-3-527-83799-1

Cover Design: ADAM DESIGN, Weinheim, Germany
Typesetting Straive, Chennai, India
Printing and Binding CPI Group (UK) Ltd, Croydon, CR0 4YY

C120398_040423

Contents

UV-Visible Photocatalysis for Clean Energy Production and Pollution Remediation: Materials, Reaction Mechanisms, and Applications – A Preface

Detlef Bahnemann[1,2]

[1] *Leibniz University Hannover, Institute for Technical Chemistry, 30167 Hannover, Germany*
[2] *Saint-Petersburg State University, Laboratory, Photoactive Nanocomposite Materials, Saint-Petersburg 198504, Russia*

First of all, I like to take this opportunity to congratulate the editors on this exceptional collection of feature and review articles assembled within their new book titled *UV-Visible Photocatalysis for Clean Energy Production and Pollution Remediation: Materials, Reaction Mechanisms, and Applications*, thus managing to bring together (almost) all major authorities in the field of photocatalysis worldwide at once. And I like to thank them for this opportunity to contribute a small preface for their excellent book.

The research field of photocatalysis has seen well over 40 years of active research now and currently appears to be in its heydays. Many industrial applications of this technology are already available on the market today such as air and water cleaning devices, self-cleaning surfaces, solar cells, and even solar fuel generators. However, unfortunately, several even of the most basic principles of photocatalysis are still far from being understood today. Nowadays, different experimental techniques are most certainly available to study some of the most important and crucial features of TiO_2 photocatalytic systems. Time-resolved analytical tools such as transient optical spectroscopy, for example, seem to be ideally suited to study the generation, trapping, and transfer of electrons and holes, that is, the "initiators" of all photocatalytic processes, in particular, when different detection methods such as transient optical spectroscopy and microwave conductivity measurements are combined. The respective data analysis is found to be very complex, and it has been shown that photoelectrochemical concepts involving band bending and particle–particle contacts need to be considered in depth [1].

While the generally accepted mechanistic picture uses a single photocatalyst particle excited by a single photon to explain the entire photocatalytic world, recent research results have created the need of more elaborate mechanisms involving, for example, three-dimensional self-assembled particle agglomerates acting as antenna and charge transfer systems for their comprehensive interpretation. It is through these ensemble and cooperative properties that even widely used effects such as photoinduced superhydrophilicity are currently being explained, even though considerable scientific controversies still exist today as to the (only) correct model [1].

One of the most important limitations for the application of photocatalytic techniques for water decomposition, carbon dioxide fixation, and ammonia synthesis from molecular nitrogen is that when employing pure water as the only reductant, all of these processes are usually found to be rather inefficient. This is generally related to the fact that the simultaneous reduction of H_2O, CO_2, and N_2 with the oxidation of water yielding O_2 (or even H_2O_2) is a complex multistep reaction involving (at least!) four electrons. Using sacrificial molecules as electron donors can remarkably improve the production of the reduction products with the holes being scavenged by these molecules, thus reducing the charge carrier recombination significantly. Furthermore, as O_2 is not produced, the back reaction to produce water is suppressed, increasing the yield of reduction products and avoiding a subsequent gas separation stage. However, it should be noted here that even the yield of the H_2, CO (or, e.g. CH_3OH or CH_4), and NH_3 formation will eventually be reduced by competing reduction reactions with the products formed upon the oxidation of the sacrificial reagents [2].

Organic compounds such as alcohols, organic acids, and hydrocarbons can act as efficient hole scavengers (that is, as electron donors) for the photocatalytic fuel generation process. In particular, methanol is frequently used as sacrificial reagent. For practical applications, the utilization of methanol will only be environmentally sensible provided that it is derived from biomass or from toxic residues that must be disposed of. Adding methanol as electron donor to react irreversibly with the photogenerated holes can enhance the photocatalytic electron/hole separation efficiency, resulting in higher quantum yields. Since electron donors are consumed in the photocatalytic reaction, their continuous addition is required to sustain the photocatalytic fuel production. Two possible mechanisms are proposed for the photocatalytic oxidation of methanol: (i) the direct oxidation by photogenerated holes and (ii) the indirect oxidation *via* interfacially formed $^{\cdot}OH$ radicals that are products of trapping valence band holes by surface –OH groups or adsorbed water molecules. It is still a challenge to distinguish between the two mechanisms in practice due to the lack of suitable probe techniques [2].

While the choice of sacrificial electron donors for studies concerning the photocatalytic formation of, e.g. molecular H_2 appears to be rather large, the sacrificial photocatalytic oxidation of water is only reported for a rather limited variety of electron acceptors. By far the vast majority of the research groups working on this topic employs silver cations, Ag^+, as electron acceptors, resulting in the fact that the photocatalytic formation of molecular oxygen is accompanied by the deposition of metallic silver nanocontacts on the semiconductor's surface. Obviously, this will lead to irreversible optical changes of these systems due to the plasmonic absorption band of the silver nanoparticles in the visible spectral region. Moreover, noble metal nanoparticles are known for their catalytic activity, resulting most likely in changes in the chemical and photochemical properties of these systems. Furthermore, it has been suggested and also experimentally verified that the role of suitable sacrificial electron acceptors such as Ag^+ is highly underestimated [3]. In particular, their possible involvement in the actual water oxidation mechanism has so far not been discussed and accepted. We are convinced that it is most certainly highly indicated to study the role of such metal cations in photocatalytic water oxidation in detail. Their catalytic role has rather recently been proposed for the first time; however, it could be part of a much more general mechanism, thus opening up new design features for photocatalytic and photoelectrochemical energy-to-fuel conversion systems. I am convinced that this is even more true for photocatalytic systems that can be activated by visible or even NIR illumination.

Finally, I like to emphasize that photocatalytic processes are inherently one-electron transfer reactions. In particular, when nanocrystalline photocatalysts are employed under solar illumination, the time interval between the absorption of two photons by one (that is, the same) photocatalyst particle is in the order of milliseconds. Hence, two free radical intermediates, that is, one reduced as well as one oxidized product of the reactions induced by the first photon, need to wait together (!!) for at least this time period on the very small surface of the same nanoparticle before having a chance to somehow form more stable products by being reduced or oxidized, respectively, again. Despite being crucial chemical species, the formation and fate of these initial free radical products are rather rarely studied [4]. The same applies to the formation as well as to further reactions of the stable reaction intermediates: it is fair to state that hardly any current (or even former) study is able to account for a full qualitative or even quantitative analysis of all reaction products formed during the photocatalytic redox processes, as these studies (almost) all just focus on the "desired" products, i.e. H_2, CO, CH_4, or NH_3 (to name but a few). This is one of the main reasons for my initial statement regarding the still rather incomplete understanding of the basic principles of photocatalysis.

In summary, it is fair to say that the impressive collection of review and feature articles combined into this really nice book clearly shows the fascinating development within the research area of visible light-driven photocatalysis during the last decade [5]. I am convinced that we will see a similar progress in the future and already look forward to the next summarizing contribution by "our" editors.

Hannover and Saint-Petersburg, 31.08.2022

Detlef Bahnemann

References

1 Schneider, J., Matsuoka, M., Takeuchi, M. et al. (2014). Understanding TiO_2 photocatalysis: mechanisms and materials. *Chem. Rev.* 114: 9919.

2 Schneider, J. and Bahnemann, D.W. (2013). Undesired role of sacrificial reagents in photocatalysis. *J. Phys. Chem. Letters* 4: 3479.

3 Jeon, T.H., Monllor-Satoca, D., Moon, G.-H. et al. (2020). Ag(I) ions working as a hole-transfer mediator in photoelectrocatalytic water oxidation on WO_3 film. *Nat. Commun.* 11: 967.

4 Haisch, C., Nunes, B.N., Schneider, J. et al. (2018). Transient absorption studies on nanostructured materials and composites: towards the development of new photocatalytic systems. *Z. Phys. Chem.* 232: 1469–1493.

5 Etacheri, V., Di Valentin, C., Schneider, J. et al. (2015). Visible-light activation of TiO_2 photocatalysts: advances in theory and experiments. *J. Photochem. Photobiol. C: Photochem. Rev.* 25: 1–29.

List of Contributors

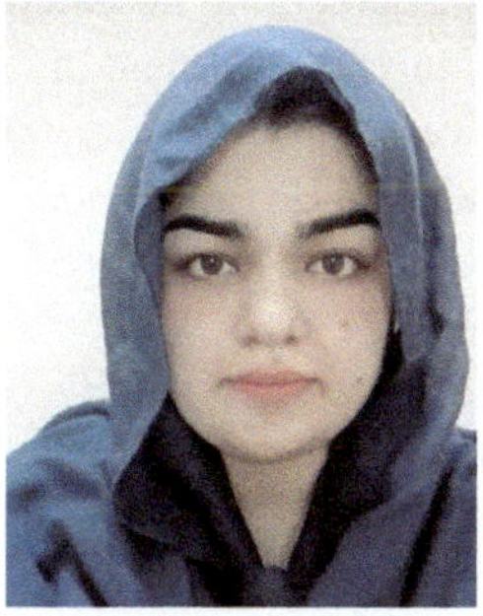

Summan Aman
University of Gujrat
Department of Chemistry
Gujrat, 50700
Pakistan

Masakazu Anpo
Osaka Metropolitan University
Graduate School of Engineering
Department of Applied Chemistry
1-1, Gakuen-cho, Naka-ku, Sakai
Osaka 599-8531
Japan

and

Fuzhou University
State Key Laboratory of Photocatalysis
on Energy and Environment
Fuzhou, Fujian 350116
P. R. China

Detlef Bahnemann
Leibniz University Hannover
Institute for Technical Chemistry
Hannover 30167
Germany

and

Saint-Petersburg State University
Laboratory, Photoactive Nanocomposite
Materials
Saint-Petersburg 198504
Russia

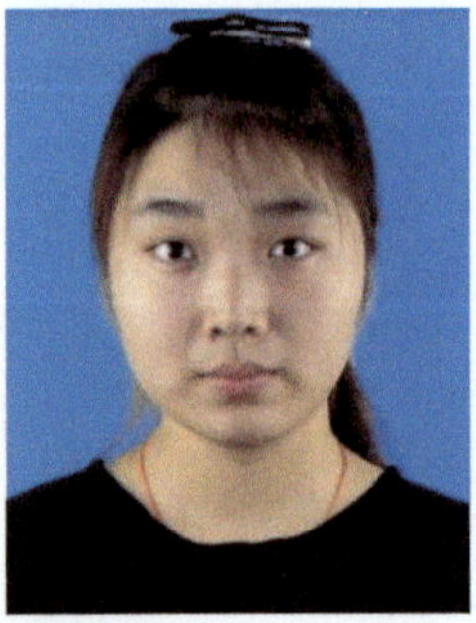

Sha Bai
Beijing University of Chemical Technology
State Key Laboratory of Chemical Resource Engineering
No.15 Beisanhuan East Road
Chaoyang District
Beijing 100029
P. R. China

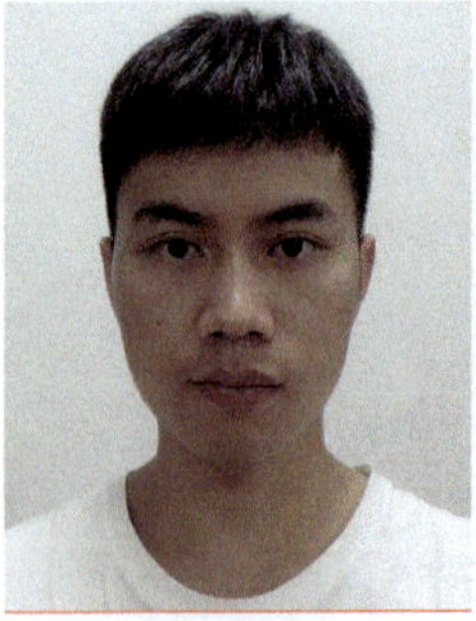

Chuanbiao Bie
China University of Geosciences
Laboratory of Solar Fuel, Faculty of Materials Science and Chemistry
388 Lumo Road
Wuhan, Hubei 430074
P. R. China

Daomei Chen
Yunnan University
Yunnan Province Engineering Research Center of Photocatalytic Treatment of Industrial Wastewater
School of Materials and Energy, School of Chemical Sciences & Technology
School of Engineering
2 Cuihu North Road
Kunming, 650091
P. R. China

Xianjie Chen
Southwest University of Science and Technology
State Key Laboratory of Environment-friendly Energy Materials, School of Materials and Chemistry
Qinglong Road
Mianyang 621010
P. R. China

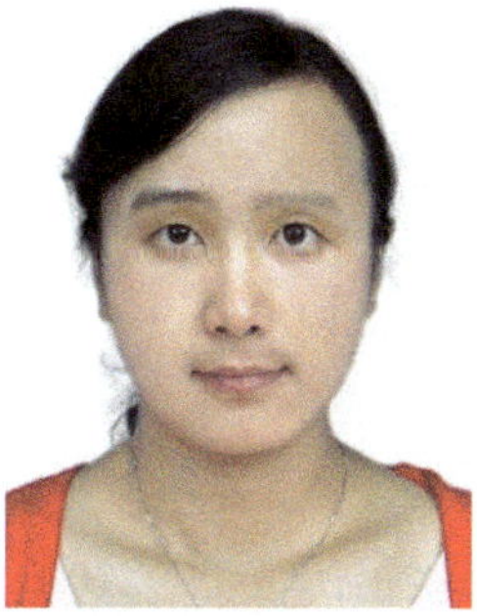

Yongjuan Chen
Yunnan University
Yunnan Province Engineering Research Center of Photocatalytic Treatment of Industrial Wastewater
School of Materials and Energy, School of Chemical Sciences & Technology
School of Engineering
2 Cuihu North Road
Kunming, 650091
P. R. China

Kazunari Domen
Shinshu University
Research Initiative for Supra-Materials
Interdisciplinary Cluster for Cutting Edge Research
4-17-1 Wakasato, Nagano-shi
Nagano 380-8553
Japan

and

The University of Tokyo
Office of University Professors
2-11-16 Yayoi, Bunkyo-ku
Tokyo 113-8656
Japan

Wenxin Dai
Fuzhou University
College of Chemistry, State Key Laboratory of Photocatalysis on Energy and Environment
2 Xueyuan Road, Minhou District
Fuzhou 350108
P. R. China

Guohui Dong
Shaanxi University of Science and Technology
School of Environmental Science and Engineering
Xi'an 710021
P. R. China

Tingting Dong
Shaanxi University of Science
and Technology
School of Environmental Science
and Engineering
Xi'an 710021
P. R. China

E.I. García-López
University of Palermo
Chemical and Pharmaceutical Sciences
and Technologies (STEBICEF)
Department of Biological
Viale delle Scienze
Palermo 90128
Italy

Xianzhi Fu
Fuzhou University
State Key Laboratory of Photocatalysis
on Energy and Environment
Building No.1, National University
Science Park of Fuzhou University
No. 2, Xueyuan Road, Minhou County
Fuzhou
Fujian 350116
P. R. China

Jiao He
Yunnan University
Yunnan Province Engineering Research
Center of Photocatalytic Treatment of
Industrial Wastewater
School of Materials and Energy, School
of Chemical Sciences & Technology
School of Engineering
2 Cuihu North Road
Kunming, 650091
P. R. China

Rana M. Irfan
Sogang University
Department of Physics
Seoul, 04107
Korea

Wei Li
Shaanxi University of Science &
Technology
College of Chemistry and Chemical
Engineering
Xi'an
Shaanxi 710021
P. R. China

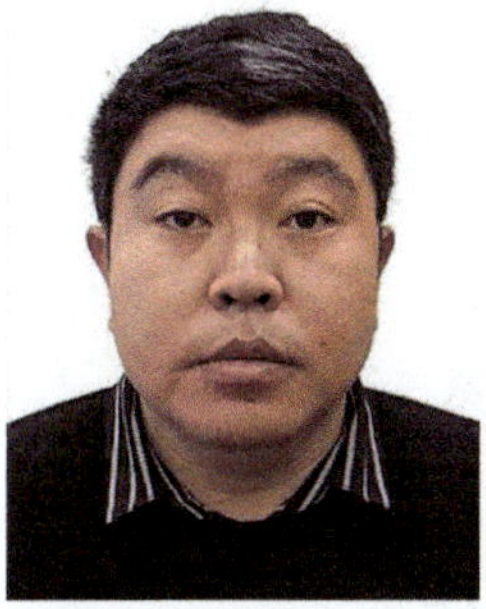

Liang Jiang
Yunnan University
Yunnan Province Engineering Research
Center of Photocatalytic Treatment of
Industrial Wastewater
School of Materials and Energy, School
of Chemical Sciences & Technology
School of Engineering
2 Cuihu North Road
Kunming, 650091
P. R. China

Xiaoman Li
Institute of Ceramics, Chinese
Academy of Sciences
State Key Laboratory of High
Performance Ceramics and Superfine
Microstructure
Shanghai 1295 Dingxi Road
Shanghai 200050
P. R. China

Kangkang Lian
Fuzhou University
State Key Laboratory of Photocatalysis on Energy and Environment
Department of Chemistry
Wulong Jiangbei Avenue
Fuzhou, Fujian
P. R. China

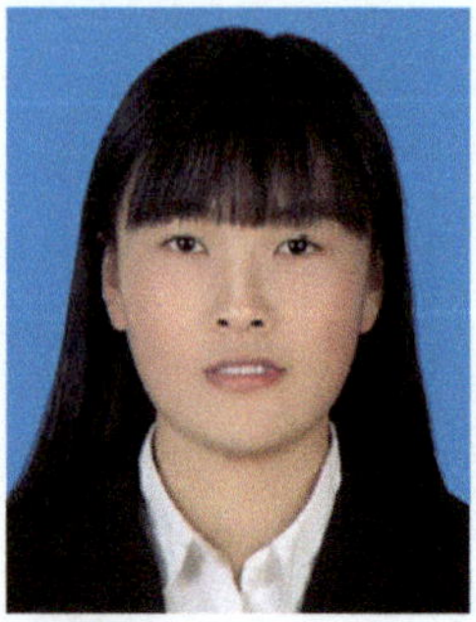

Yujie Liu
East China University of Science & Technology
School of Chemistry and Molecular Engineering
130 Meilong Road
Shanghai 200237
P. R. China

Sen Lin
Fuzhou University
State Key Laboratory of Photocatalysis on Energy and Environment
Department of Chemistry
Wulong Jiangbei Avenue
Fuzhou, Fujian
P. R. China

Xiaoqian Ma
Yunnan University
Yunnan Province Engineering Research Center of Photocatalytic Treatment of Industrial Wastewater
School of Materials and Energy, School of Chemical Sciences & Technology
School of Engineering
2 Cuihu North Road
Kunming, 650091
P. R. China

G. Marcì
University of Palermo
"Schiavello-Grillone" Photocatalysis
Group, Department of Engineering
Viale delle Scienze
Palermo, 90128
Italy

Masashi Morita
Tokyo University of Agriculture
and Technology
Graduate School of Engineering
Department of Applied Chemistry
2-24-16 Naka-cho, Koganei
Tokyo, 184-8588
Japan

Alireza Z. Moshfegh
Sharif University of Technology
Department of Physics
Azadi Avenue, 11155-9161
Tehran
Iran

and

Sharif University of Technology
Institute for Nanoscience
and Nanotechnology
Azadi Avenue, 14588-89694
Tehran, Iran

Amene Naseri
Agricultural Biotechnology Research
Institute of Iran (ABRII), Agricultural
Research, Education, and Extension
Organization (AREEO)
Department of Nanotechnology
Karaj, 3135933151
Iran

Bernaurdshaw Neppolian
SRM Institute of Science
and Technology
Department of Chemistry
Chennai, 603203
Tamil Nadu
India

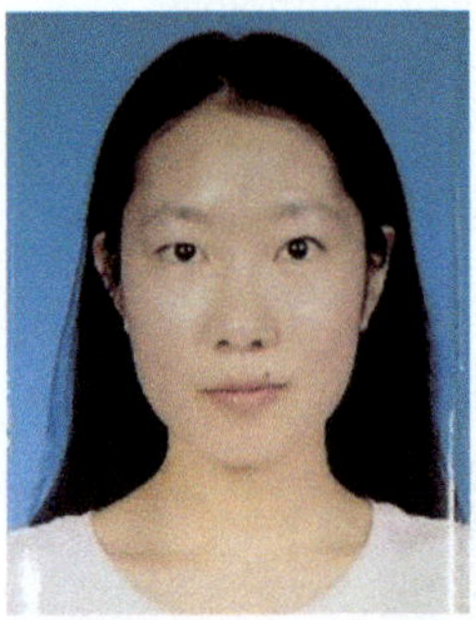

Chenjun Ning
Beijing University of Chemical
Technology
State Key Laboratory of Chemical
Resource Engineering
No.15 Beisanhuan East Road
Chaoyang District
Beijing 100029
P. R. China

Makoto Ogawa
Vidyasirimedhi Institute of Science
and Technology (VISTEC)
School of Energy Science
and Engineering
555 Moo 1 Tumbol Payupnai, Amphoe
Wangchan
Rayong 21210
Thailand

L. Palmisano
University of Palermo
"Schiavello-Grillone" Photocatalysis
Group
Department of Engineering
Viale delle Scienze
Palermo, 90128
Italy

Jing Pan
State Key Laboratory of Biogeology
and Environmental Geology
Faculty of Materials Science
and Chemistry
China University of Geosciences
Wuhan, Hubei
P. R. China

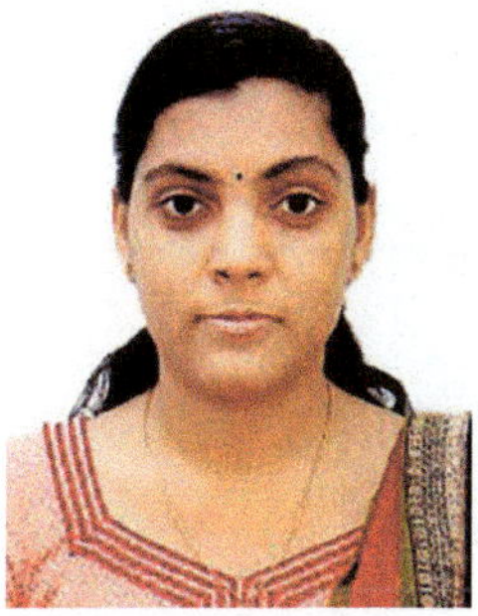

Aswathy Rajan
SRM Institute of Science
and Technology
Department of Chemistry
Chennai, 603203
Tamil Nadu
India

Morasae Samadi
Alzahra University
Department of Chemistry, Faculty of
Physics and Chemistry
Tehran, 1993893973
Iran

Oleksandr Savateev
Max Planck Institute of Colloids
and Interfaces
Department of Colloid Chemistry
Am Mühlenberg 1
Potsdam 14476
Germany

Tianyang Shen
Beijing University of Chemical Technology
State Key Laboratory of Chemical Resource Engineering
No.15 Beisanhuan East Road
Chaoyang District
Beijing 100029
P. R. China

Yu-Fei Song
Beijing University of Chemical Technology
State Key Laboratory of Chemical Resource Engineering
No.15 Beisanhuan East Road
Chaoyang District
Beijing 100029
P. R. China

Yujie Song
Hainan Provincial Key Laboratory of Fine Chemicals
Hainan University
No. 58 Renmin Avenue
Haikou, Hainan 570228
P. R. China

Ren Su
Soochow University, Soochow Institute for Energy and Materials InnovationS (SIEMIS)
Key Laboratory of Advanced Carbon Materials and Wearable Energy Technologies of Jiangsu Province
Suzhou 215006
P. R. China

and

Synfuels China Technology Co. Ltd.
Leyuan South Street II, No.1, Yanqi Economic Development Zone C#
Huairou District
Beijing 101407
P. R. China

Songmei Sun
Institute of Ceramics, Chinese Academy of Sciences
State Key Laboratory of High Performance Ceramics and Superfine Microstructure
Shanghai 1295 Dingxi Road
Shanghai 200050
P. R. China

Tsunehiro Tanaka
Kyoto University, Graduate School of Engineering
Department of Molecular Engineering
Kyoto Daigaku Katsura
Kyoto 6158510
Japan

Masato Takeuchi
Osaka Metropolitan University
Graduate School of Engineering
Department of Applied Chemistry
1-1, Gakuen-cho, Naka-ku, Sakai
Osaka 599-8531
Japan

Xiaoping Tao
Shinshu University
Research Initiative for Supra-Materials
Interdisciplinary Cluster for Cutting Edge Research
4-17-1 Wakasato, Nagano-shi
Nagano 380-8553
Japan

Muhammad Tayyab
East China University of Science & Technology
School of Chemistry and Molecular Engineering
130 Meilong Road
Shanghai 200237
P. R. China

Chuanyi Wang
Shaanxi University of Science & Technology, School of Environmental Sciences and Engineering
Shaanxi Key Laboratory of Chemical Additives for Industry
Xi'an, Shaanxi 710021
P. R. China

Kentaro Teramura
Kyoto University, Graduate School of Engineering
Department of Molecular Engineering
Kyoto Daigaku Katsura
Kyoto 6158510
Japan

Jiaqiang Wang
Yunnan University
Yunnan Province Engineering Research Center of Photocatalytic Treatment of Industrial Wastewater
School of Materials and Energy, School of Chemical Sciences & Technology
School of Engineering
2 Cuihu North Road
Kunming, 650091
P. R. China

Wenzhong Wang
Institute of Ceramics, Chinese Academy of Sciences
State Key Laboratory of High Performance Ceramics and Superfine Microstructure
Shanghai 1295 Dingxi Road
Shanghai 200050
P. R. China

and

University of Chinese Academy of Sciences
Hangzhou Institute for Advanced Study, School of Chemistry and Materials Science
1 Sub-lane Xiangshan
Hangzhou 310024
P. R. China

Xinchen Wang
Fuzhou University
State Key Laboratory of Photocatalysis on Energy and Environment
Building No.1, National University Science Park of Fuzhou University
No. 2, Xueyuan Road, Minhou County
Fuzhou, Fujian 350116
P. R. China

Zhiwen Wang
Hainan Provincial Key Laboratory of Fine Chemicals
Hainan University
No. 58 Renmin Avenue
Haikou, Hainan 570228
P. R. China

Zhonghui Wang
Fuzhou University
State Key Laboratory of Photocatalysis on Energy and Environment
Department of Chemistry
Wulong Jiangbei Avenue
Fuzhou
Fujian
P. R. China

Zhongming Wang
Fuzhou University
College of Chemistry, State Key Laboratory of Photocatalysis on Energy and Environment
2 Xueyuan Road, Minhou District
Fuzhou 350108
P. R. China

and

Anhui University of Technology
College of Energy and Environment
Department of Environmental Sciences and Engineering
59 Hudong Road, Huashan District
Ma Anshan 243002
P. R. China

Fan Xia
State Key Laboratory of Biogeology and Environmental Geology
Faculty of Materials Science and Chemistry
China University of Geosciences
Wuhan, Hubei
P. R. China

Cailin Xiao
Institute of Ceramics, Chinese Academy of Sciences
State Key Laboratory of High Performance Ceramics and Superfine Microstructure
Shanghai 1295 Dingxi Road
Shanghai 200050
P. R. China

Jiadong Xiao
Shinshu University
Research Initiative for Supra-Materials
Interdisciplinary Cluster for Cutting Edge Research
4-17-1 Wakasato, Nagano-shi
Nagano 380-8553
Japan

Xuanwen Xu
Kyoto University, Graduate School of Engineering
Department of Molecular Engineering
Kyoto Daigaku Katsura
Kyoto 6158510
Japan

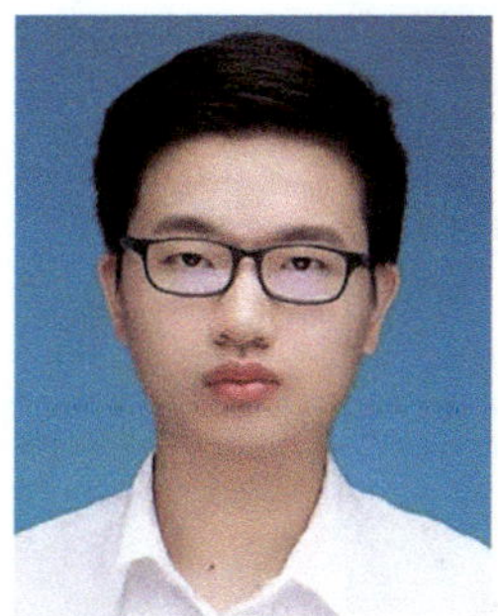

Zehong Xu
East China University of Science & Technology
School of Chemistry and Molecular Engineering
130 Meilong Road
Shanghai 200237
P. R. China

Jiaguo Yu
China University of Geosciences
Laboratory of Solar Fuel, Faculty of Materials Science and Chemistry
388 Lumo Road
Wuhan, Hubei 430074
P. R. China

Jinlong Zhang
East China University of Science & Technology
School of Chemistry and Molecular Engineering
130 Meilong Road
Shanghai 200237
P. R. China

Wenhui Yue
East China University of Science & Technology
School of Chemistry and Molecular Engineering
130 Meilong Road
Shanghai 200237
P. R. China

Liang Zhou
East China University of Science & Technology
School of Chemistry and Molecular Engineering
130 Meilong Road
Shanghai 200237
P. R. China

Yuanyi Zhou
Institute of Ceramics, Chinese Academy of Sciences
State Key Laboratory of High Performance Ceramics and Superfine Microstructure
Shanghai 1295 Dingxi Road
Shanghai 200050
P. R. China

Yongfa Zhu
Southwest University of Science and Technology
State Key Laboratory of Environment-friendly Energy Materials, School of Materials and Chemistry
Qinglong Road
Mianyang 621010
P. R. China

and

Tsinghua University
Department of Chemistry
Shuangqing Road
Beijing 100084
P. R. China

1

Introduction

Xinchen Wang, Masakazu Anpo, and Xianzhi Fu

Fuzhou University, State Key Laboratory of Photocatalysis on Energy and Environment, Building No.1, National University Science Park of Fuzhou University, No. 2, Xueyuan Road, Minhou County, Fuzhou, Fujian 350116, P.R. China

1.1 Challenges and Objectives in the Use of Solar Energy

The increase in the world population and rampant unregulated economic growth have all led to accelerated energy consumption and the unabated release of nonbiodegradable materials such as toxic agents, plastics, and industrial wastes such as solvents and by-products into the air and sea and various waterways. This has, in turn, led to pollution-related diseases and abnormal climate changes such as global warming. Thus, humanity is now facing the serious and urgent issues of the lack of natural energy resources and fossil-fuel-driven environmental destruction and pollution on a global scale. While our industries are constantly providing a variety of new products and materials based on innovative new technologies, it is now imperative to focus on recycling those materials, reducing waste, and raising awareness of the great impact that our consumerism has on our environment.

The greatest challenge for researchers is to develop environmentally harmonious, ecologically clean and safe, sustainable, and energy-efficient chemical technologies. Photocatalysis, which utilizes the inexhaustible energy of the sun, can be harnessed and converted into clean chemical or electrical energy for nontoxic, economically viable technologies. In this book, we will present not only the fundamentals of photocatalysis but also the advances in research such as the photocatalytic splitting of water to produce clean H_2, the photocatalytic reduction of CO_2 to form hydrocarbons, the photocatalytic fixation of N_2, photo-induced super hydrophilicity to design useful materials, and promising applications of photocatalytic systems for the purification of polluted water.

UV-Visible Photocatalysis for Clean Energy Production and Pollution Remediation: Materials, Reaction Mechanisms, and Applications, First Edition. Edited by Xinchen Wang, Masakazu Anpo, and Xianzhi Fu.
© 2023 WILEY-VCH GmbH. Published 2023 by WILEY-VCH GmbH.

1.2 Brief History of the Progress in Photocatalysts and Photocatalytic Reactions

Metal oxide-semiconducting photocatalysts such as TiO_2 materials have been the focus of much attention for their ability to induce efficient and effective reactions at room temperature under sunlight irradiation. In fact, as shown in Figure 1.1, in the past half-century, many scientists have studied the effect of UV light irradiation on semiconductors [1, 2]. In 1972, Honda and Fujishima reported the photosensitization effect of a TiO_2 electrode in the electrochemical photolysis of water at the TiO_2 electrode with a Pt counter electrode under UV light irradiation [3]. Schrauzer and Guth reported the photolysis of water and the photoreduction of nitrogen on TiO_2 photocatalyst in 1977 [4]. Since then, many researchers have investigated the development of inorganic semiconducting materials for the photocatalytic decomposition of water to produce hydrogen [5–7].

As a photocatalytic material, TiO_2 is cost-efficient, nontoxic, and has relatively high activity and stability even in aqueous solutions. Thus, as shown in Figure 1.1, it was first applied to the photocatalytic decomposition of water to produce hydrogen under relatively strong UV light irradiation [1, 5, 6]. In relation to these reactions, the idea of a dye-sensitized solar cell was reported by Gratzel in 1991 [8]. There is also a report on the photocatalytic degradation or mineralization of organic pollutants on inorganic semiconducting materials under relatively weak UV light irradiation [9]. Since then, various visible-light-responsive photocatalysts have been reported, and the separate evolution of H_2 and O_2 from water using a visible-light-responsive TiO_2 thin film photocatalyst was successfully carried out in 2004 [10]. Moreover, the highly efficient photocatalytic splitting of water to produce an H_2 and O_2 mixture using particulate oxynitrides and oxychalcogenides as semiconducting powdered materials under visible-light irradiation was reported by Kudo and Domen [11].

In 2009, Wang et al. reported the highly efficient production of H_2 and O_2 from water using a visible-light-responsive polymeric graphic carbon nitride (g-C_3N_4) photocatalyst [12]. Since then, as one of the most promising photocatalytic materials over metal oxides and sulfides, graphic carbon nitride (g-C_3N_4) nanomaterials have been investigated by many researchers with a focus on their design, construction, and optimization to establish the high conversion of light energy into chemical energy [7, 13]. For this material, chemical engineering technologies such as element doping have been applied. Various heterojunctions have also been reported to enhance the efficiency of the photocatalytic splitting of water to produce hydrogen [14]. Significantly, almost 100% quantum yield for the photocatalytic splitting of water using Al-doped $SrTiO_3$ with metal cocatalysts as well as the large-scale photocatalytic production of H_2 from water under sunlight irradiation has been reported by Domen and coworkers [15]. Such remarkable progress in the photocatalytic splitting of water has spurred further research into such practical applications that will lead to clean renewable energy technologies and the purification of the environment.

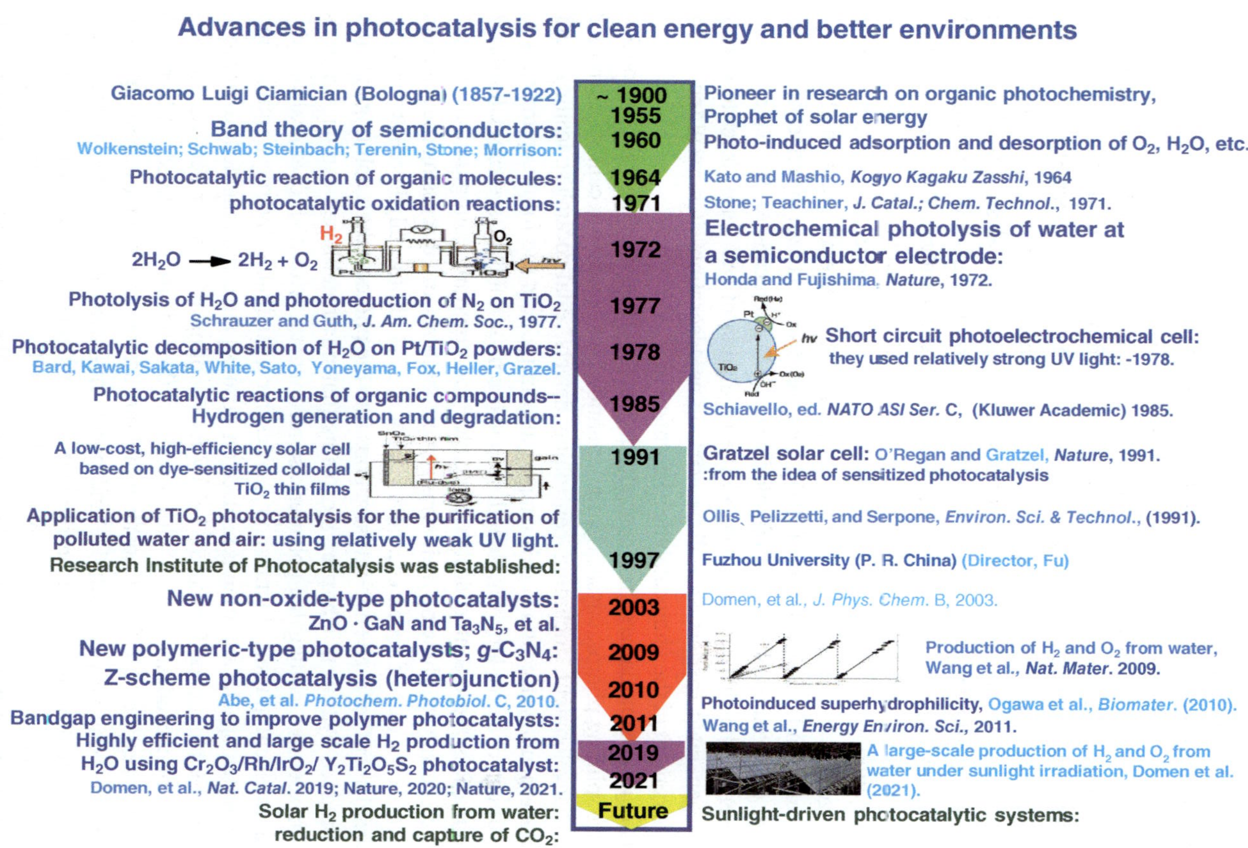

Figure 1.1 Progress in research on photocatalysts and their practical applications (past, present, and future).

1.3 Brief Introduction of the Chapters

This book consists of four categories of the fundamentals of photocatalysis which are detailed in four chapters, the photocatalytic splitting of water to produce hydrogen in seven chapters, the photocatalytic reduction of CO_2 and fixation of N_2 in six chapters, and the application of photocatalysis in four chapters.

Chapter 2 introduces dual-functioning photocatalysis to produce hydrogen from the treatment of wastewater as a promising approach to address both global environmental issues and energy problems. The fundamentals of dual-functional photocatalytic reactions via heterojunction formation and cocatalysts are also explained. **Chapter 3** summarizes the reaction mechanism of the photocatalytic oxidation of alcohol with O_2 using a carbon nitride photocatalyst. Special attention is given to strategies that improve selectivity toward carbonyl compounds and dehydrogenation employing a hydrogen-evolving cocatalyst. **Chapter 4** introduces an S-scheme heterojunction as a new concept in photocatalysis. S-scheme heterojunctions lead to an enhancement in photocatalytic reactions, including hydrogen evolution from water, the reduction of CO_2, the degradation of organic pollutants, and the production of H_2O_2. **Chapter 5** summarizes the effect of defects on the electronic and optical properties of the ZnO photocatalyst. Special attention is given to the role of the defects in the regulation of the Fermi level to enrich photogenerated carriers as well as working as adsorption sites, resulting in an enhancement of the photocatalytic performance.

Chapter 6 summarizes the recent advances in the study of visible-light-responsive water-splitting photocatalysts. Special attention is focused on recent strategies for their development, including controlling defects, dual-cocatalyst loading, and surface nanolayer coating. This chapter also introduces the future approaches to the design of active long-wavelength-responsive photocatalytic systems. **Chapter 7** summarizes the essential design principles and emerging configurations of dual redox cocatalysts, showing their synergistic operations and discussing the selection process of a pair of redox cocatalysts for a special photocatalytic redox reaction. Special attention is given to investigations into the photocatalytic splitting of water to produce hydrogen. **Chapter 8** explains that nonmetal-based carbon nitride is an inexpensive, conjugated polymeric semiconductor with high chemical stability and excellent physicochemical properties for sustainable hydrogen production from water. Special attention is focused on various preparation methods to improve its photocatalytic performance, i.e. by its coupling with other semiconductors, the use of cocatalysts, doping with metals, or the induction of defects. **Chapter 9** summarizes the significant advances in the design and photocatalytic splitting of water using perylene diimide and porphyrin-based supramolecular materials. Special attention is focused on the ongoing challenges and opportunities for the future development of supramolecular photocatalysts in high-quality nanodevices. **Chapter 10** summarizes the development of the visible-light-responsive TiO_2 thin film photocatalyst on a Ti foil substrate by an radiofrequency-magnetron sputtering (RF-MS) deposition method for the separate evolution of hydrogen and oxygen from water. The integration of this photocatalytic reaction system and an artificial light-type plant factory to realize the production of pure hydrogen and fresh vegetables using sunlight efficiently and effectively is introduced. **Chapter 11** summarizes the features of the photocatalytic evolution of hydrogen on CdS semiconductors. Special attention is focused on the merit and/or demerits of the CdS photocatalytic material as well as several improvements for enhancing the hydrogen production performance of CdS. **Chapter 12** explains the theoretical approach for the efficient photocatalytic splitting of water to produce hydrogen. Special attention is focused on band structure engineering and carrier separation of three-dimensional and two-dimensional photocatalytic materials to improve the yield of hydrogen evolution.

Chapter 13 gives an overview of the progress in the development of cocatalysts for the photocatalytic conversion of CO_2 by H_2O as an electron donor. Special attention is focused on the effects of various cocatalysts on the reduction of CO_2. **Chapter 14** summarizes the photocatalytic activity of the single-site Ti-oxo species prepared within silica or silicate for the reduction of CO_2 with H_2O. Special attention is focused on the preparation and characterization of the single-site species in nano spaces of silica and silicate materials. **Chapter 15** summarizes the potential strategy to create an efficient catalyst for the directional conversion of CO_2 through regulating the coordination modes of CO_2 on two-dimensional oxide nanosheets. **Chapter 16** summarizes the progress on layered double hydroxides (LDHs)-based nanomaterials for solar energy conversion through the reduction of CO_2 by fine-tuning the composition, coordination environment, hybrid structure, and topological transformation of LDHs. **Chapter 17** introduces the background and reaction mechanism of the photocatalytic fixation of N_2, explaining the influencing factors such as types of materials, and also provides a brief overview of its current challenges and future prospects. **Chapter 18** provides a general overview of the investigations on the photocatalytic fixation of N_2 and introduces the research highlights in these research fields.

Chapter 19 discusses the contributions of photocatalysts to the remediation of polluted environments and their reaction conditions. Detailed analytical methods of the reactants, products, and intermediates are provided. The reaction mechanisms of a few case studies using model reactions are reported. **Chapter 20** summarizes the development of photocatalytic reactors in the treatment of wastewater using biotemplated photocatalysts with structural specialty, complexity, and related unique properties. Special attention is focused on the design, assembly, and manufacturing of skid-mounted photocatalytic reactors. **Chapter 21** introduces the high surface wettability of TiO_2 thin films and TiO_2-based binary oxide thin films under UV light irradiation, and its dependence on their surface morphology, surface area, and crystal phase. **Chapter 22** introduces the advances in surfaces with super wettability and photocatalytic activity in terms of their formation and applications. Special attention is focused on the challenges and opportunities for the further development of surfaces with both super wettability and photocatalytic activity.

These chapters provide useful information on the various advances in the development of photocatalysis and photocatalytic materials, significantly, for the photocatalytic production of clean hydrogen from water as well as the photocatalytic purification of polluted water under sunlight irradiation.

1.4 Conclusion and Perspectives

Most traditional inorganic metal oxide and sulfide photocatalysts function only under UV light, which makes up less than 3–4% of the sunlight reaching the earth. Thus, visible-light-responsive inorganic photocatalysts which effectively and efficiently utilize sunlight by more than 30–35% have been intensely studied. Polymeric metal-free graphitic carbon nitride (g-C_3N_4) nanomaterials have also been developed as efficient photocatalysts operating even under visible-light irradiation. Various sciences and technologies to control the morphology of photocatalysts and the formation of a heterojunction to enhance photocatalytic activity and stability have also been developed. These investigations continue the research of Ciamician (1857–1922), the father of organic photochemistry, who conducted the first systematic studies on the behavior of organic substances toward light. Further research into the development of highly efficient and photocatalytic reaction systems will be vital in establishing artificial photosynthetic systems to convert solar energy effectively and efficiently into useful chemical or electrical energy, thus ensuring a stable energy source as well as a cleaner and more sustainable environment for future generations.

References

1 Wang, X., Anpo, M., and Fu, X. (2020). *Current Developments in Photocatalysis and Photocatalytic Materials, New Horizons in Photocatalysis*. Elsevier, Amsterdam, Netherlands and references therein.

2 Wang, B., Anpo, M., and Wang, X. (2018). Visible light-responsive photocatalysts – from TiO_2 to carbon nitrides and boron carbon nitrides. *Adv. Inorg. Chem.* 72: 49–85. and references therein.

3 Fujishima, A. and Honda, K. (1972). Electrochemical photolysis of water at a semiconductor electrode. *Nature* 238: 39–38.

4 Schrauzer, G. and Guth, T. (1977). Photolysis of water and photoreduction of nitrogen on titanium dioxide. *J. Am. Chem. Soc.* 99: 7189–7193.

5 Bard, A. (1979). Photoelectrochemistry and heterogeneous photocatalysis at semiconductors. *J. Photochem.* 10: 59–75.

6 Schneider, J., Matsuoka, M., Takeuchi, M. et al. (2014). Understanding TiO_2 photocatalysis: mechanisms and materials. *Chem. Rev.* 114: 9919–9986. and references therein.

7 Fabian, K., Zheng, Y., Schwarz, D. et al. (2017). Functional carbon nitride materials – design strategies for electrochemical devices. *Nat. Rev. Mater.* 2: 17030. and references therein.

8 O'Regan, B. and Grätzel, M. (1991). A low-cost, high-efficiency solar cell based on dye-sensitized colloidal TiO_2 films. *Nature* 353: 737–740.

9 Schiavello, M. (ed.) (1985). *Photoelectrochemistry, Photocatalysis, and Photoreactors: Fundamentals and Developments*, NATO ASI Series, C, vol. 146. Kluwer Academic, Amsterdam, Netherlands.

10 Anpo, M., Kikuchi, H., Hosoda, T. and Takeuchi, M. et al. (2004). Decomposition of H_2O with the separation of the evolved H_2 and O_2 using visible-light-responsive TiO_2 thin film photocatalysts. In: Proceedings of 13th International Congress on Catalysis (Paris), Vol. 4 (Fuels and Energy for The Future), 1942–1944 (2004) Institut Francais du Petrole (IFP), Curran Associates, Inc., Paris, France.

11 Maeda, K., Takata, T., Hara, M. et al. (2005). GaN:ZnO solid solution as a photocatalyst for visible-light-driven overall water splitting. *J. Am. Chem. Soc.* 127: 8286–8287.

12 Wang, X., Maeda, K., Thomas, A. et al. (2009). A metal-free polymeric photocatalyst for hydrogen production from water under visible light. *Nat. Mater.* 8: 76–80.

13 Lin, Z. and Xinchen Wang, X. (2013). Nanostructure engineering and doping of conjugated carbon nitride semiconductors for hydrogen photosynthesis. *Angew. Chem. Int. Ed.* 52: 1735–1738.

14 Suzuki, H., Tomita, O., Higashi, M., and Abe, R. (2015). Z-scheme water splitting into H_2 and O_2 using tungstic acid as an oxygen-evolving photocatalyst under visible light irradiation. *Chem. Lett.* 44: 1134–1136.

15 Lyu, H., Hisatomi, T., Goto, Y. et al. (2019). An Al-doped $SrTiO_3$ photocatalyst maintaining sunlight-driven overall water splitting activity for over 1000 h of constant illumination. *Chem. Sci.* 10: 3196–3201.

Part I

Fundamentals of Photocatalysis

2

Visible-Light Active Photocatalysts in Pollutant Degradation/Conversion with Simultaneous Hydrogen Production

Amene Naseri[1], *Morasae Samadi*[2], *and Alireza Z. Moshfegh*[3,4]

[1]*Agricultural Biotechnology Research Institute of Iran (ABRII), Agricultural Research, Education, and Extension Organization (AREEO), Department of Nanotechnology, Karaj, 3135933151, Iran*
[2]*Alzahra University, Department of Physical chemistry and Nanochemistry, Faculty of Chemistry, Tehran, 19938-91176, Iran*
[3]*Sharif University of Technology, Department of Physics, Azadi Avenue, 11155-9161, Tehran, Iran*
[4]*Sharif University of Technology, Institute for Nanoscience and Nanotechnology, Azadi Avenue, 14588-89694, Tehran, Iran*

2.1 Introduction

Photocatalysis is a green and sustainable technology that many researchers have studied because of its ability to utilize solar irradiation to address the energy crisis and environmental issues resulting from organic pollutants [1]. Organic pollutants possess high toxicity, carcinogenicity, and refractory degradation, causing severe risks to human health. Recently, the photocatalytic degradation of organic pollutants has been considered a promising water treatment approach [2]. Moreover, hydrogen production by photocatalytic water splitting has been an efficient technology for converting solar light energy into hydrogen energy as a fuel [3]. However, most current research in the field of photocatalysis has either focused on the decontamination of wastewater or the production of hydrogen via water splitting in a separate process.

Dual-functional photocatalytic processes can effectively improve the efficiency of light quantum utilization for photocatalytic H_2 evolution reactions (HERs) with simultaneous organic molecular pollutant degradation to less harmful products. To develop and reach this goal, a growing number of documents have been published in recent years in this active research field (Figure 2.1a, orange columns). Considering the reactions under visible irradiation, more than 50% of these studies have been conducted in this region of the solar spectrum, as shown by the blue columns in Figure 2.1a. Since visible light accounts for about 42% of solar light irradiation, designing visible light nanomaterials are an essential step toward preparing efficient photocatalysts that are highly active under solar light irradiation.

As shown on the right side of Figure 2.1b, during concurrent H_2 production and pollutant degradation, the reactive oxygen species (ROS) generated from photoinduced holes decompose organic materials into less toxic molecules. At the same

UV-Visible Photocatalysis for Clean Energy Production and Pollution Remediation: Materials, Reaction Mechanisms, and Applications, First Edition. Edited by Xinchen Wang, Masakazu Anpo, and Xianzhi Fu.
© 2023 WILEY-VCH GmbH. Published 2023 by WILEY-VCH GmbH.

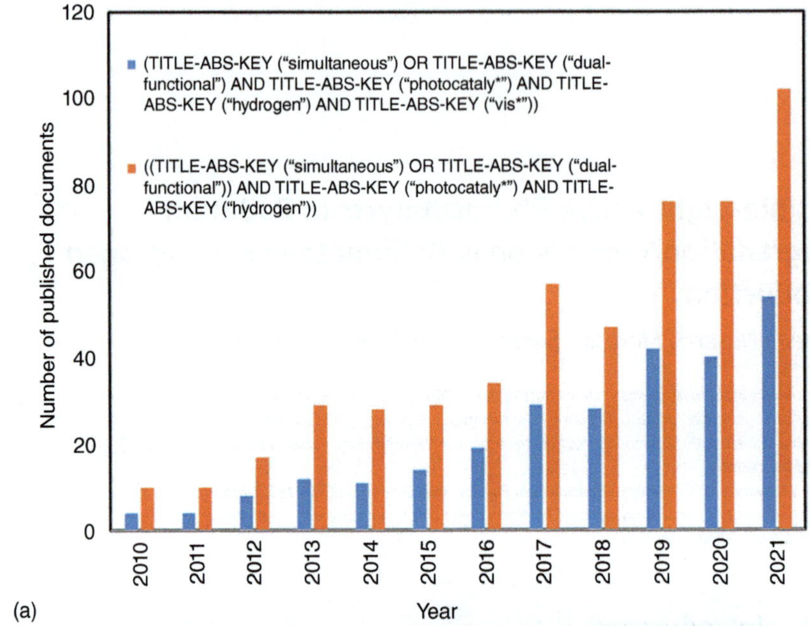

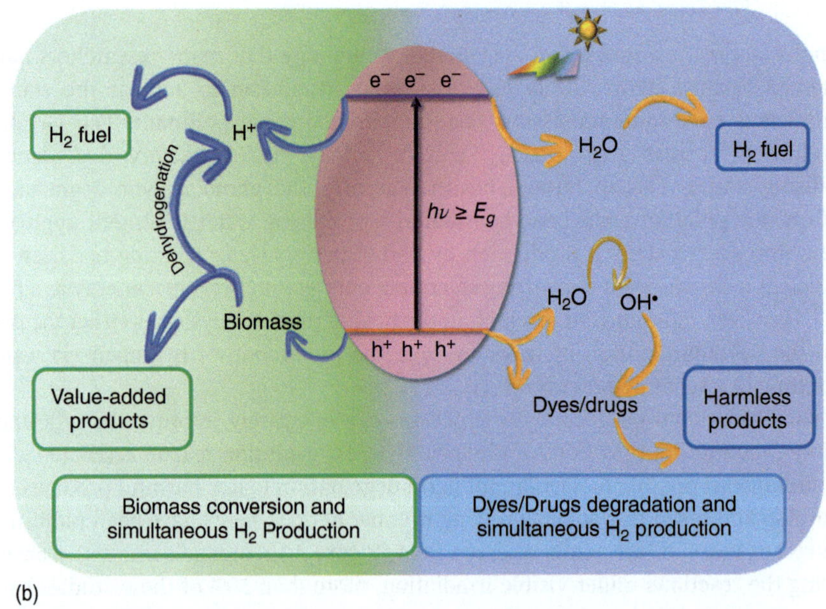

Figure 2.1 (a) Number of published documents vs. year derived from Scopus, 29 June 2022, the search query string in "TITLE-ABS-KEY" are shown in plot legend. (b) Schematic illustration of dual-functional photocatalysts for simultaneous pollutant degradation/conversion and H_2 production (right side: dyes/drugs degradation and H_2 production, left side: biomass conversion and H_2 production).

time, photogenerated electrons convert water molecules via a reduction process into H_2 fuel [4]. On the other hand, if the biomass-derived substances are consumed as a pollutant, dehydrogenating these hydrogen-containing materials as a result of their photo-oxidation process provides the proton needed to produce H_2 fuel (left side of Figure 2.1b [5, 6].

This chapter will discuss the concurrent photocatalytic reactions for H_2 production from organic water pollutants in detail.

2.2 Principles of Simultaneous Photocatalysis

Many researchers have paid attention to saving visible light energy in chemical bonds. It is known that photocatalytic pollutiondegradation for water treatment has been a significant field of study since the 1960s. Recently, energy production from organic pollutants degradation simultaneously on a dual-functional photocatalyst has been used as an efficient alternative to a single-functional one [7]. As shown in Figure 2.1b, upon light absorption with sufficient energy equal to or higher than the band gap of an appropriate semiconductor (SC) photocatalyst, electrons and holes are generated in the SC. The photogenerated holes at the valence band can be consumed in parallel photocatalytic processes to degrade various pollutants in order to treat water. At the same time, the photogenerated electrons in the conduction band can be used to produce clean energy and recover resources. Based on this approach, producing H_2 from various organic substances, including dyes, drugs, and biomass-derived chemicals, has been a hot topic in photocatalysis studies very recently [6]. Due to the technological importance of this simultaneous process, the principles of dual-functional photocatalysis and the progress in designing highly efficient photocatalysts will be discussed in Sections 2.2.1, 2.2.2, and 2.3.

2.2.1 Dual-Functional vs. Conventional Photocatalysts

Compared with conventional photocatalysts, the dual-functional ones can be more efficient for H_2 generation and organic pollutant oxidation at the same time [8]. This is due to the chemical potential of pollutant oxidation having the ability to supply chemical potential energy for the production of H_2 via water splitting [9].

Thermodynamically, only the redox couples whose energy levels are contained within the band gap of the semiconductor can occur in photocatalysis. In simultaneous reactions, the redox potential of both photo-oxidation and photo-reduction reactions should be laid within the conduction band and valence band edges of the semiconductor photocatalyst. For instance, the redox potentials of some known concurrent photocatalytic reactions allowed thermodynamically on the surface of TiO_2 as a common photocatalyst are shown in Figure 2.2a [9]. To further understand the issue, the H_2 production with simultaneous ROS generation to degrade organic pollutants, an appropriate semiconductor photocatalyst with the right band gap energy and position must be considered. The critical problem of conducting cooperative photocatalytic reactions is selectively controlling the charge transfers at the

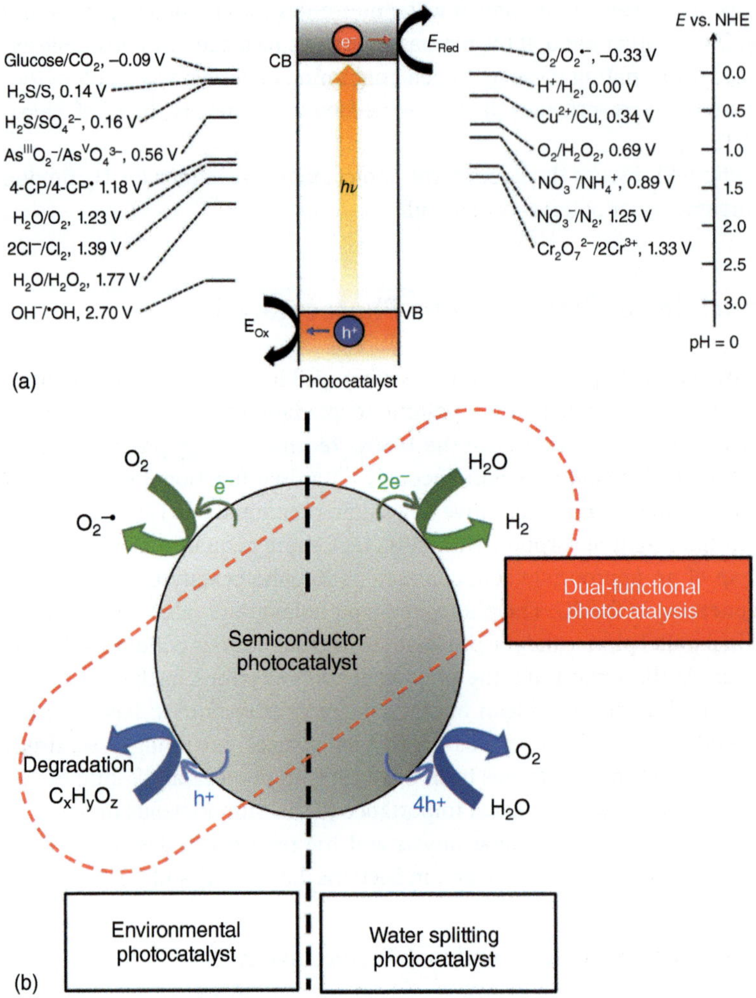

Figure 2.2 (a) Energy level diagram for TiO_2 and various redox pairs that should act as either an electron acceptor (right side) or an electron donor (left side). Source: Reprinted with permission from Jeon et al. [9]/American Chemical Society. (b) Common reactions of semiconductor photocatalysis for individual environmental remediation and water splitting. The red dashed line represents the uncommon dual-functional photocatalysis for simultaneous water treatment and H_2 production. Source: Reproduced with permission from Jeon et al. [9]/American Chemical Society.

interface. In H_2-producing water treatment reactions, the photogenerated electrons should reduce water (or protons) selectively for H_2 production via 2-electron transfer (not to reduce O_2 through a 1-electron reaction) (Figure 2.2b). The photoinduced holes should be able to generate ROS to oxidize organic substances mainly via 1-hole transfer, not to oxidize water to produce O_2 via multihole transfer. The 1-hole oxidation of water to ROS requires a highly positive potential (2.70 V vs. NHE) but occurs rapidly with little kinetic limitation [9]. The 4-hole oxidation

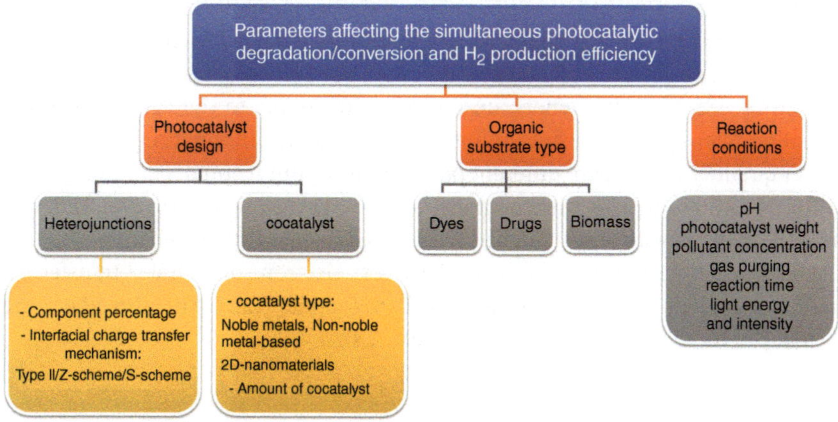

Figure 2.3 Strategies to design highly efficient dual-functional photocatalysts.

of water to O_2 needs a much lower potential (1.23 V vs. NHE) but is kinetically hindered [9]. As a result, the control over the single- vs. multi-electron/hole transfer determines the overall process pathway. The dual-functional photocatalysts should enable to conduct of both the single-hole transfer and the multi-electron transfer at the same time [10]. Various parameters have been studied to achieve this goal, including photocatalyst design, organic substrate type, and reaction conditions. Various parameters that affect the simultaneous photocatalytic reaction are shown schematically in Figure 2.3, and they will be discussed in detail later.

2.2.2 Reaction Efficiency Evaluation

For the HER, apparent quantum efficiency (AQE) is calculated based on the number of photons, N_{photons}, at a specific wavelength, incident on the front window of the reactor cell (flat parallel windows), and the number of molecules produced in the reaction, $N_{\text{molecules}}$, as described below:

$$\text{AQE} = \frac{N_{\text{molecules}}\,(\text{mol s}^{-1})\ \text{transformed/produced}}{N_{\text{photons}}(\text{Einstein s}^{-1})\ \text{incident inside reactor cell}} \tag{2.1}$$

For organic pollutant degradation, the change in concentration (absorbance) of the pollutant during the photocatalytic reaction is expressed as conversion as defined below:

$$\text{conversion}\ (\%) = \left(\frac{(C_0 - C)}{C_0}\right) \times 100 \tag{2.2}$$

where C_0 and C are concentration at the initial and the later time of illumination, in different studies, $\frac{C}{C_0}$ variations during light illumination have been used to determine the photocatalytic degradation of organic pollutants or biomass conversion in a separate study.

For the case of simultaneous H_2 production from wastewater, the rate of H_2 evolution (in mmol g^{-1} h^{-1} or mmol g^{-1}) over a photocatalyst is typically used along with the conversion/degradation percentage of the organic pollutant. It is worth noting that there is a lack of a standard measure of efficiency, which enables a clear comparison of various photocatalysts. Because the reaction rates of photocatalysts reported by different researchers are not comparable due to the variations in the irradiation intensity. As a result, AQE normalized to the illumination conditions is a better quantity to compare different photocatalysts. But it still depends on experimental conditions and measurement methods [6]. Therefore, the H_2 evolution rate and AQE with conversion percentage can be used to quantify the reaction.

Parameters determining the selectivity of electron and hole transfer in dual-functional photocatalysts (mentioned in Section 2.2.1) can also affect the efficiency of these reactions (shown in Figure 2.3).

2.3 Cooperation Photocatalysts for Organic Pollutant Degradation/Conversion and H_2 Fuel Production

As mentioned in Section 2.2 and Figure 2.3, factors affecting the selectivity and efficiency of concurrent H_2 production and organic pollutant degradation/conversion are classified into three main groups. These key factors will be discussed in Sections 2.3.1–2.3.3 in detail.

2.3.1 Photocatalyst Design

Generally, researchers have modified the surface of the photocatalysts by coupling the semiconductors and/or cocatalyst addition to assist and enhance photocatalytic H_2 evolution rate with simultaneous organic pollutant degradation to less harmful chemicals.

It is well known that the mechanism of the combined H_2 evolution and organic material conversion in nanocomposites consisting of more than one semiconductor mainly occurs through the type II heterostructure or Z-scheme mechanism. To elucidate the right pathway, the conduction and valence band potentials of the semiconductors should be determined and compared with E^o $(H^+/H_2) = 0.00$ eV for HER. Concerning this issue, both type II heterostructure and Z-scheme mechanism for $Bi_7O_9I_3/B_4C$ heterojunction were studied and depicted in Figure 2.4 [11]. The result showed that in the Z-scheme mechanism, the accumulated photoelectrons on B_4C are more negative than the proton reduction and enable H_2 production. However, in the case of type II heterojunction, the remaining photoelectrons on the $Bi_7O_9I_3$ conduction band do not have the ability to reduce protons to H_2. Therefore, the Z-scheme mechanism is the dominant mechanism.

The rate of H_2 generation via organic pollutants degradation is also affected by the photocatalyst composition and the characteristics of the organic pollutants. Shen et al. engineered the composition and percentage of compounds in 2D/2D/3D $Cu_2O/RGO/BiVO_4$ and the 7/3 mass ratio of $Cu_2O/BiVO_4$ demonstrated the best photocatalytic performance for simultaneous tetracycline (TC) degradation and H_2 production [12]. The related photocatalytic process is illustrated in Figure 2.5,

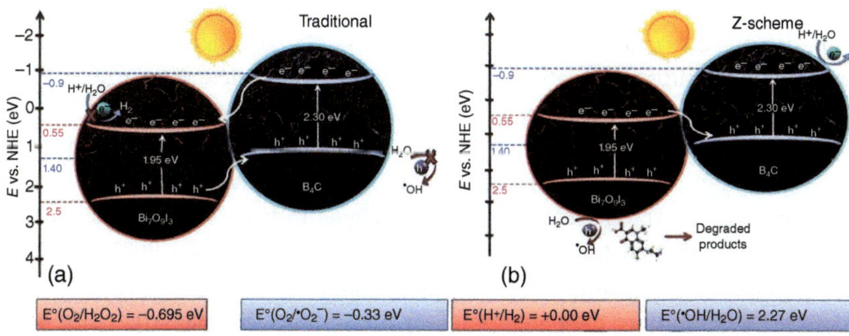

Figure 2.4 Comparison of the separation mechanisms of photogenerated electron hole on the $Bi_7O_9I_3/B_4C$ heterojunction. (a) Type II heterojunction, (b) Z-scheme mechanism. Source: Reproduced with permission from Rana et al. [11]/Elsevier.

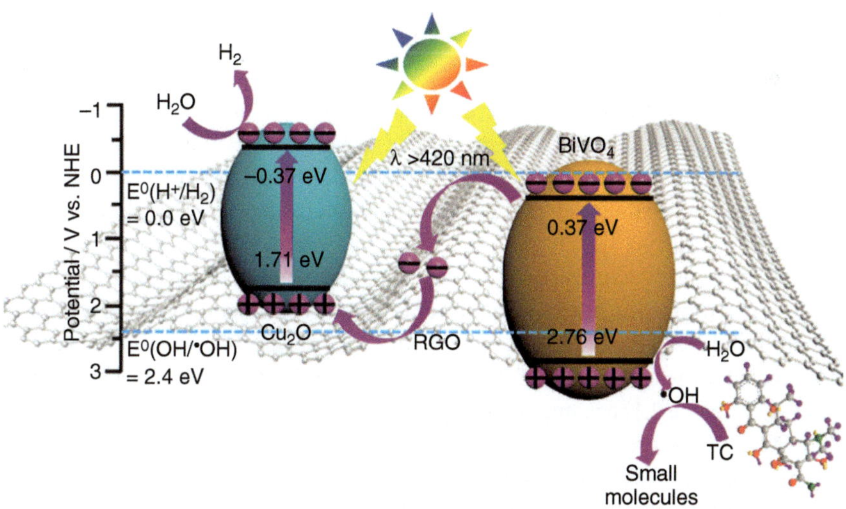

Figure 2.5 The photocatalytic process on the dual-functional characteristics of $Cu_2O/RGO/BiVO_4$ composites for H_2 production and tetracycline degradation. Source: Reproduced with permission from Shen et al. [12]/Elsevier.

which shows the Z-scheme configuration and RGO as an electron mediator [13] for efficient electron–hole recombination. The remaining electrons on the Cu_2O enforce H_2 evolution, and photogenerated holes in the $BiVO_4$ decompose TC.

The role of the cocatalyst in the modification of photocatalysts for enhanced H_2 generation is well established. The utilization of cocatalysts can facilitate photogenerated electron–hole separation due to their electron sink function. Also, it can reduce the activation energy for H^+ to H_2 conversion on the photocatalyst surface [7]. Because of the efficiency of noble metals as cocatalysts, they have been mainly reported. Among various cocatalysts, Pt has demonstrated a tremendous attraction for H_2 evolution due to its several advantages, including low overpotential for the reaction; suitable Fermi level for withdrawing the photoexcited electrons; and the ability for effective charge separation and providing several active surface sites for the photocatalytic reaction [14]. In addition, Pt has a large work function; thus, it has a strong capability to trap electrons from the CB of a semiconductor photocatalyst. Therefore, it could separate photogenerated charge carriers efficiently and, as a result, it could enhance photocatalytic H_2 evolution [15]. Wang et al. synthesized Pt/g-C_3N_4 through the photo-deposition method and studied the effect of Pt size on the conversion of biomass-derived benzaldehyde to benzoic acid and H_2 fuel [16]. The results demonstrated the order of activities as follows Pt single atom (0.2 nm) > nanocluster (1 nm) > nanoparticle (4 nm diameter) > nanoparticle (7 nm diameter). Based on their data analysis, they have found that the larger size of Pt can promote the recombination of both photon-induced charge carriers, which decreases the photocatalytic efficiency. Another determining factor in utilizing metals as a cocatalyst is the amount of deposited nanostructure. Yao et al. decorated 0.5, 1, and 2 wt% Au nanoparticles on the surface of sheaf-like TiO_2. The optimal amount of 1 wt% for coproduction of hydrogen and ibuprofen degradation showed that higher Au content enforced the recombination of photogenerated electron hole and deteriorated the efficiency [17].

Despite the advantages of Pt as a cocatalyst, it is a rare and expensive element; hence, researchers prefer to utilize other cheap and earth abundant cocatalysts for the broad application of photocatalysis. Regarding this, utilizing noble-metal-free cocatalysts has received a lot of interest recently for photocatalytic H_2 evolution [18].

Using 2D nanostructures, such as metal chalcogenides, MXenes, and carbon nitrides are a promising alternative for noble metal cocatalyst to produce hydrogen efficiently and cost-effectively. Recently, Wang et al. prepared CdS nanorods/$Ti_3C_2T_x$ nanosheets for the conversion of furfuryl alcohol (FOL) to furfural (FAL) and furic acid (FA) (Scheme 2.1) accompanying hydrogen production under a blue LED lamp [19]. FAL and FA are valuable chemicals to produce a high-impact platform for fuels, pharmaceuticals, and plastics. CdS with a 2.4 eV band gap has photocatalytic activity under visible light. Moreover, 2D $Ti_3C_2T_x$ nanosheets, known as MXenes, have excellent potential for utilization as a cocatalyst for photocatalytic hydrogen evolution. Different amounts of $Ti_3C_2T_x$ (X = 0.25, 0.50, and 1.00 wt%) were loaded on CdS nanorods with optimized activity at 0.50 wt%, which resulted in 90.2% conversion of FOL and 62.0 μmol evolution of hydrogen. In the $Ti_3C_2T_x$/CdS nanocomposite, the dehydrogenation of FOL was conducted via photogenerated holes on the CdS and hydrogen evolution occurred via transformed photoelectrons

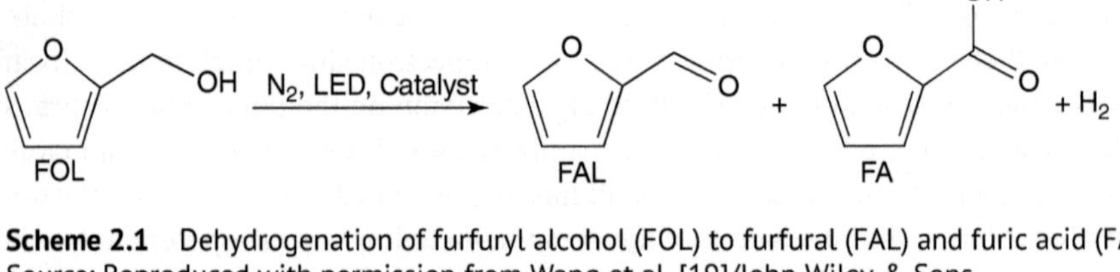

Scheme 2.1 Dehydrogenation of furfuryl alcohol (FOL) to furfural (FAL) and furic acid (FA). Source: Reproduced with permission from Wang et al. [19]/John Wiley & Sons.

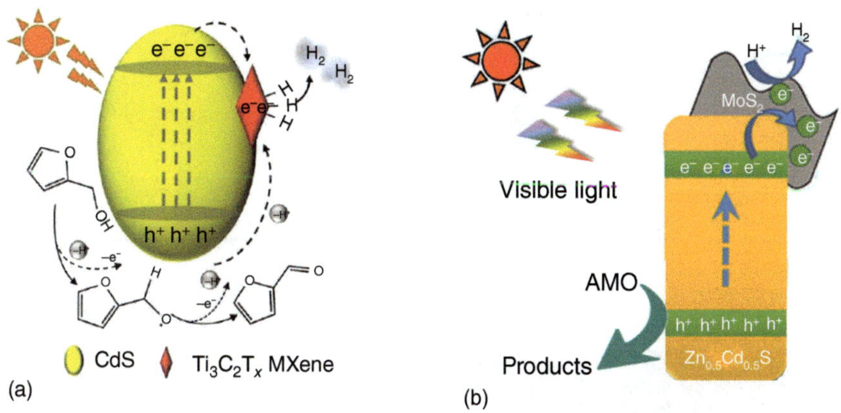

Figure 2.6 (a) Photocatalytic conversion of furfuryl alcohol (FOL) to furfural (FAL) and furic acid (FA) with simultaneous hydrogen generation over $Ti_3C_2T_x$/CdS nanocomposite under visible light. Source: Reproduced with permission from Wang et al. [19]/John Wiley & Sons. (b) Simultaneous amoxicillin (AMO) degradation and hydrogen production over 8%MoS_2@$Zn_{0.5}Cd_{0.5}S$. Source: Reproduced with permission from Samadi et al. [22]/Elsevier.

on the $Ti_3C_2T_x$ [20] (Figure 2.6a). Therefore, the performance of 2D MXenes as cocatalysts can improve the separation of electrons and holes, leading to higher photocatalytic H_2 evolution.

In addition to Mxenes, other 2D transition metal dichalcogenides like MoS_2, WS_2, $MoSe_2$, and WSe_2 have been utilized as cocatalyst [21]. To elucidate the role of MoS_2 as a cocatalyst, different amounts were loaded on $Zn_xCd_{1-x}S$ to prepare y%MoS_2@$Zn_xCd_{1-x}S$ ($y = 0, 5, 8, 10, 15$) [22]. The best result for simultaneous amoxicillin (AMO) degradation and hydrogen production was reported for 8%MoS_2@$Zn_{0.5}Cd_{0.5}S$. Higher amounts of MoS_2 could block light absorption by $Zn_xCd_{1-x}S$ and thus decrease the photocatalytic efficiency. As shown in Figure 2.6b, the role of MoS_2 is attributed to the efficient separation of photogenerated electron–hole pairs, and the hydrogen is generated on the MoS_2 cocatalyst.

2.3.2 Organic Substrate Type

It is well known that the presence of sacrificial agents (here, organic pollutants including dyes, drugs, and biomass) can increase the HER efficiency. For example, H_2 production over the Pt/TiO_2 photocatalyst increases initially in the early hours in the presence of acid orange 7 (AO7), basic blue 41, and basic red 46 azo dyes [23].

Regarding, Yan et al. investigated the dual-purpose performance of resorcinol (RC), TC, and bisphenol A (BPA) on urchin-like oxygen doped MoS_2/$ZnIn_2S_4$ (OMS/ZIS) composite during simultaneous photodegradation and H_2 production [24]. Figure 2.7a shows that BPA has superior hydrogen production compared to the others. Utilizing density functional theory (DFT) to calculate the energy levels of LUMO and HOMO for TC, RC, and BPA shows that the higher HOMO energy level of BPA is more favorable for its direct oxidation via photogenerated holes. Moreover, more significant adsorption of TC on the photocatalyst surface can result

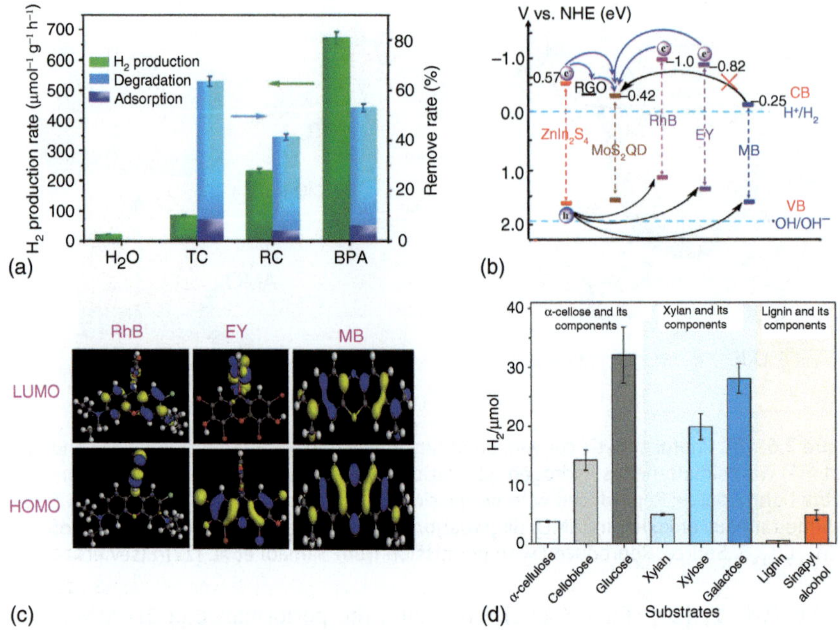

Figure 2.7 (a) Photocatalytic H_2 production in the presence of water, tetracycline (TC), resorcinol (RC), and bisphenol A (BPA) on urchin-like oxygen doped $MoS_2/ZnIn_2S_4$ after four hours of reaction. Source: Reproduced with permission from Yan et al. [24]/Springer Nature. (b) The mechanism of charge carrier separation, including the band structure diagram of MoS_2 QDs@$ZnIn_2S_4$@RGO and the HOMO and LUMO energies of rhodamine B (RhB), eosin Y (EY), and methylene blue (MB). Source: Reproduced with permission from Zhang et al. [25]/Elsevier. (c) Frontier electron densities of LUMO and HOMO of RhB, EY, MB. Source: Reproduced with permission from Zhang et al. [25]/Elsevier. (d) Photocatalytic H_2 production in the presence of different lignocellulose components over cyanamide-functionalized carbon nitride under AM 1.5G irradiation for 24 hours at 25 °C. Source: Reproduced with permission from Kasap et al. [26]/American Chemical Society.

in a higher degradation rate of TC. But it can still compete for H adsorption, which leads to lower H_2 production.

In the other study, the LUMO and HOMO levels of rhodamine B (RhB), eosin Y (EY), and methylene blue (MB) were calculated to elucidate the mechanism of dye degradation and H_2 production over the MoS_2QDs@$ZnIn_2S_4$@RGO photocatalyst (Figure 2.7b) [25]. A higher amount of H_2 was generated in the presence of RhB, and no H_2 evolution was produced in the presence of MB. As depicted in Figure 2.7b, the accumulated photoelectrons on the MoS_2QDs can convert H^+ to H_2 fuel, but, in the case of MB as a photosensitizer, the generated electrons in its LUMO cannot transfer to the conduction band of MoS_2QDs due to the more positive potential to generate hydrogen. Furthermore, based on the DFT calculations in Figure 2.7c, the HOMO and LUMO of RhB, EY, and MB show the overlapping of the electron cloud in the LUMO and HOMO of MB accelerates its electron and hole recombination, leading to lower hydrogen production.

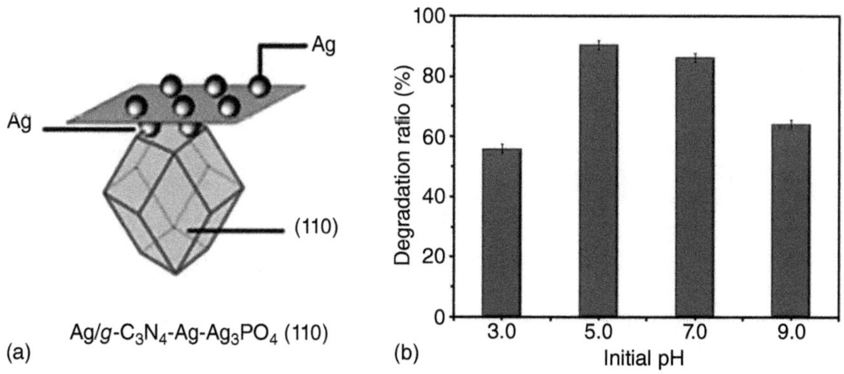

Figure 2.8 (a) Schematic of Ag/*g*-C_3N_4-Ag-Ag_3PO_4 (110) composite structure. (b) The effect of pH values on the photocatalytic degradation of levofloxacin on the Ag/*g*-C_3N_4-Ag-Ag_3PO_4 surface. Source: Reproduced with permission from Li et al. [27]/Elsevier.

Moreover, the effect of different kinds of biomass on photoreforming into H_2 was investigated. Figure 2.7d shows the photocatalytic H_2 generation in the presence of lignocellulose solutions of α-cellulose, cellobiose, glucose, xylan, xylose, galactose, lignin, and sinapyl alcohol over cyanamide-functionalized carbon nitride [26]. The result shows that lignocellulose, which can quench photogenerated holes effectively, demonstrates higher H_2 production.

Furthermore, the type of organic pollutants in H_2 production was investigated. RhB, methyl orange (MO), and levofloxacin (LEV) were compared through photodegradation on the surface of the Ag/g-C_3N_4-Ag-Ag_3PO_4 (110) photocatalyst system (Figure 2.8a). In this study, Li et al. investigated the photodegradation of different organic pollutants and hydrogen production on g-C_3N_4-Ag-Ag_3PO_4 decorated with Ag nanoparticles as a cocatalyst [27]. The order of hydrogen production amounts through LEV, RhB, and MO was 218.87 > 156.33 > 98.24 μmol, respectively. The results showed that the positively charged LEV molecule could react more efficiently with negatively charged photocatalysts compared to RhB and MO; therefore, it led to more hydrogen production. Moreover, they studied the effect of photocatalyst amounts on hydrogen production, demonstrating an optimal value of $1.0\,g\,l^{-1}$. With a higher photocatalyst dosage, a shielding effect was observed, resulting in lower light absorption for the photocatalytic process. In addition, higher H^+ ions are produced by decreasing pH in favor of the HER. But as depicted in Figure 2.8b, there is an optimum pH value regarding LEV photocatalytic degradation. This is due to the surface charge changes of LEV molecules and photocatalysts at different pH values. At pH = 5, LEV and the Ag/g-C_3N_4-Ag-Ag_3PO_4 (110) composite have positive and negative charges, respectively. Thus, this process leads to improved efficient interaction and enhanced photocatalytic degradation.

2.3.3 Reaction Conditions

Different reaction parameters, including the amount of photocatalyst, concentration of organic pollutants, and pH, can influence the reaction rate. These parameters

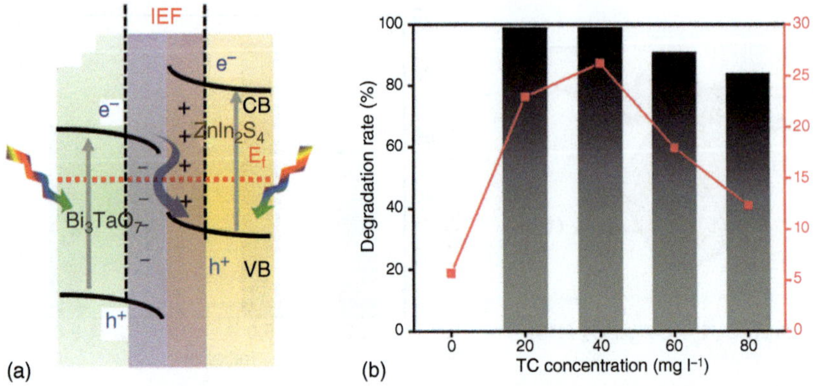

Figure 2.9 (a) The mechanism of electron–hole separation on $Bi_3TaO_7/ZnIn_2S_4$ hybrid with S-scheme heterojunction, internal electric field (IEF) induces charge transfer mechanism of S-scheme heterojunction under solar light irradiation. (b) Photocatalytic hydrogen evolution over $Bi_3TaO_7/ZnIn_2S_4$ hybrid with 1% Pt cocatalyst in tetracycline (TC) solution with different concentrations. Source: Reproduced with permission from Wang et al. [28]/Elsevier.

can influence the H_2 photocatalytic production rate and organic pollutant degradation/conversion activity.

Regarding the organic pollutant amounts for hydrogen production efficiency, Wang et al. examined the degradation of TC with different concentrations on 0D Bi_3TaO_7/3D $ZnIn_2S_4$ nanocomposite, which exhibited an S-scheme mechanism (Figure 2.9a) [28]. Figure 2.9b shows that increasing TC has direct correlation with H_2 evolution amounts up to 40 mg l^{-1}. In the case of higher TC concentrations, they have found that consumption of more photogenerated holes of $ZnIn_2S_4$ leads to disrupting the S-scheme mechanism and lower photocatalytic H_2 production measured on the CB of $ZnIn_2S_4$.

In the other detailed study, to obtain the optimal conditions for formaldehyde (HCHO) conversion and simultaneous H_2 evolution over $ZnIn_2S_4$-NiO/$BiVO_4$ heterojunction, the variation of pH and photocatalyst weight, as well as the concentration of oxygen (O_2) purging, were investigated very recently [29]. The authors analyzed their data and demonstrated that in the absence of O_2 purging into the reaction medium, formaldehyde was polymerized to polyformaldehyde, and no hydrogen evolved. In contrast, at 80% O_2 concentration, the photocatalytic conversion of formaldehyde to formic acid (HCOOH) and H_2 production were obtained. Higher amounts of oxygen covered the surface of the photocatalyst and reduced the conversion efficiency. The mechanism of the photocatalytic process could be a type II heterostructure or Z-scheme mechanism at the interface between $ZnIn_2S_4$ and NiO/$BiVO_4$. Utilizing different scavengers and also the adequate thermodynamic potential for H_2 production, the Z-scheme heterojunction was confirmed as the main route of the reaction, as depicted in Figure 2.10.

Concerning the effect of pH on the degree of photocatalytic conversion of xylose to H_2 and lactic acid, the results showed that the addition of higher KOH concentrations could lead to the generation of more reactive oxidation species and, as a result,

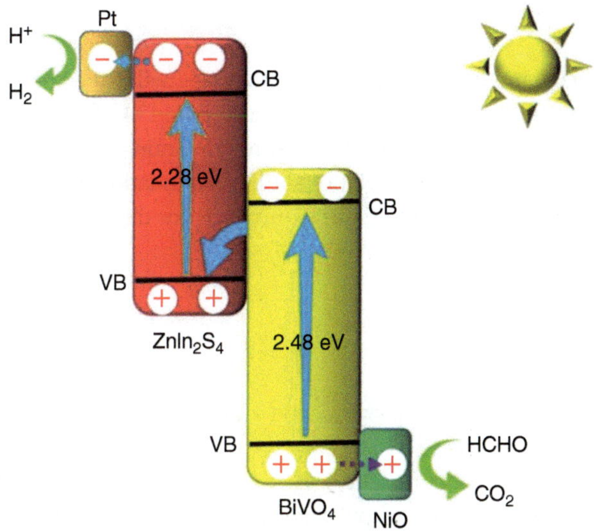

Figure 2.10 Possible direct Z-scheme mechanism for the formaldehyde conversion and simultaneous H_2 evolution over $ZnIn_2S_4$-NiO/$BiVO_4$ heterojunction. Source: Reproduced with permission from Yang et al. [29]/Royal Society of Chemistry.

the efficiency of the reaction was increased [30]. Therefore, due to the diverse effects of acidity and alkalinity in the different studies, it is necessary to further study the effect of pH changes on photodegradation reaction and hydrogen production rate in more detail in the future.

It should be noted that the parameters mentioned above, i.e. photocatalyst design, organic pollutant substance, and reaction conditions, are not independent. Therefore, many researchers have been studying the effects of different variables to achieve the highest photocatalytic efficiency of dual-functional photocatalytic systems with efficient hydrogen production in order to understand the reaction process.

2.4 Conclusions

Simultaneous photocatalytic reactions for H_2 production from a wide range of organic pollutants in water have been a promising and efficient approach to addressing various environmental and energy issues facing our society. Thus, many researchers have recently attempted to design highly efficient dual-functional photocatalysts toward this goal. In this context, heterojunction design, tuning the percentage of the components, and interfacial charge transfer mechanisms has been examined. Furthermore, the effect of the organic substrate type and reaction conditions has also been studied to determine the highest pollutant photodegradation activity and hydrogen production rate. Despite various attempts, there is a need to conduct more investigations to accomplish higher AQE and rate of H_2 production from wastewater treatment under visible-light irradiation on a pilot scale.

Acknowledgment

The authors would like to thank the Research and Technology Council of the Sharif University of Technology. The author Alireza Z. Moshfegh also would like to thanks the Iran National Science Foundation for the support through Grant No. 940009 and the author Amene Naseri appreciates Iran National Elites Foundation for their financial support. Morasae Samadi thanks to Alzahra University for their financial support.

References

1 Eisenstein, M. (2015). Pesticides: seeking answers amid a toxic debate. *Nature* 521 (7552): S52–S55.

2 Naseri, A., Samadi, M., Ebrahimi, M. et al. (2020). Heterogeneous photocatalysis by organic materials: from fundamentals to applications. In: *Current Developments in Photocatalysis and Photocatalytic Materials* (ed. X. Wang, M. Anpo and X. Fu), 457–473. Elsevier.

3 Naseri, A., Samadi, M., Pourjavadi, A. et al. (2017). Graphitic carbon nitride (g-C_3N_4)-based photocatalysts for solar hydrogen generation: recent advances and future development directions. *J. Mater. Chem. A* 5 (45): 23406–23433.

4 Kampouri, S., Nguyen, T.N., Spodaryk, M. et al. (2018). Concurrent photocatalytic hydrogen generation and dye degradation using MIL-125-NH_2 under visible light irradiation. *Adv. Funct. Mater.* 28 (52): 1806368.

5 Akhundi, A., Naseri, A., Abdollahi, N. et al. (2022). Photocatalytic reforming of biomass-derived feedstock to hydrogen production. *Res. Chem. Intermed.* 48: 1793–1811.

6 Qi, M.Y., Conte, M., Anpo, M. et al. (2021). Cooperative coupling of oxidative organic synthesis and hydrogen production over semiconductor-based photocatalysts. *Chem. Rev.* 121 (21): 13051–13085.

7 Naseri, A., Asghari Sarabi, G., Samadi, M. et al. (2022). Recent advances on dual-functional photocatalytic systems for combined removal of hazardous water pollutants and energy generation. *Res. Chem. Intermed.* 48 (3): 911–933.

8 Akhundi, A., Moshfegh, A.Z., Habibi-Yangjeh, A., and Sillanpää, M. (2022). Simultaneous dual-functional photocatalysis by g-C_3N_4-based nanostructures. *ACS ES&T Eng.* 2 (4): 564–585.

9 Jeon, T.H., Koo, M.S., Kim, H., and Choi, W. (2018). Dual-functional photocatalytic and photoelectrocatalytic systems for energy-and resource-recovering water treatment. *ACS Catal.* 8 (12): 11542–11563.

10 Kim, J., Monllor-Satoca, D., and Choi, W. (2012). Simultaneous production of hydrogen with the degradation of organic pollutants using TiO_2 photocatalyst modified with dual surface components. *Energy Environ. Sci.* 5 (6): 7647–7656.

11 Rana, A., Kumar, A., Sharma, G. et al. (2021). Pharmaceutical pollutant as sacrificial agent for sustainable synergistic water treatment and hydrogen production

via novel Z-scheme $Bi_7O_9I_3/B_4C$ heterojunction photocatalysts. *J. Mol. Liq.* 343: 117652.

12 Shen, H., Wang, M., Zhang, X. et al. (2020). 2D/2D/3D architecture Z-scheme system for simultaneous H_2 generation and antibiotic degradation. *Fuel* 280: 118618.

13 Liu, Y., Naseri, A., Li, T. et al. (2022). Shape-controlled photochemical synthesis of noble metal nanocrystals based on reduced graphene oxide. *ACS Appl. Mater. Interfaces* 14 (14): 16527–16537.

14 Kampouri, S. and Stylianou, K.C. (2019). Dual-functional photocatalysis for simultaneous hydrogen production and oxidation of organic substances. *ACS Catal.* 9 (5): 4247–4270.

15 Zhu, Y., Wang, T., Xu, T. et al. (2019). Size effect of Pt co-catalyst on photocatalytic efficiency of g-C_3N_4 for hydrogen evolution. *Appl. Surf. Sci.* 464: 36–42.

16 Wang, L., Tang, R., Kheradmand, A. et al. (2021). Enhanced solar-driven benzaldehyde oxidation with simultaneous hydrogen production on Pt single-atom catalyst. *Appl. Catal. B Environ.* 284: 119759.

17 Yao, X., Hu, X., Liu, Y. et al. (2020). Simultaneous photocatalytic degradation of ibuprofen and H_2 evolution over Au/sheaf-like TiO_2 mesocrystals. *Chemosphere* 261: 127759.

18 Deng, F., Zou, J.-P., Zhao, L.-N. et al. (2019). Nanomaterial-based photocatalytic hydrogen production. In: *Nanomaterials for the Removal of Pollutants and Resource Reutilization*, 59–82. Elsevier.

19 Wang, J., Liu, X., and Li, Z. (2021). Acceptorless photocatalytic dehydrogenation of furfuryl alcohol (FOL) to furfural (FAL) and furoic acid (FA) over $Ti_3C_2T_x$/CdS under visible light. *Chem. Asian J.* 16 (19): 2932–2938.

20 Huang, K., Li, C., Li, H. et al. (2020). Photocatalytic applications of two-dimensional Ti_3C_2 MXenes: a review. *ACS Appl. Nano Mater.* 3 (10): 9581–9603.

21 Samadi, M., Sarikhani, N., Zirak, M. et al. (2018). Group 6 transition metal dichalcogenide nanomaterials: synthesis, applications, and future perspectives. *Nanoscale Horiz.* 3 (2): 90–204.

22 Wei, Z., Xu, M., Liu, J. et al. (2020). Simultaneous visible-light-induced hydrogen production enhancement and antibiotic wastewater degradation using $MoS_2@Zn_xCd_{1-x}S$: solid-solution-assisted photocatalysis. *Chin. J. Catal.* 41 (1): 103–113.

23 Patsoura, A., Kondarides, D.I., and Verykios, X.E. (2006). Enhancement of photoinduced hydrogen production from irradiated Pt/TiO_2 suspensions with simultaneous degradation of azo-dyes. *Appl. Catal. B Environ.* 64 (3): 171–179.

24 Yan, T., Yang, Q., Feng, R. et al. (2022). Highly effective visible-photocatalytic hydrogen evolution and simultaneous organic pollutant degradation over an urchin-like oxygen-doped $MoS_2/ZnIn_2S_4$ composite. *Front. Environ. Sci. Eng.* 16 (10): 131.

25 Zhang, S., Wang, L., Liu, C. et al. (2017). Photocatalytic wastewater purification with simultaneous hydrogen production using MoS_2 QD-decorated hierarchical assembly of $ZnIn_2S_4$ on reduced graphene oxide photocatalyst. *Water Res.* 121: 11–19.

26 Kasap, H., Achilleos, D.S., Huang, A., and Reisner, E. (2018). Photoreforming of lignocellulose into H_2 using nanoengineered carbon nitride under benign conditions. *J. Am. Chem. Soc.* 140 (37): 11604–11607.

27 Li, S., Zhang, M., Qu, Z. et al. (2020). Fabrication of highly active Z-scheme Ag/g-C_3N_4-Ag-Ag_3PO_4 (110) photocatalyst for visible light photocatalytic degradation of levofloxacin with simultaneous hydrogen production. *Chem. Eng. J.* 382: 122394.

28 Wang, K., Shao, X., Zhang, K. et al. (2022). 0D/3D Bi_3TaO_7/$ZnIn_2S_4$ heterojunction photocatalyst towards the degradation of antibiotics coupled with simultaneous H_2 evolution: in situ irradiated XPS investigation and S-scheme mechanism insight. *Appl. Surf. Sci.* 596: 153444.

29 Yang, R., Chen, Q., Ma, Y. et al. (2021). Highly efficient photocatalytic hydrogen evolution and simultaneous formaldehyde degradation over Z-scheme $ZnIn_2S_4$-NiO/$BiVO_4$ hierarchical heterojunction under visible light irradiation. *Chem. Eng. J.* 423: 130164.

30 Yang, X., Ma, J., Sun, S. et al. (2022). K/O co-doping and introduction of cyano groups in polymeric carbon nitride towards efficient simultaneous solar photocatalytic water splitting and biorefineries. *Green Chem.* 24 (5): 2104–2113.

3

Selective Oxidation of Alcohols Using Carbon Nitride Photocatalysts

Oleksandr Savateev

Max Planck Institute of Colloids and Interfaces, Department of Colloid Chemistry, Am Mühlenberg 1, Potsdam 14476, Germany

3.1 Introduction

In 1962, Schwab et al. described the oxidation of methanol in the gas phase with O_2 over UV-sensitized ZnO [1]. It may therefore be considered the first reported example of alcohol oxidation mediated by semiconductor photocatalysis. Eventually, the oxidation of alcohols to carbonyl compounds became a benchmark reaction that has been used for more than 50 years to evaluate the performance of new materials in photocatalysis.

Despite the relative simplicity of alcohol oxidation to carbonyl compounds – it requires a light source, a photocatalyst, and O_2, or alternatively, a cocatalyst capable of facilitating the evolution of H_2, the photocatalytic approach holds promise as a convenient and atom-efficient method in preparative organic chemistry (Figure 3.1).

Indeed, selective oxidation of alcohols to carbonyl compounds on a preparative scale in general requires an excess of chemical oxidants, such as pyrridinium chlorochromate, iodosobenzene diacetate, and TEMPO, which are not only converted into chemical waste, decreasing the overall atom efficiency of the process, but also complicating purification of the target compound (Figure 3.1). Therefore, by utilizing one of the most abundant oxidants, O_2, the sustainability of carbonyl compound synthesis from alcohols may be greatly improved. Nevertheless, challenges, such as low selectivity in the case of electron-rich substrates and substrates bearing multiple functional groups, remain.

Due to a number of features, among which are suitable potentials of the conduction and valence bands, high chemical stability, and weakly basic characteristics that originate from the conjugated structure composed of alternating C=N and C—N bonds, heptazine-based graphitic carbon nitrides have been studied in visible-light organic photocatalysis, including the synthesis of carbonyl compounds from alcohols [2]. In 2010, Wang and coworkers reported the first method of benzylic and aliphatic alcohol oxidation mediated by mesoporous graphitic carbon nitride (mpg-CN) upon illumination with $\lambda > 420\,\text{nm}$ [3].

UV-Visible Photocatalysis for Clean Energy Production and Pollution Remediation: Materials, Reaction Mechanisms, and Applications, First Edition. Edited by Xinchen Wang, Masakazu Anpo, and Xianzhi Fu.
© 2023 WILEY-VCH GmbH. Published 2023 by WILEY-VCH GmbH.

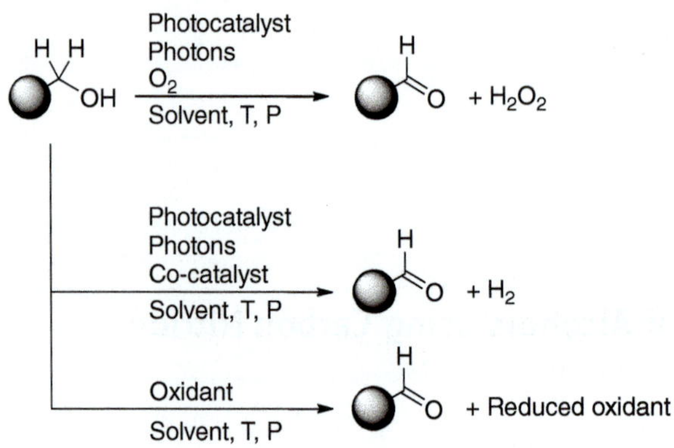

Figure 3.1 Oxidation of primary alcohols to aldehydes is performed photocatalytically and using chemical oxidants.

In this chapter, graphitic carbon nitride materials that were employed in the selective oxidation of alcohols are highlighted. The mechanism of alcohol oxidation to carbonyl compounds and side processes that decrease the selectivity of this reaction are discussed. The approaches employed in carbon nitride photocatalysis to suppress side reactions and improve selectivity toward carbonyl compounds are discussed.

3.2 Heptazine-Based Graphitic Carbon Nitrides

In general terms, carbon nitrides are materials composed of alternating carbon and nitrogen atoms. Depending on their local geometry, periodic structures may be divided into β-carbon nitride and graphitic carbon nitrides. In β-carbon nitride, C is sp^3-hybridized (tetrahedral) and N is sp^2-hybridized (trigonal planar). In graphitic carbon nitrides, both C and N atoms are trigonal planar. Given that the valence of

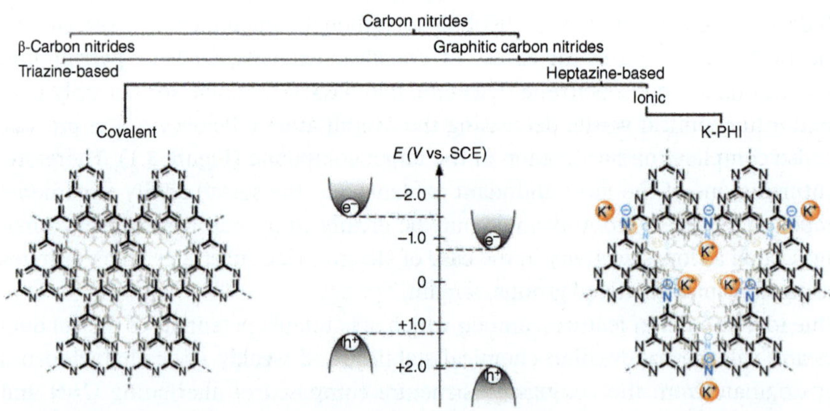

Figure 3.2 Ideal structure of covalent and ionic carbon nitrides represented by perfectly condensed heptazine-based graphitic carbon nitride (left) and K-PHI (right). Potentials of valence and conduction bands in mpg-CN and K-PHI vs. SCE are shown on the left and right sides of the scale respectively. NHE scale was converted to SCE scale according to the equation: $E_{SCE} = E_{NHE} - 0.25$. [5].

carbon is 4, whereas nitrogen has a valence of 3, in graphitic carbon nitrides, C and N atoms may only be linked in a way to create a network of either 1,3,5-triazine or heptazine units [4]. Furthermore, heptazine-based graphitic carbon nitrides may be divided into covalent structures and ionic structures. Covalent graphitic carbon nitrides contain only polar covalent C—N and C=N bonds. The bespoken mpg-CN may thus be assigned to covalent structures. Ionic carbon nitrides contain in their structure deprotonated nitrogen atoms, such as bridging imide groups between neighboring heptazine units (Figure 3.2). In order to maintain electric neutrality, the excessive negative charge is compensated for by alkali metal cations. Ionic carbon nitrides include potassium poly(heptazine imide) (K-PHI, shown in Figure 3.2) [6, 7] and cyanimide-functionalized graphitic carbon nitride (NCN-CNx) [8].

Potentials of valence and conduction bands in mpg-CN (+1.15 and −1.55 V vs. SCE, respectively) [3] and K-PHI (+1.95 and −0.75 V vs. SCE) [6] are shown in Figure 3.2, which define primary optical gap of 2.7 eV.

3.3 Mechanism of Alcohols Oxidation by Carbon Nitrides

Benzyl alcohol is often used as a model substrate, with O_2 as an electron acceptor. In carbon nitride photocatalysis, the products of their interaction are benzaldehyde and hydrogen peroxide [3]. Thermochemical calculations suggest that this reaction is exergonic, in other words, spontaneous, and characterized by the Gibbs free energy change under the standard conditions (ΔG^o) of −22.9 kcal mol^{-1} [9]. However, as indicated by numerous control experiments, the reaction does not proceed at a significant rate in the dark and in the absence of a photocatalyst, which is due to high activation energy. A photocatalyst absorbs electromagnetic radiation and utilizes the energy of incident photons stored temporarily in the form of electron-hole pairs to overcome the activation barrier. In other words, carbon nitride mediates the flow of electrons (and protons) from alcohol to oxygen upon excitation with visible photons. Taking into account reduction potential of O_2 to superoxide radical of −0.89 V vs. SCE in acetonitrile [6], thermodynamic driving force for a photoinduced electron transfer (PET) from the photoexcited mpg-CN conduction band to the electron acceptor is −0.66 eV and as such feasible (step b in Figure 3.3). Superoxide radical is registered as a DMPO-O_2H adduct by EPR spectroscopy [3].

Similar analysis is also applicable to another elementary step – one-electron oxidation of benzyl alcohol. Given that oxidation potential of benzyl alcohol is > +2.2 V vs. SCE in acetonitrile [10], thermodynamic driving force for PET from the substrate to the valence band of the photoexcited mpg-CN is 1.05 eV, which renders direct oxidation of the substrate very challenging. Therefore, in a series of elementary steps, the role of mpg-CN as the photocatalyst is to activate oxygen to $O_2^{\cdot-}$, which in turn facilitates oxidation of benzyl alcohol.

In step c, $O_2^{\cdot-}$ abstracts a proton from benzyl alcohol, which gives $HO_2^{\cdot}$ and $PhCH_2O^-$, while the negative charge is compensated by the hole resting in the carbon nitride [3]. The resultant alkoxide is a stronger reductant compared to benzyl alcohol. Therefore, its oxidation to the alkoxyl radical by the hole remaining

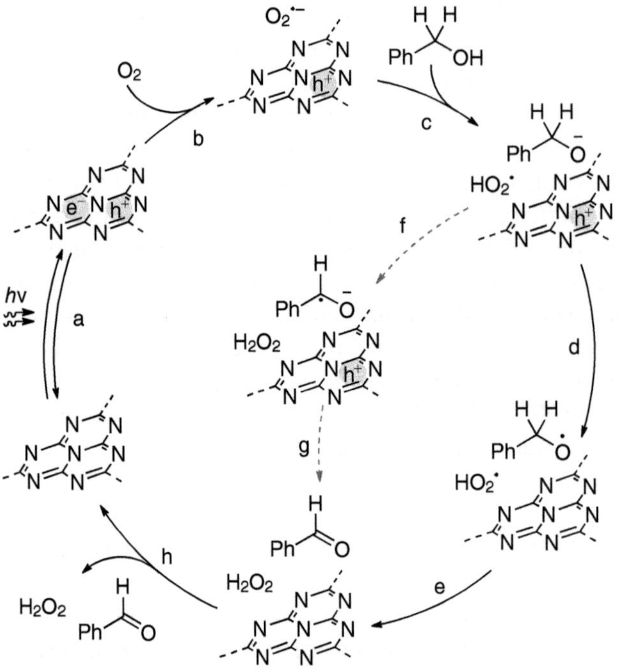

Figure 3.3 A schematic mechanism of benzyl alcohol oxidation by mpg-CN photocatalyst using O_2 as the oxidant.

in the mpg-CN is more feasible (step d). Alternatively, steps c–d could merge into a single proton-coupled electron transfer (PCET) reaction – electron is transferred to the mpg-CN, while proton to $O_2^{\cdot-}$. Hydrogen abstraction by hydroperoxyl radical gives surface-bound benzaldehyde and H_2O_2 (step e).

Although it has been discussed in the literature [3], abstraction of hydrogen atoms by $HO_2^{\cdot}$ from $PhCH_2O-$ is less thermodynamically feasible because it produces a highly reactive radical anion with an oxidation potential of −1.93 V vs. SCE (step f) [11].

Desorption of benzaldehyde and hydrogen peroxide from the surface of carbon nitride, which is typically a weakly endergonic process, recovers the photocatalyst (step h). Using deuterium labeled benzyl alcohol, it was demonstrated that the abstraction of benzylic hydrogen atom in step e (or step f) is rate limiting [3].

In this reaction, cyclopropylphenyl carbinol, the radical clock, gave cyclopropylphenyl ketone, which implies the reaction will proceed via steps d–e rather than f–g. On the other hand, it could mean that the reaction in step g proceeds much faster than the cyclopropyl ring opening ($k \sim 10^6\ s^{-1}$) [12].

The formation of one equivalent H_2O_2 per benzaldehyde molecule is required by the stoichiometry of this reaction. However, the amount of H_2O_2 is always lower compared to the aldehyde, which is explained by the fact that H_2O_2 itself acts as an oxidant and, therefore, it is consumed, at least partially, in the reaction [13]. As will be shown below, the high reactivity of H_2O_2 generated *in situ* upon photocatalytic oxidation of primary alcohols is the origin of the low selectivity toward carbonyl compounds.

The scope of alcohols is represented mainly by molecules possessing benzylic moieties, bearing electron donating and electron withdrawing groups. For a comprehensive list of substrates and carbon nitride-based photocatalysts, readers are suggested to refer to a review in Savateev and Antonietti [14]. The larger scope of benzylic substrates compared to aliphatic ones reported in literature is likely due to the stabilizing effect of the aromatic ring in the structure of the intermediates, which facilitates this reaction. When aliphatic and benzylic alcohol moieties are present in one substrate, oxidation proceeds selectively at the benzylic site. Alternatively, when two reagents are added into the reactor, benzylic alcohol is oxidized exclusively to carbonyl compounds. Therefore, intrinsically higher reactivity of benzylic alcohols compared to aliphatic ones allows for selective oxidation of this type of substrates. Examples of more complex molecules bearing multiple functional groups are scarce.

Hydrogen peroxide is characterized by low, 43.7 kcal mol^{-1} (or 1.9 eV) [15], O—O bond dissociation free energy (BDFE, ΔG). Homolytic cleavage of O—O bond in H_2O_2 generates two reactive $HO^{\cdot}$ radicals. Alternatively, one-electron reduction of H_2O_2 by the photogenerated electrons in carbon nitride produces HO^- and $HO^{\cdot}$ radicals. $HO^{\cdot}$ radical is capable of abstracting hydrogen atoms from other organic molecules in solution and, as a result, initiating chain radical reactions. In Figure 3.4, radical $X^{\cdot}$ derived from X–H compounds standing on the right side (higher BDFE value) can abstract H· from all compounds standing on the left (lower BDFE value).

For example, carbonyl C-H BDFE in aldehydes is 81.3 kcal mol^{-1} [17]. Therefore, $HO^{\cdot}$ radical is capable of abstracting hydrogen atoms from benzaldehyde and generating an acyl radical, which, followed by trapping O_2, gives perbenzoic acid. Homolytic cleavage of the O—O bond in perbenzoic acid followed by hydrogen atom transfer gives benzoic acid – a major by-product in photocatalytic oxidation of benzyl alcohol using O_2.

When the reaction is carried out in methanol, the carbon nitride photocatalyst, which is rich in acidic sites, selectively oxidizes benzyl alcohol to methyl benzoate [18]. In this case, photocatalytic oxidation of hemiacetal as the intermediate was postulated. Reported external quantum efficiency (EQE) of benzyl alcohol oxidation to either benzaldehyde or methylbenzoate over various carbon nitride photocatalysts are presented in Table 3.1.

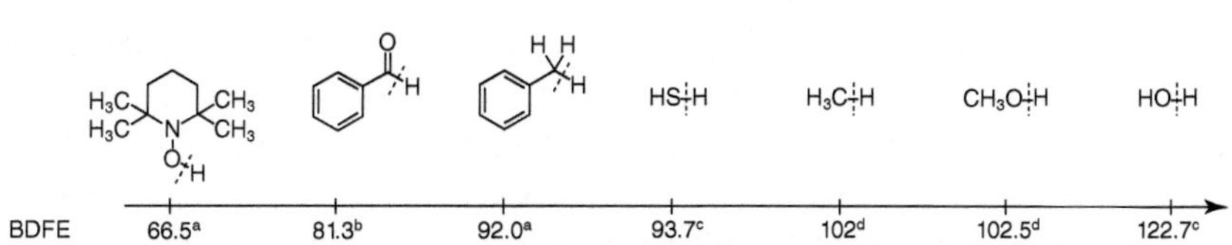

Figure 3.4 X–H BDFE values (kcal mol^{-1}) in organic and inorganic compounds. a – in MeCN; b – in gas phase; c – in water; d – in DMSO. Source: Warren et al. [16]/American Chemical Society.

Table 3.1 EQE of benzyl alcohol photocatalytic oxidation by carbon nitrides using O_2.

Carbon nitride photocatalyst	Product	λ (nm)	EQE (%)	Reference
Carbon nitride thin film	Benzaldehyde	400	1.8	[19]
mpg-CN	Benzaldehyde	470	1.6	[20]
Mesoporous carbon nitride modified with thiophene moieties	Benzaldehyde	420	8.6	[21]
Acidified carbon nitride	Methyl benzoate	420	0.41	[18]

3.4 Improving Selectivity of Alcohols Oxidation

Oxidation of secondary alcohols generates ketones, which, unlike aldehydes, do not possess labile C—H bonds next to the carbonyl group. Therefore, ketones are chemically more stable under the employed photocatalytic conditions, which results in overall higher selectivity of secondary alcohol oxidation. All approaches that were developed to terminate oxidation of *primary* alcohol at the step of aldehyde formation and therefore improve selectivity may be divided into several groups:

- Optimizing reaction time and conversion of alcohol.
- Substituting O_2 by the oxidant that in the photocatalytic cycle is reduced to a product more stable than H_2O_2.
- Using carbon nitride photocatalysis in combination with H_2-evolving catalyst under anaerobic conditions.
- Employing photo-chargeable ionic carbon nitrides under anaerobic conditions.

3.4.1 Optimizing Reaction Time and Conversion of Alcohol

This approach is highly relevant to photocatalytic oxidation of alcohols in water. The choice of water as a solvent in organic photocatalysis is controversial. On the one hand, it is abundant and environmentally friendly. On the other hand, the solubility of organic compounds in water generally is low, which naturally narrows the scope of the substrates. Despite water is characterized by one of the highest O–H BDFE values of 122.7 kcal mol^{-1} [16] photocatalysts with highly positive potential of the valence band, such as TiO_2 (valence band potential +2.7 V vs. NHE [3] or +2.45 V vs. SCE [5]) oxidize HO^- to $HO^\cdot$ radical ($E(HO^-/HO^\cdot) = +1.89$ V vs. NHE [16]). On the other hand, potential of the valence band in mpg-CN, +1.15 V vs. SCE, does not permit oxidation of HO^- to $HO^\cdot$ radical. From this standpoint, carbon nitride photocatalysis minimizes side reactions and therefore improves selectivity by preventing the generation of excessive amounts of $HO^\cdot$ radicals in the reaction medium. Nevertheless, as shown in Figure 3.5, selectivity toward 4-methoxybenzaldehyde gradually decreases as conversion of 4-methoxybenzylalcohol approaches 100%. The yield of aldehyde in an acidic environment is higher compared to an alkaline environment, which is

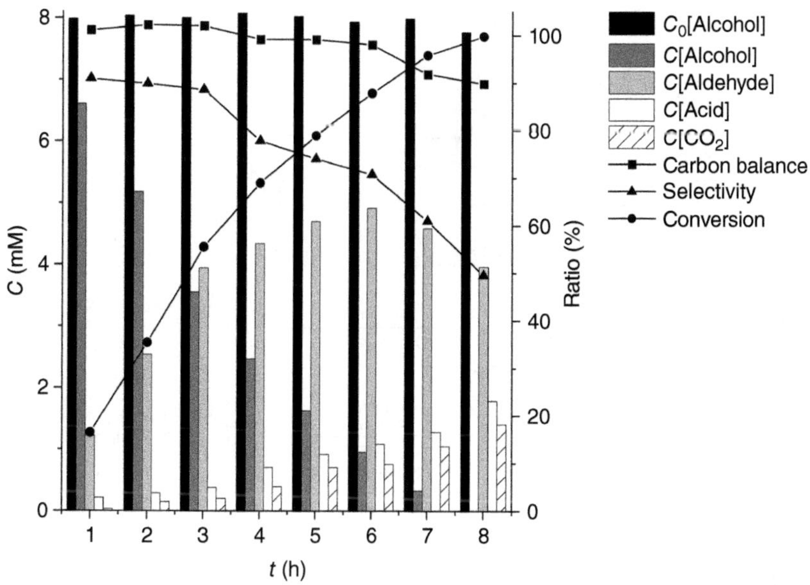

Figure 3.5 Kinetic profile of 4-methoxybenzyl alcohol oxidation in water mediated by mpg-CN in the presence of O_2. Conditions: mpg-CN (50 mg); 4-methoxybenzyl alcohol (0.8 mmol); water (20 ml); light ($\lambda > 420$ nm); O_2 (0.8 MPa); T 60 °C. Source: Reproduced with permission from Long et al. [13]/John Wiley & Sons.

explained by the enhanced rate of PCET from mpg-CN to O_2 (step c in Figure 3.3). When the reaction is carried out in 1M HCl, mpg-CN can oxidize aliphatic alcohols such as cyclohexanol and diethylcarbinol to the corresponding ketones [13].

Flow technology complements organic photochemistry greatly [22]. Bajada et al. designed a triphasic (gas–liquid–solid) packed column photoreactor with mpg-CN immobilized inside the transparent fluorinated ethylene propylene tube ($L = 75$ cm, I.D. = 2 mm) [20]. A solution of alcohol in acetonitrile was divided into subunits by means of an O_2 gas spacer and passed through the tube (Figure 3.6a). Using 1-naphthalenemethanol as the substrate, the corresponding aldehyde was obtained with 90% selectivity at complete conversion of the substrate when the residence time was set to 90 minutes (Figure 3.6b). The developed photoreactor can convert up to 170 μmol of 1-naphthalenemethanol per hour, or 0.64 g per day. The developed method allows conducting the reaction at 55 °C (due to heat generated by the LEDs) instead of 100 °C and maintaining O_2 pressure slightly above 1 bar to compensate for the backpressure compared to 8 bar used in batch [3].

Mazzanti et al. coated common lab glass vials and quartz microfluidic cells with 111 nm carbon nitride thin film using chemical vapor deposition (CVD, Figure 3.6c) and used fabricated photoreactors for oxidation of benzylic alcohols [19]. After 24 hours of reaction mixture illumination in visible batch wall reactor (Vis-BWR), benzoic acid was obtained as the sole product (Figure 3.6d). Using visible flow wall reactor (Vis-FWR) with recirculation of the reaction mixture, benzaldehydes

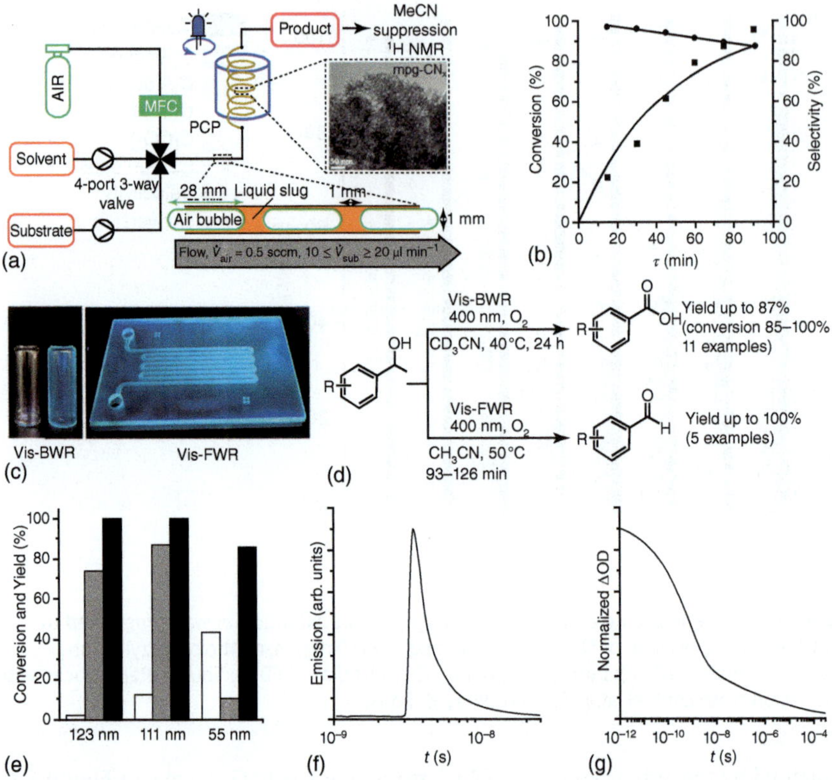

Figure 3.6 Photocatalysis in flow applied to selective oxidation of alcohols by stationary carbon nitrides. (a) Schematic diagram of the triphasic packed column photoreactor (PCP). MFC: mass flow controller. (b) Conversion of benzyl alcohol and selectivity toward benzaldehyde vs. residence time. Source: Bajada et al. [20]/reproduced with permission from American Chemical Society. (c) Photographs of visible batch wall reactor (Vis-BWR) under ambient light and illumination with UV-light (left) and visible (microfluidic) flow wall reactor (Vis-FWR) under UV light (right). (d) Oxidation of benzylic alcohols to acids in Vis-BWR and alcohols to aldehydes in Vis-FWR. (e) Conversion of benzyl alcohol (black bars); yield of benzaldehyde (white bars) and benzoic acid (gray bars) vs. carbon nitride film thickness in Vis-BWR. (f) Time-resolved fluorescence intensity in semilogarithmic scale acquired for 111 nm thick carbon nitride film in Vis-BWR upon excitation at 375 nm. (g) Decay of absorption of 111 nm thick carbon nitride films in Vis-BWR in semilogarithmic scale. Source: Mazzanti et al. [19]/reproduced with permission from American Chemical Society.

were obtained selectively with 86–100% yield by setting the overall residence time to 93–126 minutes. CVD allows controlling the film thickness from 55 to 123 nm, which correlates with the performance of the device (Figure 3.6e). The advantages of the developed approach consist not only in the immobilization of the photocatalyst at the surface of photoreactors and semitransparency. Spectroscopic data indicate that excited state dynamics of carbon nitride thin film is altered, when it is assembled in thin film compared to conventional carbon nitride powder prepared in a crucible. Time-resolved spectroscopy indicates several pathways for carbon nitride excited state relaxation back to the ground state. One pathway is relaxation with emission of photons, characterized by a lifetime 0.6 ns (Figure 3.6f). As indicated by transient absorption spectroscopy, 10% of the excited state population survives after 1 μs (Figure 3.6g). Such rather long lifetime of presumably triplet-excited states is beneficial for energy transfer to generate singlet oxygen – in this case, the active specie in the oxidation of benzylic alcohols.

3.4.2 Substituting O_2 by Other Oxidants

The main by-product of alcohol oxidation discussed in Section 3.4.1 that cannot be easily separated from the reaction mixture is benzoic acid. In order to consider photocatalytic oxidation with O_2 as a viable synthetic procedure, it must yield an aldehyde with 100% selectivity at full conversion of the alcohol. Despite O_2 being the most abundant and sustainable reagent, selectivity of alcohol oxidation is greatly improved while maintaining full conversion of the alcohol by using oxidants which in the photocatalytic cycle are reduced to stable products.

When H_2O_2 is added in exact stoichiometric quantity vs. alcohol, upon illumination under anaerobic conditions it is reduced by carbon nitride to HO^- and $HO^\cdot$. The latter species enables selective oxidation of primary alcohols to aldehydes (Figure 3.7a) [23]. The ultimate product of H_2O_2 reduction, water, is not involved in the downstream chemistry.

Savateev et al. successfully replaced O_2 with α-octaatomic sulfur (S_8) in the synthesis of benzaldehyde from benzyl alcohol by K-PHI photocatalyst (Figure 3.7b) [24]. The reason for high (>99%) selectivity toward benzaldehyde at close to 100% conversion of the alcohol is that the product of S_8 reduction, H_2S, is less reactive compared to H_2O_2. S–H BDFE in H_2S is 93.7 kcal mol^{-1} (Figure 3.4) – significantly higher compared to O–O BDFE in H_2O_2 (43.7 kcal mol^{-1}). Nevertheless, they pointed to a trace amount of the by-product, presumably thiobenzoic acid. To further increase selectivity toward benzaldehyde, ZnO was used as H_2S scavenger.

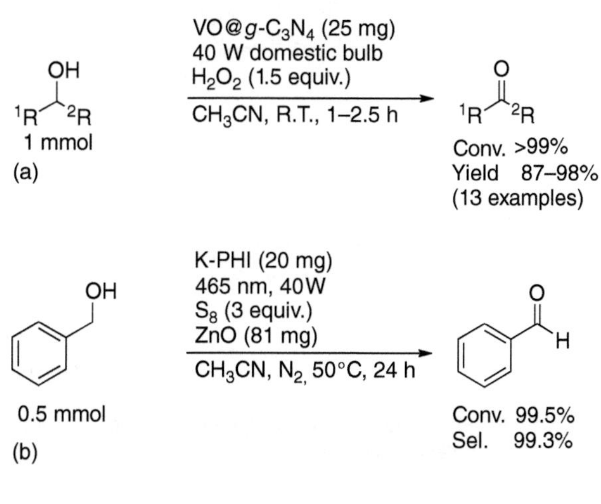

Figure 3.7 Photocatalytic oxidation of alcohols over carbon nitrides using oxidants different from O_2. (a) H_2O_2 and carbon nitride photocatalyst modified with vanadium oxides, VO@*g*-C_3N_4. (b) S_8 and K-PHI.

3.4.3 Combining Carbon Nitride Photocatalyst with H_2 Evolving Catalyst

Under anaerobic conditions (Figure 3.8) [25], Kasap et al. designed and employed a system composed of NCN-CNx and Ni cocatalyst. In this approach, subsequent oxygenation of benzaldehyde to benzoic acid, which otherwise proceeds readily in O_2 atmosphere, is avoided. As a result, two value-added products, benzaldehyde and H_2, are obtained selectively from benzylic alcohols. NiP TOF $31.1 \pm 3.1\ h^{-1}$ was achieved with full-solar spectrum irradiation.

Hydrogen evolution reaction (HER), commonly studied in carbon nitride photocatalysis, requires a sacrificial electron donor, typically triethanolamine or simple alcohols [26]. Under anaerobic conditions, carbon nitride photocatalyst featuring photochemically deposited Pt nanoparticles oxidizes selectively benzyl alcohol to benzaldehyde accompanied by H_2 evolution [27].

3.4.4 Employing Photo-Chargeable Ionic Carbon Nitrides Under Anaerobic Conditions

Band gap excitation of inorganic semiconductors, such as TiO_2, as well as microporous ionic carbon nitrides, K-PHI, in the presence of sacrificial electron donors, such as alcohols, leads to an accumulation of electrons and protons in the material (Figure 3.9) [28].

A molecule of alcohol as the electron donor provides two electrons and two protons to be stored in the semiconductor. The specific capacity (δ_{max}) of

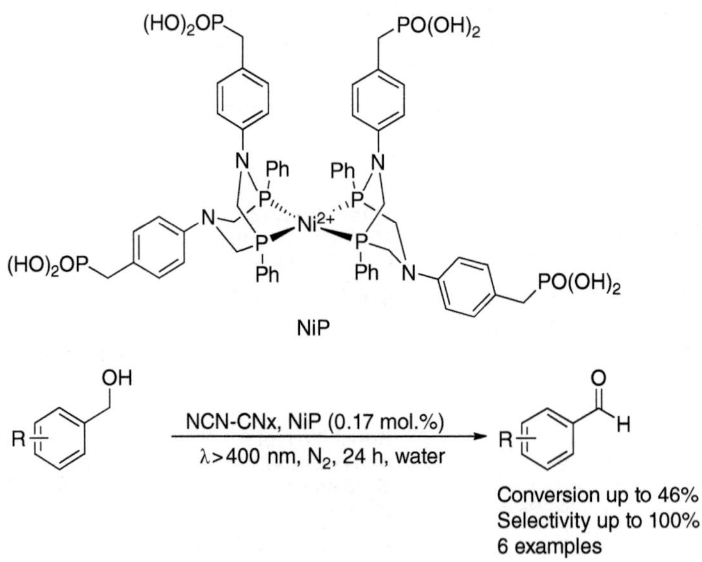

Figure 3.8 Photocatalytic dehydrogenation of benzylic alcohols coupled with H_2 evolution assisted by molecular NiP catalyst.

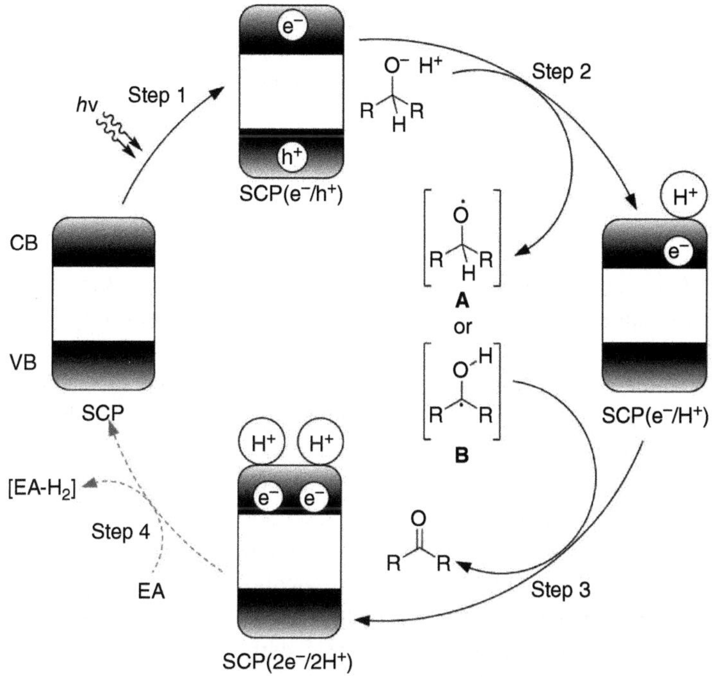

Figure 3.9 A schematic mechanism of semiconductor particle photo-charging using alcohol as donor of electrons and protons. SCP denotes semiconductor particle; CB – conduction band; VB – valence band; h^+ – photo-generated hole; H^+ – proton; EA – electron acceptor (oxidant); $[EA\text{-}H_2]$ – reduced (hydrogenated) electron acceptor. Source: Reproduced with permission from Savateev [28]/John Wiley & Sons.

photo-chargeable materials depends most strongly on the structure of the semiconductor, diameter of particles, type of electron donor, and counter ion, and ranges from 3×10^{-6} to $2 \times 9{\cdot}10^{-4}$ mol[e⁻] per gram of the material [28]. For ionic carbon nitrides δ_{max} reaches few mmol[e⁻] g^{-1} [29]. Therefore, by adjusting the ratio between a semiconductor and an alcohol, it is possible to convert selectively the given amount of alcohol into an aldehyde. For example, in the absence of NiP (Figure 3.8) and upon illumination of the reaction mixture under anaerobic conditions for 24 hours, 5 mg of NCN-CNx selectively produces 7.3 μmol of *p*-tolualdehyde along with a negligible amount of H_2 [25]. Assuming that each 4-methylbenzyl alcohol provides two electrons and two protons, in this experiment, 14.6 μmol of electrons and charge-compensating protons are transferred and stored in the material, while the specific density of electrons in photo-charged NCN-CNx is 2.9 mmol[e⁻] g^{-1}.

By refilling the photoreactor with O_2 in dark, allows production of H_2O_2 (denoted as $[EA\text{-}H_2]$ in Figure 3.9) upon scavenging electrons from the photo-charged carbon nitride [30]. This approach allows for the selective synthesis of benzaldehyde and H_2O_2.The method has the potential to replace or at least mimic industrial anthraquinone processes for H_2O_2 synthesis, though further optimization of reaction conditions is required. It has been observed by many groups that the yield of aldehydes in the control experiments, under anaerobic conditions is nonzero, which could be explained by photo-charging of semiconductors [20]. Nevertheless, to make such conclusion, it is imperative to eliminate even traces of O_2 or other oxidants from the reaction mixture.

3.5 Conclusion

Carbon nitride photocatalysis employs electromagnetic radiation in the visible range as an energy source and O_2 as the most abundant oxidant to convert alcohols to carbonyl compounds. The selectivity of this process is, however, limited by the subsequent overoxidation of aldehyde to carboxylic acid mediated by hydroxyl radicals, which are formed either upon H_2O_2 homolysis or one-electron reduction of H_2O_2 by the photoexcited carbon nitride. Strategies to increase selectivity of alcohol oxidation are primarily built around decreasing the detrimental effect of $HO^{\cdot}$ radicals. These strategies include: (i) limiting conversion of alcohol by decreasing the reaction time in batch and optimizing residence time in flow photoreactor. (ii) Substitution of O_2 with the stoichiometric vs. the alcohol amount H_2O_2, which is reduced to H_2O, or α-octaatomic sulfur, which is reduced to less reactive H_2S. (iii) Combination of carbon nitride photocatalysis with H_2-evolving catalysis, either molecular or Pt nanoparticles. The approach allows for the synthesis of two value-added products – carbonyl compound and H_2. (iv) Coupling of alcohols oxidation with photocharging of semiconductors, which allows decoupling dehydrogenation of alcohol to aldehyde during the light phase from H_2O_2 generation upon purging photocharged ionic carbon nitride with O_2 in dark.

The combination of high-performing carbon nitride photocatalysts with the abovementioned strategies allows for selective synthesis of carbonyl compounds from the corresponding aldehydes. Further research directed toward application of carbon nitride photocatalysis for synthesis of more complex molecules, bearing functional groups and industrially relevant molecules especially under natural outdoor solar light, will facilitate development of a more sustainable approach in synthetic organic chemistry.

References

1 Schwab, G.-M., Noller, H., Steinbach, F., and Venugopalan, M. (1962). Oxidation of carbon monoxide and methyl alcohol photosensitized in the gas phase by zinc oxide. *Nature* 193: 774–775.

2 Savateev, A., Ghosh, I., König, B., and Antonietti, M. (2018). Photoredox catalytic organic transformations using heterogeneous carbon nitrides. *Angew. Chem. Int. Ed.* 57: 15936–15947.

3 Su, F., Mathew, S.C., Lipner, G. et al. (2010). mpg-C_3N_4-catalyzed selective oxidation of alcohols using O_2 and visible light. *J. Am. Chem. Soc.* 132: 16299–16301.

4 Miller, T.S., Jorge, A.B., Suter, T.M. et al. (2017). Carbon nitrides: synthesis and characterization of a new class of functional materials. *Phys. Chem. Chem. Phys.* 19: 15613–15638.

5 Pavlishchuk, V.V. and Addison, A.W. (2000). Conversion constants for redox potentials measured versus different reference electrodes in acetonitrile solutions at 25 °C. *Inorg. Chim. Acta* 298: 97–102.

6 Savateev, A., Tarakina, N.V., Strauss, V. et al. (2020). Potassium poly(heptazine imide): transition metal-free solid-state triplet sensitizer in cascade energy transfer and [3+2]-cycloadditions. *Angew. Chem. Int. Ed.* 59: 15061–15068.

7 Schlomberg, H., Kröger, J., Savasci, G. et al. (2019). Structural insights into poly(heptazine imides): a light-storing carbon nitride material for dark photocatalysis. *Chem. Mater.* 31: 7478–7486.

8 Lau, V.W.-h., Klose, D., Kasap, H. et al. (2017). Dark photocatalysis: storage of solar energy in carbon nitride for time-delayed hydrogen generation. *Angew. Chem. Int. Ed.* 56: 510–514.

9 Thermochemical data were taken from https://webbook.nist.gov/chemistry/.

10 Mayeda, E.A., Miller, L.L., and Wolf, J.F. (1972). Electrooxidation of benzylic ethers, esters, alcohols, and phenyl epoxides. *J. Am. Chem. Soc.* 94: 6812–6816.

11 Roth, H.G., Romero, N.A., and Nicewicz, D.A. (2016). Experimental and calculated electrochemical potentials of common organic molecules for applications to single-electron redox chemistry. *Synlett* 27: 714–723.

12 Bowry, V.W., Lusztyk, J., and Ingold, K.U. (1990). Evidence for reversible ring-opening of the α-cyclopropylbenzyl radical. *J. Chem. Soc., Chem. Commun.* 923–925.

13 Long, B., Ding, Z., and Wang, X. (2013). Carbon nitride for the selective oxidation of aromatic alcohols in water under visible light. *ChemSusChem* 6: 2074–2078.

14 Savateev, A. and Antonietti, M. (2018). Heterogeneous organocatalysis for photoredox chemistry. *ACS Catal.* 8: 9790–9808.

15 Bach, R.D., Ayala, P.Y., and Schlegel, H.B. (1996). A reassessment of the bond dissociation energies of peroxides. An ab initio study. *J. Am. Chem. Soc.* 118: 12758–12765.

16 Warren, J.J., Tronic, T.A., and Mayer, J.M. (2010). Thermochemistry of proton-coupled electron transfer reagents and its implications. *Chem. Rev.* 110: 6961–7001.

17 Simões, J.A.M. and Griller, D. (1989). Enthalpy of formation of the benzoyl radical by photoacoustic calorimetry. *Chem. Phys. Lett.* 158: 175–177.

18 Wang, C., Wan, Q., Cheng, J. et al. (2021). Efficient aerobic oxidation of alcohols to esters by acidified carbon nitride photocatalysts. *J. Catal.* 393: 116–125.

19 Mazzanti, S., Manfredi, G., Barker, A.J. et al. (2021). Carbon nitride thin films as all-in-one technology for photocatalysis. *ACS Catal.* 11: 11109–11116.

20 Bajada, M.A., Vijeta, A., Savateev, A. et al. (2020). Visible-light flow reactor packed with porous carbon nitride for aerobic substrate oxidations. *ACS Appl. Mater. Interfaces* 12: 8176–8182.

21 Chen, Y., Zhang, J., Zhang, M., and Wang, X. (2013). Molecular and textural engineering of conjugated carbon nitride catalysts for selective oxidation of alcohols with visible light. *Chem. Sci.* 4: 3244–3248.

22 Cambié, D., Bottecchia, C., Straathof, N.J.W. et al. (2016). Applications of continuous-flow photochemistry in organic synthesis, material science, and water treatment. *Chem. Rev.* 116: 10276–10341.

23 Verma, S., Baig, R.B.N., Nadagouda, M.N., and Varma, R.S. (2016). Selective oxidation of alcohols using photoactive VO@g-C_3N_4. *ACS Sustain. Chem. Eng.* 4: 1094–1098.

24 Savateev, A., Dontsova, D., Kurpil, B., and Antonietti, M. (2017). Highly crystalline poly(heptazine imides) by mechanochemical synthesis for photooxidation of various organic substrates using an intriguing electron acceptor – elemental sulfur. *J. Catal.* 350: 203–211.

25 Kasap, H., Caputo, C.A., Martindale, B.C.M. et al. (2016). Solar-driven reduction of aqueous protons coupled to selective alcohol oxidation with a carbon nitride–molecular Ni catalyst system. *J. Am. Chem. Soc.* 138: 9183–9192.

26 Wang, X., Maeda, K., Thomas, A. et al. (2009). A metal-free polymeric photocatalyst for hydrogen production from water under visible light. *Nat. Mater.* 8: 76–80.

27 Wang, H., Zhang, J., Jin, X. et al. (2021). General surface grafting strategy-derived carbon-modified graphitic carbon nitride with largely enhanced visible light photocatalytic H_2 evolution coupled with benzyl alcohol oxidation. *J. Mater. Chem. A* 9: 7143–7149.

28 Savateev, O. (2022). Photocharging of semiconductor materials: database, quantitative data analysis, and application in organic synthesis. *Adv. Energy Mater.* 12: 2200352.

29 Markushyna, Y., Lamagni, P., Teutloff, C. et al. (2019). Green radicals of potassium poly(heptazine imide) using light and benzylamine. *J. Mater. Chem. A* 7: 24771–24775.

30 Ou, H., Tang, C., Chen, X. et al. (2019). Solvated electrons for photochemistry syntheses using conjugated carbon nitride polymers. *ACS Catal.* 9: 2949–2955.

4

Application of S-Scheme Heterojunction Photocatalyst

Chuanbiao Bie and Jiaguo Yu

China University of Geosciences, Laboratory of Solar Fuel, Faculty of Materials Science and Chemistry, 388 Lumo Road, Wuhan, Hubei 430074, P.R. China

4.1 Introduction

Multi-component photocatalyst is developed, as single-component photocatalyst cannot meet the demand for some photocatalytic reactions. Due to structure and composition differences, the heterojunction interface is formed between two different components in the multi-component photocatalyst. The heterojunction interface is the channel for carrier separation and transfer inside the multi-component photocatalyst and largely determines the activity of the photocatalyst [1]. The well-defined heterojunctions are initially classified into three types based on the energy band arrangement: straddling band arrangement (type I), staggered band arrangement (type II), and broken band arrangement (type III). The type II heterojunction is one of the most widely discussed heterojunctions in photocatalysis.

In the type II heterojunction, photogenerated electrons are thought to transfer from the conduction band (CB) of the reduction photocatalyst (RP) to the CB of the oxidation photocatalyst (OP), while the photogenerated holes migrate in the opposite direction. This type II heterojunction mechanism has been commonly used to explain the enhanced photocatalytic performance for a long time. However, an in-depth analysis reveals the defects of type II heterojunction (Figure 4.1). First, the final retained photogenerated carriers reduce the redox ability of the entire system. Second, the photogenerated carrier transfer between two semiconductors is difficult due to electrostatic repulsion. Third, the step-down transfer of photogenerated electrons leads to a partial energy loss, reducing photocatalytic efficiency. The same is true for the step-up transfer of photogenerated holes. Therefore, type II heterojunction is incorrect in thermodynamics, kinetics, and energy in photocatalysis.

Bard proposed the traditional Z-scheme heterojunction, which is utilized in a liquid-phase system with a redox ion couple as a carrier migration bridge, to tackle the low redox ability of type II heterojunction [2]. Under light illumination, photogenerated electrons transfer from the CB of OP to the valence band (VB) of RP via the redox ion couple (Figure 4.2a). However, the electron transfer mechanism

UV-Visible Photocatalysis for Clean Energy Production and Pollution Remediation: Materials, Reaction Mechanisms, and Applications, First Edition. Edited by Xinchen Wang, Masakazu Anpo, and Xianzhi Fu.
© 2023 WILEY-VCH GmbH. Published 2023 by WILEY-VCH GmbH.

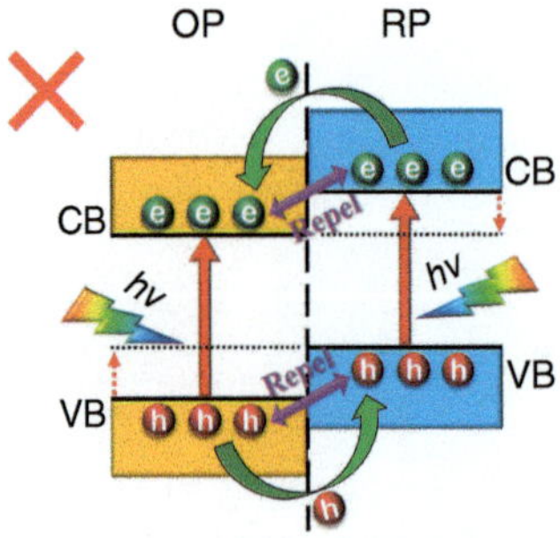

Figure 4.1 The irrationality of type II heterojunction in photocatalysis.

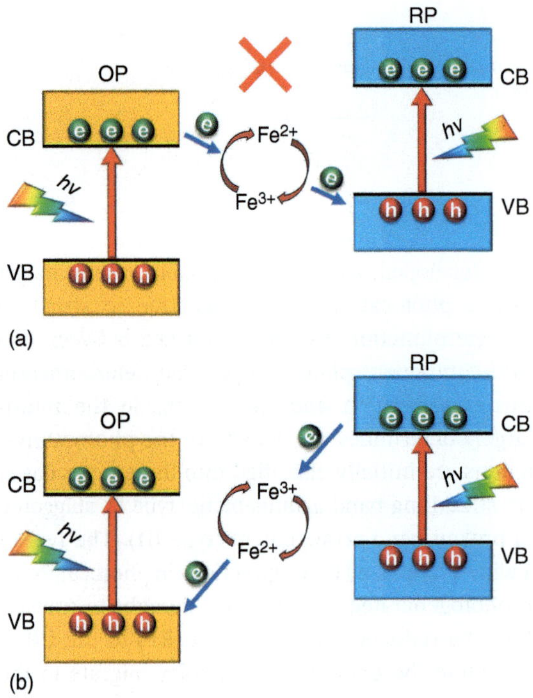

Figure 4.2 (a) Supposed and (b) real traditional Z-scheme heterojunction.

of this traditional Z-scheme heterojunction is wrong from thermodynamic and electrochemical viewpoints because the electrons are more likely to transfer from the CB of RP to the VB of OP (Figure 4.2b). This will result in a reduced redox ability of the system. Also, the application of the traditional Z-scheme system is mainly limited in the liquid-phase environment. Further, the redox ion couples have colors and are susceptible to pH.

In 2006, an all-solid-state Z-scheme heterojunction, which connected two semiconductors using noble metal nanoparticles (such as Pt and Au) as electronic conductors, was proposed to overcome the shortcomings of the traditional Z-scheme system (Figure 4.3a) [3]. Under light illumination, photogenerated electrons in the CB of OP are transported to the noble metal and finally recombine with photogenerated holes in the VB of RP. The metal acts as a bridge to realize the Z-scheme migration of charge carriers. However, the electron transfer mechanism of all-solid-state Z-scheme heterojunction is also incorrect from thermodynamic

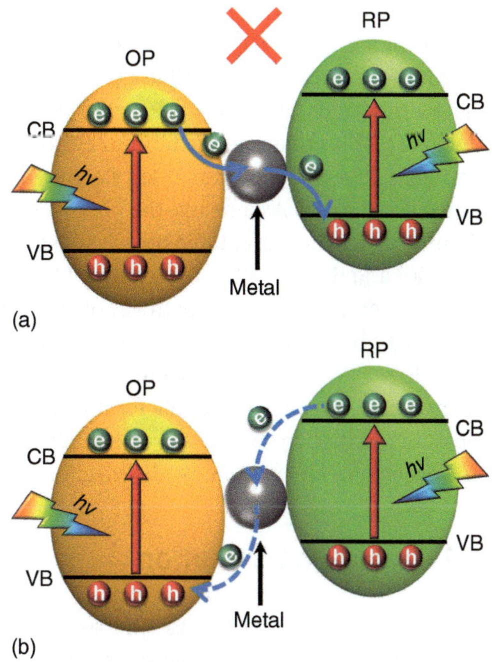

Figure 4.3 (a) Presumed and (b) actual all-solid-state Z-scheme heterojunction.

and electrochemical viewpoints because the electrons are more likely to transfer from the CB of RP and the VB of OP (Figure 4.3b). In addition, it is difficult to place the noble metal nanoparticle precisely between two semiconductors during fabrication. Furthermore, the work function of noble metals (such as Pt and Au) is generally larger than that of semiconductors. The Schottky junction formed between the noble metal and the semiconductor also inhibits the transfer of electrons from the CB of OP to the VB of RP [4, 5].

Afterward, a direct Z-scheme heterojunction was reported in 2001 [6]. Direct Z-scheme heterojunction operates without redox ion couple and metal medium. However, the mechanism of direct Z-scheme heterojunction was not well discussed.

In 2019, a new S-scheme heterojunction was proposed to overcome the problems of type II, traditional Z-scheme system, and all-solid-state Z-scheme heterojunction [7]. As shown in Figure 4.4, the requirement for constructing an S-scheme

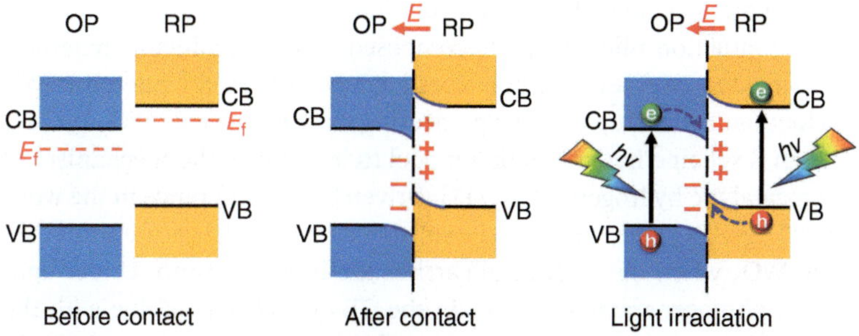

Figure 4.4 Schematic of S-scheme heterojunction. OP and RP are oxidation and reduction photocatalysts, respectively.

heterojunction is that the CB bottom, Fermi level, and VB top positions of RP are all higher than those of OP. Charge redistribution results in a built-in electric field at the interface directed from RP to OP when RP and OP are combined. Meanwhile, the energy bands of RP and OP at the interface bend up and down, respectively. Under light illumination, the built-in electric field, band bending, and electrostatic attraction drive the photogenerated electrons in OP to recombine with the photogenerated holes in RP. As a result, the photogenerated electrons with strong reduction ability and the photogenerated holes with strong oxidation ability are retained in the CB of RP and the VB of OP, respectively [4]. The S-scheme heterojunction is named after the step-like migration pathway of electrons from the CB of OP to the CB of RP under light illumination. Recent studies have shown that S-scheme heterojunction is not limited to two n-type semiconductors [5]. The universality of the S-scheme heterojunction makes it suitable for a wide range of applications in photocatalysis, including hydrogen evolution, carbon dioxide reduction, pollutant degradation, hydrogen peroxide production, disinfection and sterilization, and organic synthesis.

4.2 Hydrogen Evolution

Energy is a vital guarantee for the survival and development of human society. Hydrogen energy has attracted extensive attention due to its advantages, such as cleanliness, abundant resources, high combustion calorific value, no energy decay, and various utilization forms. Furthermore, hydrogen production is significant in promoting the energy consumption revolution and building a clean, low-carbon, and efficient energy system. Hydrogen energy is divided into gray hydrogen (a large amount of CO_2 emissions during its preparation process), blue hydrogen (a small amount of CO_2 emissions during its preparation process), and green hydrogen (almost no CO_2 emissions during its preparation process). Only green hydrogen can truly contribute to the energy transition and carbon neutrality.

Photocatalytic hydrogen production is a promising way to produce green hydrogen as it can convert solar energy to hydrogen without any carbon emissions. Numerous studies have shown that photocatalysts are crucial for efficient photocatalytic hydrogen production [8]. Therefore, various photocatalysts have been designed, prepared, and modified to improve photocatalytic hydrogen production efficiency. The S-scheme heterojunction photocatalysts developed in recent years are outstanding in photocatalytic hydrogen evolution.

S-scheme heterojunction photocatalysts composed of semiconductor materials, such as various oxides, sulfides, and polymers have been used for photocatalytic hydrogen production. For example, in the pioneering work of Fu et al., WO_3/g-C_3N_4 was chosen as an S-scheme heterojunction model to investigate the mechanism of enhanced photocatalytic hydrogen activity [7]. Driven by the difference in the work function, the WO_3/g-C_3N_4 heterojunction establishes a built-in electric field directed from g-C_3N_4 to WO_3 when the interfacial carriers are in equilibrium. Under light illumination, the photogenerated electrons in the CB of WO_3 recombine with the photogenerated holes in the VB of g-C_3N_4. Meanwhile, the photogenerated holes in the VB of WO_3 and the photogenerated electrons in the CB of g-C_3N_4 are retained.

Thus, the system obtains the strongest redox ability. Moreover, the S-scheme heterojunction facilitates the separation and transfer of photogenerated carriers. Therefore, the photocatalytic hydrogen evolution activity of the WO_3/g-C_3N_4 S-scheme heterojunction photocatalyst is greatly improved. In the follow-up study, WO_3 is often combined with another reduction semiconductor to form an S-scheme heterojunction, such as WO_3/TiO_2 [9], due to its oxidation-type energy band structure. This design idea of constructing S-scheme heterojunction photocatalysts by combining a reduction semiconductor and an oxidation semiconductor has been rapidly developed.

Cheng et al. reported an inorganic/organic S-scheme heterojunction photocatalyst for hydrogen evolution [10]. The S-scheme heterojunction is constructed from a pyrene-based conjugated polymer (labeled as PT) on which CdS nanocrystals are loaded (Figure 4.5a–c). The surface defects of PT can anchor the CdS nanocrystals

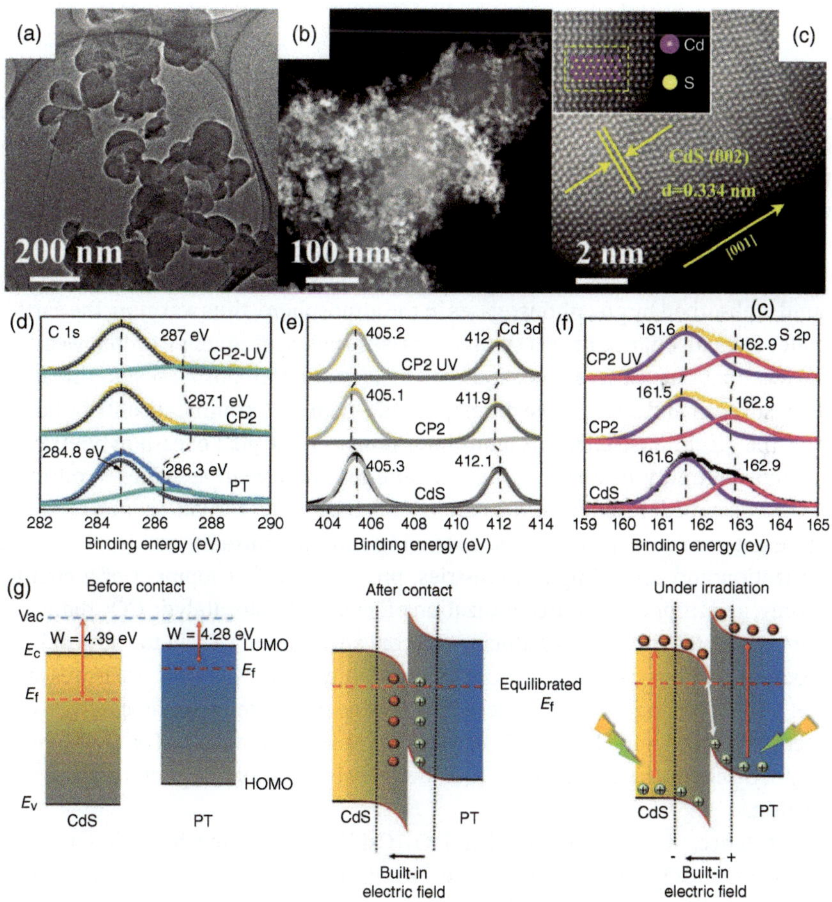

Figure 4.5 Transmission electron microscopy (TEM) images of (a) PT, (b) CdS/PT composite, and (c) CdS. ISIXPS of (d) C 1s, (e) Cd 3d, and (f) S 2p. (g) Schematic of S-scheme carrier transfer mechanism in CdS/PT composite. Source: Cheng et al. [10]/reproduced with permission from John Wiley & Sons.

to form tight contact between CdS and PT. The CdS/PT composite containing 2 wt% PT (labeled as CP2) exhibits the best photocatalytic hydrogen production activity (9.28 mmol h^{-1} g^{-1}), which is approximately eight times that of pure CdS. *In situ* irradiation Kelvin probe force microscopy (ISIKPFM) shows a built-in electric field directed from PT to CdS in the CdS/PT composite without light illumination. Interestingly, the surface potential of the CdS/PT composite changes significantly under light illumination: the potentials of CdS and PT increase and decrease, respectively. This change is consistent with the S-scheme carrier transfer mechanism. *In situ* irradiated X-ray photoelectron spectroscopy (ISIXPS) further reveals that CdS acts as an electron donor to facilitate the S-scheme charge transfer mechanism between PT and CdS under light illumination (Figure 4.5d–g). This work demonstrates that the combination of π-conjugated polymer with the inorganic semiconductor is a promising strategy for designing high-performance S-scheme heterojunction photocatalysts.

4.3 Carbon Dioxide Reduction

CO_2 emission increases day by day with the expansion of human activities and the rapid development of industry after the Industrial Revolution. Meanwhile, deforestation reduces the absorption of CO_2, leading to various environmental problems. The greenhouse effect is one of the hazards caused by excess CO_2 in the atmosphere and is the main culprit of global warming. Specifically, the short-wave radiation from sunlight is emitted to the ground. Then, the long-wave radiation released by the ground is absorbed by greenhouse gases in the atmosphere, thus warming the earth's climate. Climate change brings significant risks such as frequent extreme weather, species extinction, sea-level rise, and crop yield reduction, posing severe threats to human survival and sustainable development. Given the prominent contribution of CO_2 to the greenhouse effect, humans have proposed the goal of carbon neutrality.

Carbon neutrality means that the CO_2 emitted by human activities is offset by the CO_2 absorbed through afforestation and industrial carbon sequestration to achieve net-zero carbon emissions. Achieving carbon neutrality can accelerate the optimization and upgrading of industries, promote the development of a circular economy, and improve resource utilization efficiency. Photocatalytic CO_2 reduction is a promising technology for reducing the concentration of CO_2 in the atmosphere. It proceeds without carbon emissions. Instead, it converts CO_2 into treasures such as CH_4, CH_3OH, and CO [11]. Much effort has been devoted to photocatalytic CO_2 reduction for producing hydrocarbon with CO_2 as a feedstock [12]. The strong redox ability of S-scheme heterojunctions provides new inspiration for photocatalytic CO_2 reduction.

For instance, Xu et al. reported a $TiO_2/CsPbBr_3$ S-scheme heterojunction for photocatalytic CO_2 reduction [13]. The $TiO_2/CsPbBr_3$ S-scheme heterojunction is composed of TiO_2 nanofibers with $CsPbBr_3$ quantum dots (Figure 4.6a,b). The TiO_2 nanofiber network framework, consisting of anatase and rutile phases (Figure 4.6c), is beneficial for anchoring $CsPbBr_3$ quantum dots and mass transfer. Under light irradiation, the products in the system are mainly H_2, O_2, and CO (Figure 4.6d–f).

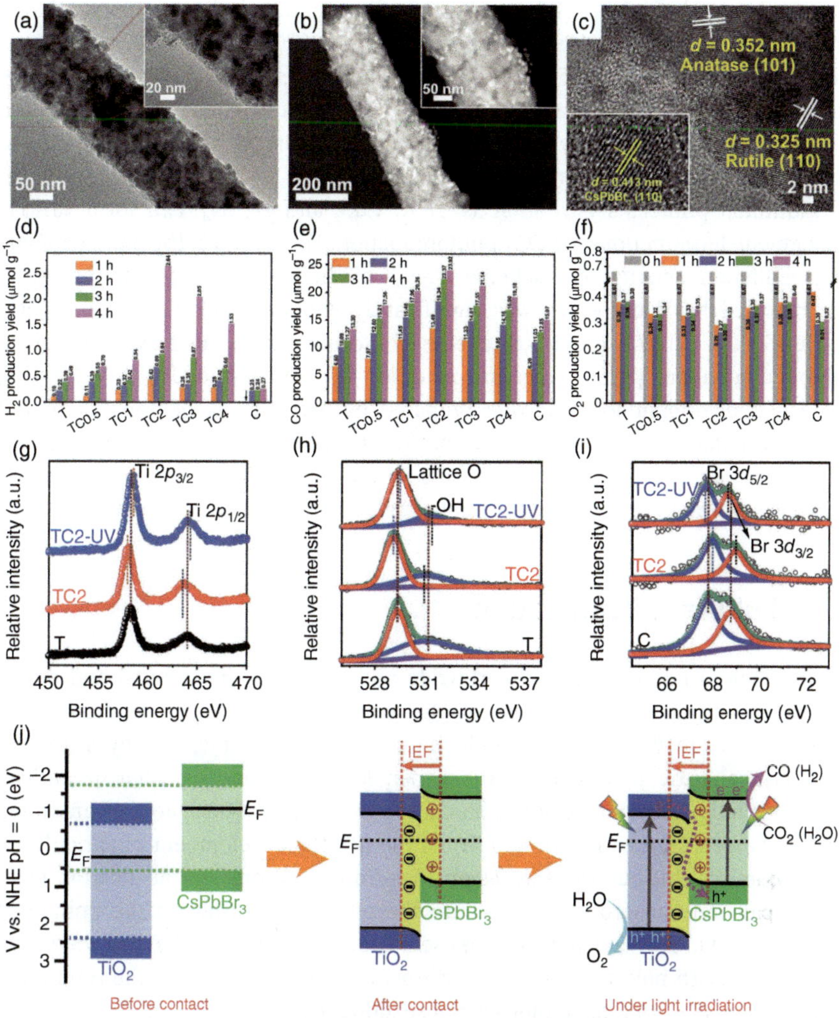

Figure 4.6 (a) TEM, (b) scanning transmission electron microscopy (STEM), and (c) high-resolution transmission electron microscopy (HRTEM) images of $TiO_2/CsPbBr_3$ composite. The evolution of (d) H_2, (e) CO, and (f) O_2 during photocatalytic CO_2 reduction over $TiO_2/CsPbBr_3$. T, C, and TC*x* refer to TiO_2, $CsPbBr_3$, and the $TiO_2/CsPbBr_3$ composite containing *x* wt% $CsPbBr_3$, respectively. ISIXPS of (g) Ti 2p, (h) O 1s, and (i) Br 3d. (j) Schematic of S-scheme charge transfer mechanism in $TiO_2/CsPbBr_3$ composite. Source: Xu et al. [13]/reproduced with permission from Springer Nature.

The CO_2-photoreduction rate of $TiO_2/CsPbBr_3$ (9.02 μmol g^{-1} h^{-1}) is higher than that of pristine TiO_2 nanofibers (4.68 μmol g^{-1} h^{-1}). The CO is demonstrated to originate from CO_2 feedstock and not from other carbonaceous impurities using isotope-labeled $^{13}CO_2$. Theoretical calculations suggest that the difference in the work function leads to the transfer of electrons from $CsPbBr_3$ to TiO_2 in the darkness and a built-in electric field in the same direction as the electron transfer. ISIXPS reveals that the photogenerated electrons in TiO_2 recombine with the photogenerated holes in $CsPbBr_3$ under light illumination, indicating the S-scheme mechanism in the $TiO_2/CsPbBr_3$ heterojunction (Figure 4.6g–i). The built-in electric field in the $TiO_2/CsPbBr_3$ S-scheme heterojunction promotes the recombination of carriers with low redox ability and facilitates the efficient photoreduction of CO_2 (Figure 4.6j). This work provides a typical example of applying S-scheme photocatalysts in CO_2 photoreduction.

Traditional photocatalysts, such as TiO_2, CdS, and g-C_3N_4, can form various S-scheme heterojunctions for CO_2 photoreduction. For example, the photocatalytic CO_2 reduction activity of S-scheme TiO_2/CdS hollow microsphere photocatalyst reported by Wang et al. was significantly enhanced relative to pure TiO_2 and CdS [14]. A TiO_2/g-C_3N_4 S-scheme photocatalyst designed by Wang et al. exhibited enhanced performance in converting CO_2 to CO and CH_4 [15]. Besides, the TiO_2/polydopamine composite prepared by Meng et al. demonstrated that S-scheme heterojunctions could also be formed between TiO_2 and organics for CO_2 photoreduction [16]. Therefore, the S-scheme heterojunction is a versatile and effective strategy for enhancing the photocatalytic CO_2 reduction activity.

4.4 Pollutant Degradation

Environmental pollution exists worldwide and threatens the survival and development of human beings. Most of the pollutants that cause environmental deterioration are difficult to degrade naturally. For example, azo dyes are the most critical toxic pollutants in the paper, leather, cosmetics, pharmaceutical, and other industries. Water pollution caused by the direct discharge of untreated dyes endangers human health and limits the healthy development of agriculture, forestry, animal husbandry, and fishery. Traditional adsorption purification technology can only adsorb pollutants into the adsorbent. Although the pollutants are transferred, they are not decomposed. More importantly, the absorbent can no longer adsorb pollutants after adsorption saturation. Therefore, the traditional adsorption purification technology is not sustainable.

The emerging photocatalytic pollutant degradation technology can degrade the organic pollutants, including dyes in water. Photocatalytic pollutant degradation, which decomposes organic pollutants through chemical reactions, is distinctly different from adsorption purification. If the photocatalyst is stable, pollutants can be continuously decomposed. Under light illumination, the photocatalyst converts the surrounding oxygen and water molecules into reactive oxygen species such as $OH^{\cdot}$ and $O_2^{\cdot -}$ radicals. These reactive oxygen species can completely decompose organic matter into nontoxic CO_2 and water through oxidation reactions. The strong redox ability of S-scheme heterojunction photocatalysts can easily generate the reactive oxygen species to mineralize organic pollutants. For example, a Bi_2O_3/TiO_2 S-scheme heterojunction fabricated by He et al. possessed enhanced photocatalytic phenol degradation activity [17]. A g-C_3N_4/SnO_2 S-scheme heterojunction reported by Pham et al. was effective in the photodegradation of nitrogen oxide [18].

Besides, the S-doped g-C_3N_4/TiO_2 S-scheme photocatalyst prepared by Wang et al. exhibited enhanced performance in photocatalytic degradation of Congo red [19]. Transmission electron microscopy (TEM) images show that the distinct heterojunction interface is formed between amorphous S-doped g-C_3N_4 nanosheets and crystalline TiO_2 fibers (Figure 4.7a,b). ISIXPS and theoretical calculations verify that electrons transfer from S-doped g-C_3N_4 to TiO_2 when S-doped g-C_3N_4

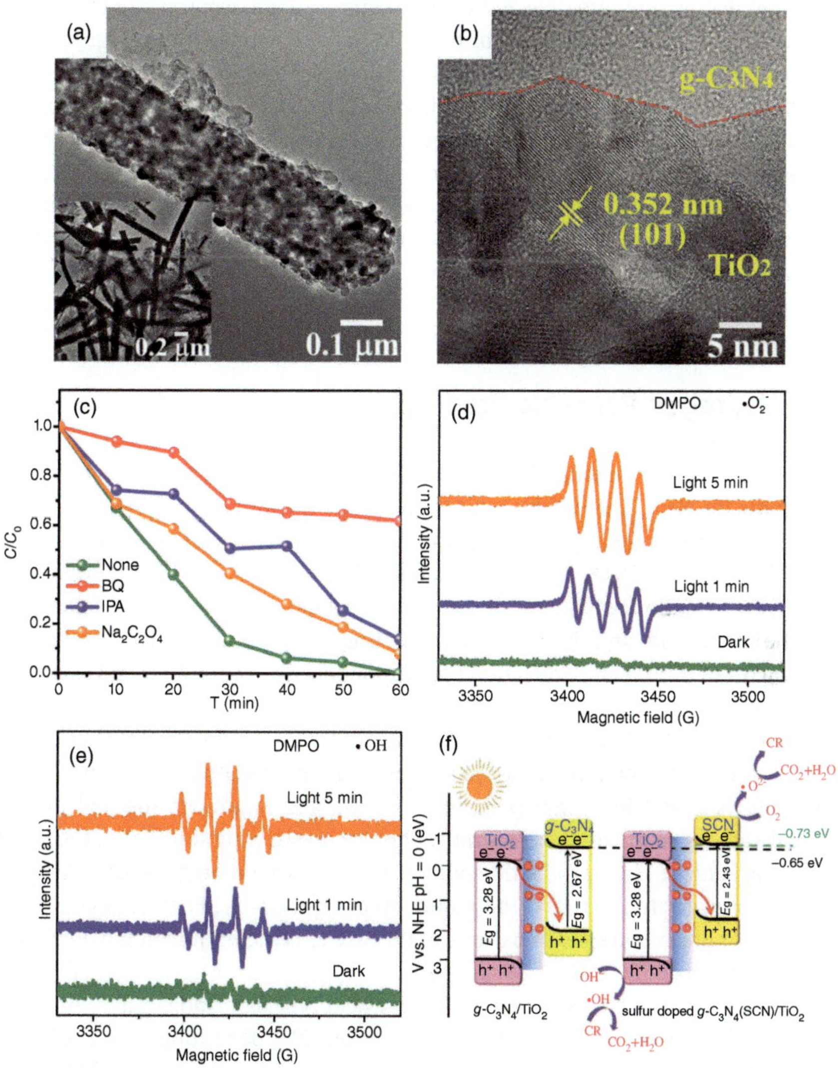

Figure 4.7 (a) TEM and (b) HRTEM images of the S-doped g-C_3N_4/TiO_2 S-scheme heterojunction. Electron paramagnetic resonance spectra of (c) $O_2^{\cdot-}$ and (d) $OH^{\cdot}$ over S-doped g-C_3N_4/TiO_2 composite. (e) Photocatalytic degradation activity of Congo Red with different scavengers. BQ and IPA are benzoquinone and isopropanol, respectively. (f) Schematic of S-scheme mechanism over S-doped g-C_3N_4/TiO_2 for photocatalytic Congo Red degradation. Source: Wang et al. [19]/reproduced with permission from Elsevier.

combines with TiO_2. Meanwhile, the energy bands of S-doped g-C_3N_4 and TiO_2 bend upward and downward, respectively. Moreover, a built-in electric field directed from S-doped g-C_3N_4 to TiO_2 is generated. Under light illumination, the built-in electric field, band bending, and Coulomb attraction synergistically promote the recombination of photogenerated electrons in TiO_2 with photogenerated holes in S-doped g-C_3N_4. Thus, the S-doped g-C_3N_4/TiO_2 S-scheme photocatalyst possesses strong redox ability. The active species that play a major role in photocatalytic Congo red degradation are detected using sodium oxalate, isopropanol (IPA), and benzoquinone (BQ) as scavengers for h^+, $OH^{\cdot}$, and $O_2^{\cdot-}$, respectively (Figure 4.7c). The results show that $O_2^{\cdot-}$ and $OH^{\cdot}$ are the main active oxide species for the photocatalytic degradation of Congo red. The signals of $O_2^{\cdot-}$ and $OH^{\cdot}$ are enhanced under irradiation, as observed in electron paramagnetic resonance spectra (Figure 4.7d,e). The S-doped g-C_3N_4/TiO_2 S-scheme heterojunction facilitates the generation of $O_2^{\cdot-}$ and $OH^{\cdot}$, thus accelerating the photocatalytic degradation of Congo red (Figure 4.7f). This work enriches the understanding of S-scheme heterojunction photocatalysts and provides a promising strategy to mitigate environmental pollution.

4.5 Hydrogen Peroxide Production

H_2O_2 is an important chemical widely used in oxide preparation, fabric bleaching, semiconductor cleaning, food bleaching, and anticorrosion. Diluted H_2O_2 solution is a good disinfectant and germicide. And high-purity H_2O_2 can be used as rocket fuel and propellant. H_2O_2 is an ideal oxidant that converts itself into oxygen and water in the oxidation reaction without introducing impurities into the reaction system. The strong oxidizing ability and nonpolluting properties make H_2O_2 a potential chemical in environmental protection applications such as wastewater and waste gas treatment.

At present, the H_2O_2 production technology is dominated by the anthraquinone process (Riedl-Pfleiderer method) [20]. However, the hazardous solvents, noble metal catalysts, and high-energy input make the anthraquinone method uneconomical and environmentally unfriendly. Photocatalysis using sunlight as the input energy can achieve H_2O_2 production, which includes two production pathways: the oxidation of water to H_2O_2 and the reduction of O_2 to H_2O_2. Photocatalytic H_2O_2 production is green and safe. In particular, the S-scheme heterojunction contributes to efficient photocatalytic H_2O_2 synthesis.

For instance, Han et al. reported an inverse opal ZnO@polydopamine S-scheme heterojunction for photocatalytic H_2O_2 production [21]. Benefiting from the Bragg diffraction and the slow photon effect of the inverse opal structure, the light absorption of ZnO@polydopamine is enhanced. Meanwhile, the S-scheme heterojunction promotes the effective separation and transfer of photoexcited carriers. Therefore, the photocatalytic H_2O_2 production performance of ZnO@polydopamine is significantly enhanced. In addition, metal–organic frameworks (MOFs) can be transformed into porous metal oxides. Liu et al. fabricated a ZnO/g-C_3N_4 S-scheme heterojunction by combining MOF-derived ZnO with g-C_3N_4 [22]. The ZnO/g-C_3N_4 photocatalyst has a hierarchical mesoporous structure that provides many active sites and effectively promotes mass transfer. Under the effect of the built-in electric field, the recombination of useful photogenerated carriers in the ZnO/g-C_3N_4 S-scheme heterojunction is significantly suppressed, resulting in an efficient photocatalytic activity for H_2O_2 production. Besides, Yang et al. constructed a TiO_2/In_2S_3 S-scheme photocatalyst to facilitate H_2O_2 synthesis [23]. TiO_2 nanofibers, on which In_2S_3 nanosheets are grown, exhibit improved photocatalytic H_2O_2 production activity and stability (Figure 4.8a–d). ISIXPS and theoretical calculations reveal the

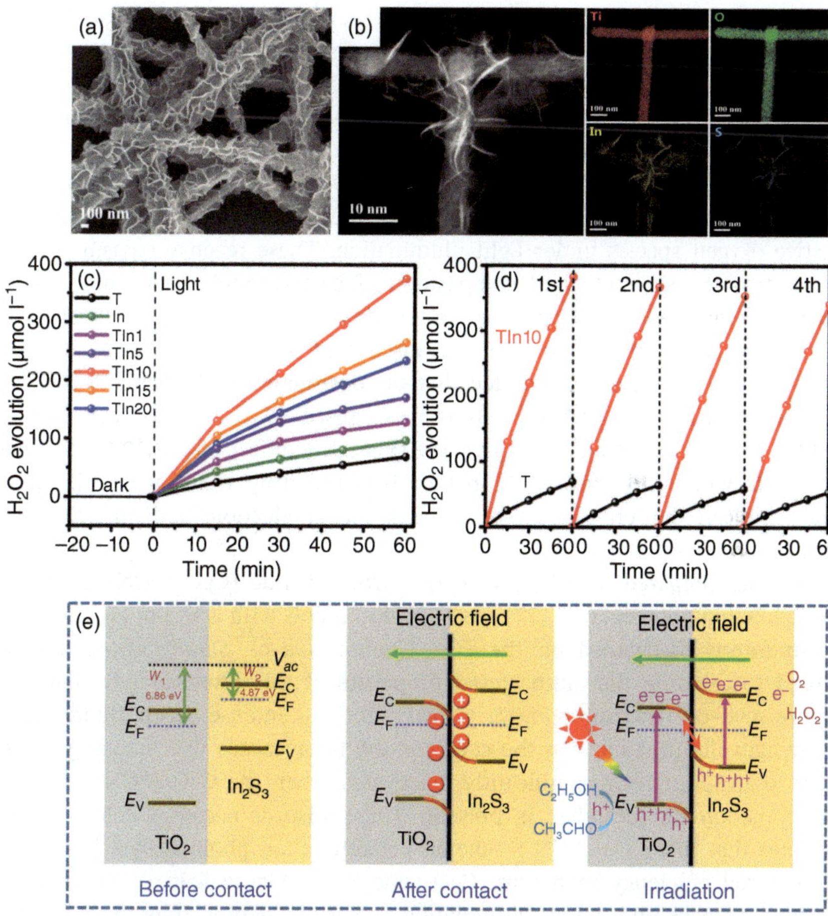

Figure 4.8 (a) Field emission scanning electron microscopy (FESEM) and (b) STEM images and corresponding elemental mapping of TiO_2/In_2S_3 composite. (c) Photocatalytic H_2O_2 production activities over samples. The T, In, and TIn*x* refer to TiO_2, In_2S_3, and the TiO_2/In_2S_3 composite containing *x* mol% In_2S_3, respectively. (d) The stability for photocatalytic H_2O_2 production over TIn10. (e) Schematic of S-scheme mechanism over TiO_2/In_2S_3 composite for photocatalytic H_2O_2 production. Source: Yang et al. [23]/reproduced with permission from Springer Nature.

S-scheme electron transfer mechanism in the TiO_2/In_2S_3 composite. The electron spin resonance spectrum indicates that the H_2O_2 production over TiO_2/In_2S_3 S-scheme photocatalyst is a stepwise single-electron reduction of O_2. Under light illumination, In_2S_3 provides the primary photogenerated electrons for O_2 reduction and strongly interacts with O_2, enhancing the photocatalytic H_2O_2 synthesis (Figure 4.8e). This work suggests that S-scheme heterojunction provides new insight for efficient photocatalytic H_2O_2 production.

4.6 Disinfection and Sterilization

Pathogens and bacteria put human health at risk. Efficient disinfection and sterilization technology are the guarantees for the safe development of human beings. Antibiotics are one of the most used sterilizing substances. However, the overuse of antibiotics is making viruses and bacteria increasingly resistant. Therefore, it is urgent to find alternative sterilization methods.

Photocatalysis can economically and effectively achieve disinfection and sterilization through advanced oxidation processes. Specifically, photocatalysts produce reactive oxygen species under light illumination. These reactive oxygen species can destroy the structure and composition of bacteria and viruses and finally decompose them non-selectively into carbon dioxide and water. The photocatalytic inactivation of microbial cells was first discovered by Matsunaga et al. in 1985 [24]. Later, Elahifard et al. reported that the destruction of *Escherichia coli* cell walls by $Ag/AgBr/TiO_2$ photocatalysts was caused by reactive oxygen species [25]. In addition, Li et al. found that the synergistic effect of g-C_3N_4/Bi_2S_3 photocatalysts with antibiotics could reduce bacterial drug resistance [26]. These studies indicate that photocatalysis has promising applications in disinfection and sterilization.

Xia et al. prepared a CeO_2/polymeric carbon nitride (CeO_2/PCN) S-scheme heterojunction photocatalyst [27]. CeO_2 quantum dots with a size of around 5 nm are uniformly distributed on the PCN (Figure 4.9a–c). *Staphylococcus aureus* is used to examine the antibacterial properties of this photocatalytic material (Figure 4.9d–o). The surviving *S. aureus* emits an intense green fluorescence under light illumination. On the contrary, the fluorescence dye mapping image shows that *S. aureus* is significantly inactivated when the CeO_2/PCN S-scheme heterojunction photocatalyst is present. The statistical bacterial colony count indicates that this CeO_2/PCN S-scheme heterojunction photocatalyst has better antibacterial efficiency than pure CeO_2 and PCN (Figure 4.9p,q). The surface potential changes under light illumination powerfully demonstrate the S-scheme electron transfer mechanism in CeO_2/PCN (Figure 4.9r–w). The built-in electric field promotes the retention of photogenerated holes in CeO_2 and photogenerated electrons in PCN. Therefore, the S-scheme heterojunction endows the CeO_2/PCN composite with strong redox ability and promotes antimicrobial efficiency.

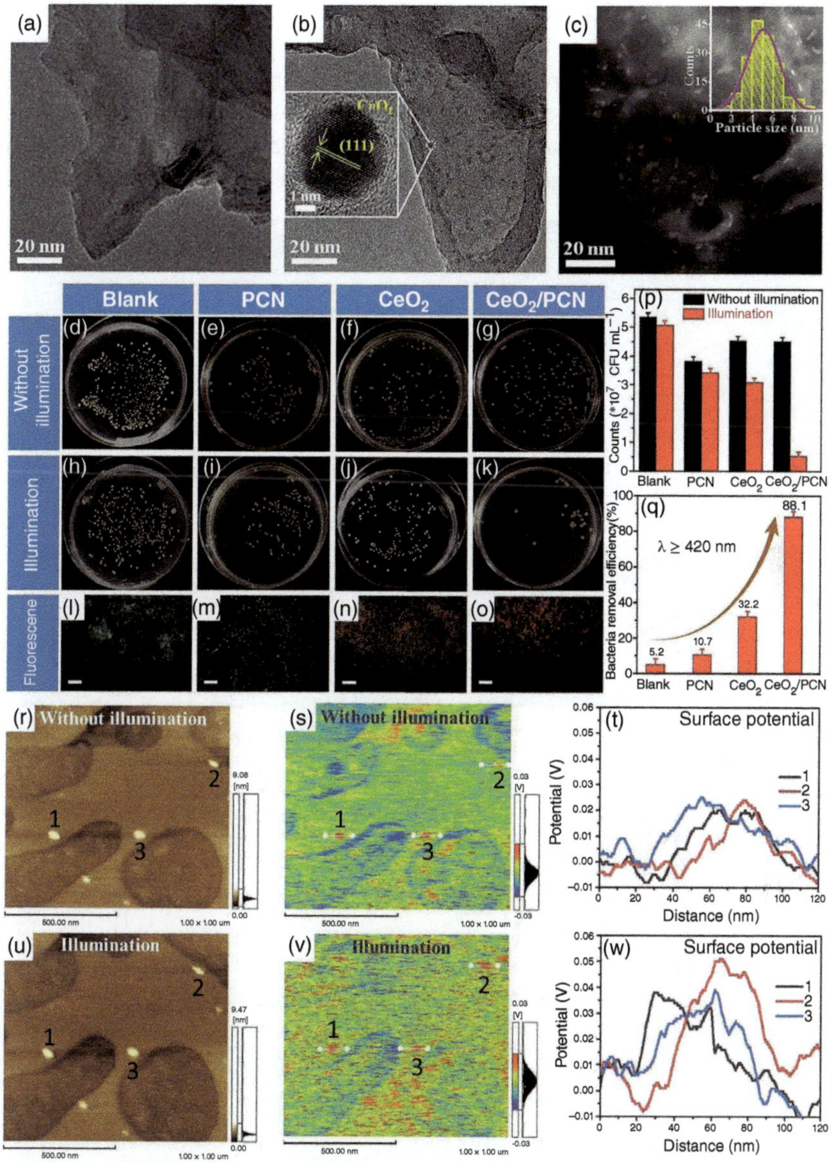

Figure 4.9 HRTEM images of (a) PCN and (b) CeO_2/PCN. (c) High-angle annular dark-field TEM image of CeO_2/PCN. (d–g) Photographs of the bacterial colony without light illumination. (h–k) Photographs of the bacterial colony under light illumination. (l–o) Fluorescence dye mapping images. (p) Colony statistics under light illumination and without light illumination. (q) The bacterial removal efficiency of samples. (r–w) Atomic force microscopy images, surface potential images, and line-scanning potentials of CeO_2/PCN without and under light illumination. Source: Xia et al. [27]/reproduced with permission from John Wiley & Sons.

4.7 Organic Synthesis

Organic synthesis is usually performed under severe conditions (e.g. high temperature and high pressure) with strong redox reagents. These traditional synthesis processes are energy-intensive and environmentally hazardous, which is inconsistent with the development of green chemistry. In recent years, photocatalytic organic

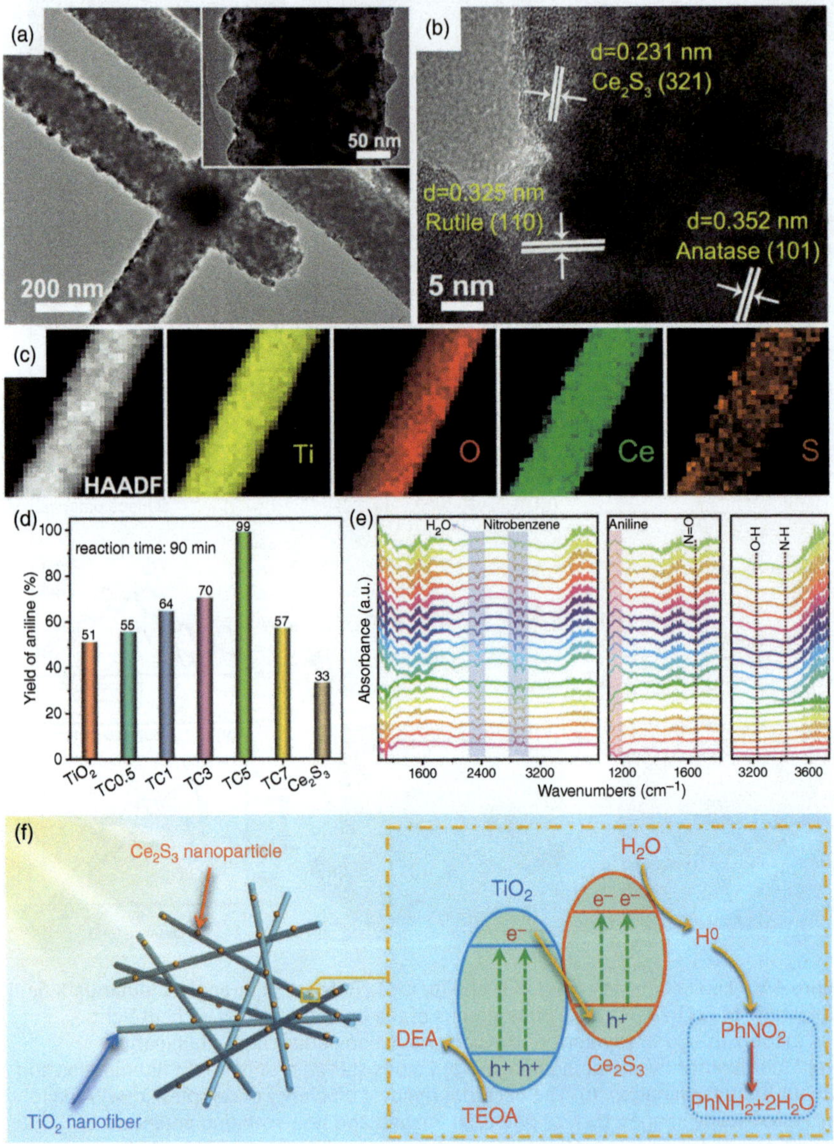

Figure 4.10 (a) TEM, (b) HRTEM, and (c) elemental mapping images of TiO_2/Ce_2S_3. (d) Aniline yield after 90 minutes photocatalytic reaction. TC*x* refers to the TiO_2/Ce_2S_3 composite containing *x* mol% Ce_2S_3. (e) *In situ* infrared spectroscopy over TC5. (f) Schematic of S-scheme mechanism over TiO_2/Ce_2S_3 composite for photocatalytic aniline synthesis. Source: Xu et al. [28]/reproduced with permission from American Chemical Society.

synthesis has progressed and shown the potential to replace traditional synthetic processes. Photocatalytic organic synthesis has many advantages over the traditional synthetic processes: (i) Photocatalytic synthesis is both economical and environmentally friendly. (ii) Photocatalysis provides a safe industrial production environment. (iii) Photocatalytic organic synthesis is driven by photogenerated carriers and energy transfer, enabling potential bond-breaking strategies that are difficult or impossible to achieve by traditional methods. (iv) The controllable input of photons (including duration, wavelength, and intensity) in the photocatalytic reaction can precisely control the reaction process and product selectivity. (v) Photocatalysis eliminates the dependence on the strong redox agents, allowing organic reactions to proceed under mild conditions. (vi) Photocatalytic reactions can be inserted into traditional processes to shorten the synthetic routes. These advantages facilitate upgrading traditional processes to more environmentally friendly and economical synthesis routes.

The development of photocatalytic organic synthesis benefits from the innovative design of photocatalysts and the new photocatalytic strategies. Recently, S-scheme heterojunction photocatalysts have demonstrated creative applications in organic synthesis. Xu et al. designed and prepared a TiO_2/Ce_2S_3 S-scheme heterojunction, revealing the photocatalytic mechanism for the hydrogenation of nitrobenzene to aniline [28]. Ce_2S_3 nanoparticles are grown on TiO_2 nanofibers composed of rutile and anatase phases (Figure 4.10a–c). The TiO_2/Ce_2S_3 S-scheme heterojunction exhibits excellent activity for synthesizing aniline from nitrobenzene. After 90 minutes of photocatalytic reaction, the yield of aniline can reach 99% (Figure 4.10d). *In situ* infrared spectroscopy shows that water is consumed with the formation of aniline (Figure 4.10e). Isotopic experiments further confirm water as a proton source, contributing to the hydrogenation of nitrobenzene for the aniline synthesis. Density functional theory calculations and ISIXPS reveal that TiO_2/Ce_2S_3 S-scheme heterojunction separates photogenerated carriers and promotes the hydrogenative reduction of nitrobenzene (Figure 4.10f). This work demonstrates the promising application of S-scheme heterojunction in organic synthesis.

4.8 Conclusion and Outlook

In summary, the carrier transfer mechanisms of type II heterojunction, traditional Z-scheme heterojunction, and all-solid-state Z-scheme heterojunction are incorrect. Thus, the new S-scheme mechanism is proposed in 2019. The S-scheme heterojunction has shown potential applications in various photocatalytic research fields, including hydrogen evolution, carbon dioxide reduction, pollutant degradation, hydrogen peroxide production, disinfection and sterilization, and organic synthesis. Moreover, S-scheme heterojunction is rapidly developing and is gradually recognized and corroborated by researchers. At the same time, there are both opportunities and challenges for S-scheme heterojunctions.

First, given that organic semiconductors, such as organic polymers, covalent organic frameworks, and MOFs, have inhomogeneous electron distribution, the research on intramolecular S-scheme heterojunctions may be interesting.

Second, S-scheme heterojunctions have been investigated in fields other than photocatalysis. For example, recently, an Fe_2O_3/C_3N_4 S-scheme heterojunction has been reported to play a positive role in a lithium–oxygen battery [29]. Therefore, applying S-scheme heterojunctions in other research fields, such as photovoltaic devices, is worth exploring.

Third, it is worth noting that S-scheme heterojunction photocatalysts are generally discussed in stand-alone powders. It may not be applicable in some systems where external circuits are applied, such as photoelectrocatalysis.

Fourth, the semiconductor is in a nonequilibrium state under light conditions. At this point, the quasi-Fermi levels at the S-scheme heterojunction interface may not be a simple alignment of the Fermi levels in two semiconductors. The nonequilibrium state of the Fermi level at the interface needs to be explored.

Fifth, S-scheme heterojunctions can be characterized by ISIXPS, ISIKPFM, electron paramagnetic resonance, and selective deposition of carrier acceptors. In addition, the recombination mode (radiative or nonradiative) between photogenerated electrons in OP and photogenerated holes in RP in S-scheme heterojunctions needs to be revealed. In the case of the radiative recombination, the S-scheme heterojunction can be verified by the fluorescence emitted when electrons and holes are recombined.

References

1 Low, J., Yu, J., Jaroniec, M. et al. (2017). Heterojunction photocatalysts. *Adv. Mater.* 29: 1601694.

2 Bard, A.J. (1979). Photoelectrochemistry and heterogeneous photo-catalysis at semiconductors. *J. Photochem.* 10: 59–75.

3 Tada, H., Mitsui, T., Kiyonaga, T. et al. (2006). All-solid-state Z-scheme in CdS–Au–TiO_2 three-component nanojunction system. *Nat. Mater.* 5: 782–786.

4 Xu, Q., Zhang, L., Cheng, B. et al. (2020). S-scheme heterojunction photocatalyst. *Chem* 6: 1543–1559.

5 Zhang, L., Zhang, J., Yu, H., and Yu, J. (2022). Emerging S-scheme photocatalyst. *Adv. Mater.* 34: 2107668.

6 Grätzel, M. (2001). Photoelectrochemical cells. *Nature* 414: 338–344.

7 Fu, J., Xu, Q., Low, J. et al. (2019). Ultrathin 2D/2D WO_3/g-C_3N_4 step-scheme H_2-production photocatalyst. *Appl. Catal. B Environ.* 243: 556–565.

8 Gao, D., Xu, J., Wang, L. et al. (2022). Optimizing atomic hydrogen desorption of sulfur-rich NiS_{1+x} cocatalyst for boosting photocatalytic H_2 evolution. *Adv. Mater.* 34: 2108475.

9 He, F., Meng, A., Cheng, B. et al. (2020). Enhanced photocatalytic H_2-production activity of WO_3/TiO_2 step-scheme heterojunction by graphene modification. *Chin. J. Catal.* 41: 9–20.

10 Cheng, C., He, B., Fan, J. et al. (2021). An inorganic/organic S-scheme heterojunction H_2-production photocatalyst and its charge transfer mechanism. *Adv. Mater.* 33: 2100317.

11 Sayed, M., Xu, F., Kuang, P. et al. (2021). Sustained CO_2-photoreduction activity and high selectivity over Mn, C-codoped ZnO core-triple shell hollow spheres. *Nat. Commun.* 12: 4936.

12 Yang, C., Wang, Y., Yu, J., and Cao, S. (2021). Ultrathin 2D/2D graphdiyne/Bi_2WO_6 heterojunction for gas-phase CO_2 photoreduction. *ACS Appl. Energy Mater.* 4: 8734–8738.

13 Xu, F., Meng, K., Cheng, B. et al. (2020). Unique S-scheme heterojunctions in self-assembled TiO_2/$CsPbBr_3$ hybrids for CO_2 photoreduction. *Nat. Commun.* 11: 4613.

14 Wang, Z., Chen, Y., Zhang, L. et al. (2020). Step-scheme CdS/TiO_2 nanocomposite hollow microsphere with enhanced photocatalytic CO_2 reduction activity. *J. Mater. Sci. Technol.* 56: 143–150.

15 Wang, L., Fei, X., Zhang, L. et al. (2022). Solar fuel generation over nature-inspired recyclable TiO_2/g-C_3N_4 S-scheme hierarchical thin-film photocatalyst. *J. Mater. Sci. Technol.* 112: 1–10.

16 Meng, A., Cheng, B., Tan, H. et al. (2021). TiO_2/polydopamine S-scheme heterojunction photocatalyst with enhanced CO_2-reduction selectivity. *Appl. Catal. B Environ.* 289: 120039.

17 He, R., Liu, H., Liu, H. et al. (2020). S-scheme photocatalyst Bi_2O_3/TiO_2 nanofiber with improved photocatalytic performance. *J. Mater. Sci. Technol.* 52: 145–151.

18 Van Pham, V., Mai, D.-Q., Bui, D.-P. et al. (2021). Emerging 2D/0D g-C_3N_4/SnO_2 S-scheme photocatalyst: new generation architectural structure of heterojunctions toward visible-light-driven NO degradation. *Environ. Pollut.* 286: 117510.

19 Wang, J., Wang, G., Cheng, B. et al. (2021). Sulfur-doped g-C_3N_4/TiO_2 S-scheme heterojunction photocatalyst for Congo Red photodegradation. *Chin. J. Catal.* 42: 56–68.

20 Campos-Martin, J.M., Blanco-Brieva, G., and Fierro, J.L.G. (2006). Hydrogen peroxide synthesis: an outlook beyond the anthraquinone process. *Angew. Chem. Int. Ed.* 45: 6962–6984.

21 Gaowei Han, F.X.B.C.Y.L.J.Y.L.Z. (2022). Enhanced photocatalytic H_2O_2 production over inverse opal ZnO@polydopamine S-scheme heterojunctions. *Acta Phys. Chim. Sin.* 38: 2112037.

22 Liu, B., Bie, C., Zhang, Y. et al. (2021). Hierarchically porous ZnO/g-C_3N_4 S-scheme heterojunction photocatalyst for efficient H_2O_2 production. *Langmuir* 37: 14114–14124.

23 Yang, Y., Cheng, B., Yu, J. et al. (2021). TiO_2/In_2S_3 S-scheme photocatalyst with enhanced H_2O_2-production activity. *Nano Res.* https://doi.org/10.1007/s12274-021-3733-0.

24 Matsunaga, T., Tomoda, R., Nakajima, T., and Wake, H. (1985). Photoelectrochemical sterilization of microbial cells by semiconductor powders. *FEMS Microbiol. Lett.* 29: 211–214.

25 Elahifard, M.R., Rahimnejad, S., Haghighi, S., and Gholami, M.R. (2007). Apatite-coated Ag/AgBr/TiO_2 visible-light photocatalyst for destruction of bacteria. *J. Am. Chem. Soc.* 129: 9552–9553.

26 Li, Y., Liu, X., Tan, L. et al. (2019). Eradicating multidrug-resistant bacteria rapidly using a multi functional g-C_3N_4@Bi_2S_3 nanorod heterojunction with or without antibiotics. *Adv. Funct. Mater.* 29: 1900946.

27 Xia, P., Cao, S., Zhu, B. et al. (2020). Designing a 0D/2D S-scheme heterojunction over polymeric carbon nitride for visible-light photocatalytic inactivation of bacteria. *Angew. Chem. Int. Ed.* 59: 5218–5225.

28 Xu, F., Meng, K., Cao, S. et al. (2022). Step-by-step mechanism insights into the TiO_2/Ce_2S_3 S-scheme photocatalyst for enhanced aniline production with water as a proton source. *ACS Catal.* 12: 164–172.

29 Zhu, Z., Lv, Q., Ni, Y. et al. (2022). Internal electric field and interfacial bonding engineered step-scheme junction for a visible-light-involved lithium–oxygen battery. *Angew. Chem. Int. Ed.* 61: e202116699.

5

The Role of the Defects on the Photocatalytic Reactions on ZnO

Zhongming Wang[1,2], Wenxin Dai[1], and Xianzhi Fu[1]

[1] *Fuzhou University, College of Chemistry, State Key Laboratory of Photocatalysis on Energy and Environment, 2 Xueyuan Road, Minhou District, Fuzhou 350108, China*
[2] *Anhui University of Technology, College of Energy and Environment, Department of Environmental Sciences and Engineering, 59 Hudong Road, Huashan District, Ma Anshan 243002, China*

5.1 Introduction

Wide bandgap semiconductors have been extensively studied as functional materials for a wide range of applications related to sustainable development. These applications include photocatalytic energy conversion, environmental governance, and high-performance electronics. Zinc oxide is a typical wide bandgap semiconductor (~3.49 eV) with unique optical and electrical properties; it is widely used as an energy storage material, sensing material, for low-temperature photocatalytic CO_2 reduction, etc. Compared with other semiconductor materials, ZnO has also attracted much attention and has been used in industrial applications because of its easy synthesis, relatively better affordability, and non-toxicity.

However, researchers found that pure ZnO has some intrinsic limitations in terms of its optical and electrical properties, and these limitations hinder its application and development. There are main two different kinds of disadvantages. First, the wide bandgap of ZnO limits its light absorption in the ultraviolet (UV) region. However, UV light only accounts for 4% of the solar spectrum [1]. Not absorbing visible light limits the performance of ZnO as a photocatalyst, disinfectant, and photoelectrode; this limitation is a common disadvantage of wide-gap band semiconductors. Second, there is low quantum efficiency, and the lifetime of the photogenerated electron–hole pairs is short; this results in lower photocatalytic and photoelectrocatalytic efficiency [2]. Also, the low conductivity of pure ZnO is not conducive to its wider applications in various electronic products. To overcome these limitations, researchers have designed and developed methods to amplify the efficiency of zinc oxide-based photocatalysts. Many approaches have been developed, including doping/co-doping with nonmetals or metals, surface modifications using noble metals, surface sensitization with organic or inorganic materials, and forming composites with other semiconductors [3].

UV-Visible Photocatalysis for Clean Energy Production and Pollution Remediation: Materials, Reaction Mechanisms, and Applications, First Edition. Edited by Xinchen Wang, Masakazu Anpo, and Xianzhi Fu.
© 2023 WILEY-VCH GmbH. Published 2023 by WILEY-VCH GmbH.

Recently, researchers have found that the vacancy formation energy of ZnO is much lower than that of other metal oxides [4]. It is easy to grow various intrinsic defects as a result of the influence of the catalyst composition and growth conditions used during the preparation process. In particular, the controlled introduction of these intrinsic defects into metal oxides can be used to adjust donor densities and band structures to control the optical, magnetic, and electronic properties; this then affects the photocatalytic performance. Therefore, a novel strategy has been developed to modify the electronic and optical properties of ZnO nanomaterials via the creation of intrinsic defects. In this chapter, we describe the types of surface defects and electrical structures, controllable preparation, and characterization of surface defects, as well as the role of defect chemistry on photocatalytic reaction behavior.

5.2 Types of Surface Defects and Their Electrical Structure

In ZnO, the valence band (VB) is mainly formed by O2p orbitals, whereas the conduction band (CB) is composed of the Zn4s orbital. ZnO usually has a chemically stable hexagonal wurtzite structure in which each zinc atom binds to four surrounding O atoms in the form of a tetrahedron. The results of many research studies show that intrinsic defects in ZnO mainly include oxygen vacancies (V_O), zinc vacancies (V_{Zn}), interstitial oxygen (O_i), and interstitial zinc (Zn_i).

5.2.1 Oxygen Vacancies

Oxygen vacancies are a significant intrinsic defect that is observed in ZnO. The formation of V_O is relatively easier in oxygen-poor conditions [5]. First, the absence of a lattice oxygen atom in the lattice leads to electron rearrangement, and a certain amount of positive charges become enriched in the vacancy center; there may be three degenerate states: V_O^{2+}, V_O^{+}, and V_O^{0}. V_O is a positive center and has a negative coulomb attraction potential, which makes the CB level enter the bandgap to form a donor level [6]. Then the level of V_O is 0.90 eV above the VB in ZnO, resulting in a bandgap of 3.49 eV [7] (Figure 5.1). Moreover, the Zn atoms that are around V_O are attracted to V_O and form inward lattice relaxation. Lattice relaxation may affect the Zn—O bond length and further affect the stability of the material.

5.2.2 Zinc Vacancies

Similarly, V_{Zn} forms relatively more easily in oxygen-rich conditions. Also, the absence of a zinc atom in the lattice leads to electron rearrangement and negative charge enrichment in the vacancy center, and there are three degenerate states: V_{Zn}^{2-}, V_{Zn}^{-}, and V_{Zn}^{0}. A negative center makes the VB level move up into the bandgap to form an acceptor level, and the level of V_{Zn} is 0.30 eV above the VB

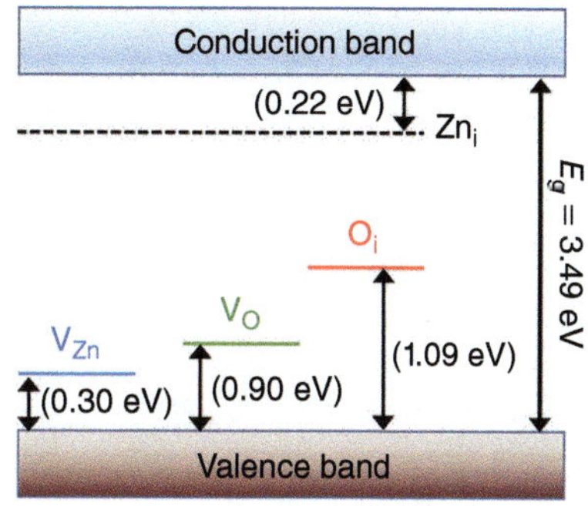

Figure 5.1 Scheme showing energy-level schematic diagram portraying the defect levels in ZnO.

(Figure 5.1). Then, the O atoms that are around V_{Zn} exhibit outward relaxations of the equilibrium bond length of Zn–O, and this is the opposite of V_O.

5.2.3 Interstitial Oxygen and Zinc

Oxygen atoms that enter the interstitial lattice form interstitial oxygen defects. O_i may appear individually in ZnO, or the O ions in the lattice may be displaced because of the relaxation that is caused by other vacancies. O_i has negative centers, and there may be three degenerate states: O_i^{2-}, O_i^{-}, and O_i^{0}; these cause the VB level to move up into the bandgap to form an acceptor level [8]. In particular, the level of O_i is 1.09 eV above the VB in ZnO (Figure 5.1).

Similarly, zinc atoms that enter the interstitial lattice form interstitial zinc defects. Zn_i may appear individually in ZnO, or the Zn ions in the lattice may be displaced because of the relaxation that is caused by other vacancies. Zn_i is a positive center, and there may be three degenerate states: Zn_i^{2+}, Zn_i^{+}, and Zn_i^{0}. These cause the CB level to enter the bandgap to form a donor level. Specifically, the level of Zn_i is 0.22 eV below the CB in ZnO (Figure 5.1).

The V_O and V_{Zn} at surface sites have lower formation energy than other defects [9]. Therefore, surface V_O and V_{Zn} are considered to be the most important defects, and they can also be present in many metal oxides. Many theoretical and experimental studies on V_O and V_{Zn} in ZnO have been conducted, and it has been proved that various physical and chemical properties of ZnO are closely related to defects; these properties include electronic structure, charge transport, and optical and surface properties. For instance, vacancy defects usually result in defect levels between the VB and CB, and this significantly enhances optical absorption and the separation and transfer of photogenerated electrons/holes. Moreover, charges are enriched around vacancies to form a local built-in electric field (Figure 5.2), and they act as active sites and gas adsorption sites [10]. These sites participate in the process of surface adsorption and reactions. By identifying the types of defects and understanding the impacts that they have on performance, researchers can artificially introduce required vacancy defects (V_O or V_{Zn}) and accurately control the electronic structure of ZnO for specific applications.

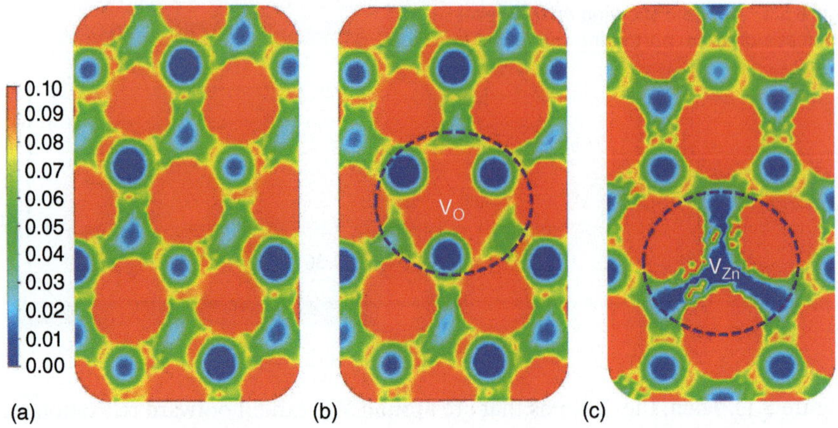

Figure 5.2 The electron localization functions of the valence electronic states (VELF) plots for (a) a perfect ZnO surface, (b) an oxygen vacancy on the surface, and (c) a zinc vacancy on the surface. Source: Adapted from Wang et al. [10].

5.3 Controllable Preparation and Characterization of Surface Vacancy Defects

5.3.1 Controllable Preparation of Surface Vacancy Defects

5.3.1.1 Formation of Vacancy Defects via Annealing at Different Conditions

In the preparation process, calcining ZnO in different conditions can provide different environments and energy needed for growing different surface defects. The dominant growth of different defects can be achieved, and this can be used as a simple and practical way to artificially control surface defects (Figure 5.3a). First, V_O can be created via the posttreatment of as-prepared pure ZnO at high temperatures. The following equilibrium describes V_O formation at elevated temperatures: $O_{lat} \rightarrow V_O + 1/2O_2$, where O_{lat} denotes lattice oxygen.

The concentration of V_O and V_{Zn} is related to the partial pressure of O_2. Specifically, the formation energy of V_O is lower in an oxygen-poor (or reducing) atmosphere, whereas the generation of V_{Zn} is favored in an oxygen-rich atmosphere. Yi and coworkers [14] synthesized ZnO nanoparticles via annealing at 250 °C in different atmospheres (pure N_2, air, and O_2). During the annealing process, the annealing atmosphere modulates the chemical nature of ZnO layers. In particular, it modulates the number of V_Os in ZnO. As the composition of O_2 gas is decreased in the annealing atmosphere, a notable increase of Vo in ZnO is observed. Lo and coworkers [15] demonstrated a strategic two-step annealing process to create a reliable structural configuration in ZnO nanorods during the first round of annealing at 800 °C under a vacuum; this created V_{Zn}s in the second step of annealing in an oxygen-rich atmosphere in which oxygen flow rates were increased. It is worth noting that the main variables to control in this method are temperature, atmosphere, and pressure.

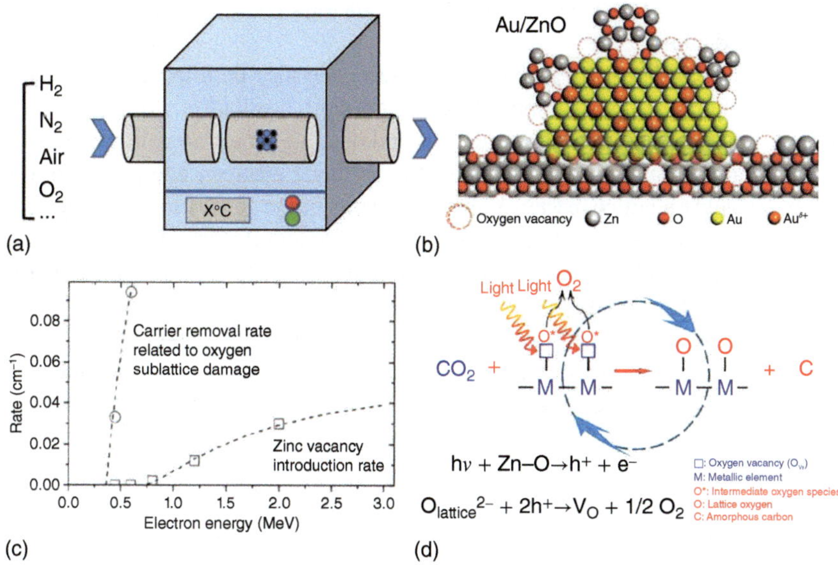

Figure 5.3 (a) Formation of ZnO with vacancy defects via annealing at different conditions; (b) schematic illustration of the influences that oxygen vacancies have on the generation of an Au/ZnO interface; (c) a square represents the zinc vacancy introduction rate as a function of electron irradiation energy; a circle represents the carrier removal rate as a function of electron irradiation energy; (d) a photocorrosion mechanism is used to directly split CO_2 into carbon and O_2. Source: Adapted from Refs. [11–13].

5.3.1.2 Formation of Vacancy Defects via Metal and Nonmetal Doping

It is well known that ZnO has a hexagonal close-packed lattice in which Zn ions occupy half of the tetrahedral sites and have vacant octahedral sites. Therefore, there is a vast number of ZnO sites that are available to be filled by dopants and/or defects. One of the most prominent ways to induce vacancy defects in ZnO is to substitute lattice Zn or O with metal (Au and Cu) and nonmetal elements (C and N) via doping. Introducing the third element breaks the equilibrium state of the electronic structure and causes an uneven distribution of electrons. This results in a rearrangement of electrons and the formation of vacancy defects. Li and coworkers [11] synthesized Au/ZnO catalysts via annealing. During the annealing process of the Au/ZnO interface, gold was inserted into the ZnO lattice, and this caused V_Os to form as a result of charge mismatches (Figure 5.3b). In addition, Krishna and coworkers [16] concluded that dopant metal ions can be mainly divided into acceptor centers and donor centers. The charge of the former is less than that of Zn^{2+}, and the latter is more than that of Zn^{2+}. In the case of ZnO that is doped with an acceptor, the extra negative charge promotes the formation of V_O or interstitial cation defects. Conversely, if ZnO is doped with a donor center, the extra positive charge promotes the formation of V_{Zn} or interstitial anion defects. This method allows a third element to promote the formation of vacancy defects and also serve as a new reaction site in the reaction.

5.3.1.3 Formation of ZnO with Vacancy Defects via High-Energy Electrons and Light Irradiation

Except for defects in the as-grown state, defects can also be introduced in many other ways, such as high-energy electrons or light irradiation. Kuznetsov and coworkers [12] constructed vacancy defects and regulated their concentration using electron irradiation at variable energies. In the lower electron energy range, the damage is solely on lattice oxygen atoms and results in the generation of V_O defects. With a further increase in electron energy, defects gradually appear on lattice zinc atoms, and V_{Zn} eventually becomes the dominant defect (Figure 5.3c). Adjusting the applied electron energy enables this method to be used to induce specific defects. Wang et al. [13] used the photocorrosion reaction ($h\nu + \text{M–O} \rightarrow h^+ + e^-$) of metal oxides under light irradiation to activate lattice oxygen through photogenerated holes (${O_{lattice}}^{2-} + 2h^+ \rightarrow V_O + 1/2O_2$), and then V_O defects were produced (Figure 5.3d).

Also, except for the above methods, rapid crystallization, atomic layer deposition, pulsed laser ablation, and other methods are powerful ways to control defects. Thus, researchers need to select appropriate methods on the basis of laboratory conditions and catalyst materials.

5.3.2 Characterization of Surface Vacancy Defects

A wide range of defects has been studied extensively. None of these results would have been possible without advanced measurement and characterization technologies. For researchers to have a clear understanding of various defects, we must summarize the results of different methods that are used to study defects.

5.3.2.1 Raman Spectroscopy

A common and useful method for evaluating defects in materials is Raman spectroscopy. From group theory, the predictable Raman active phonon modes of wurtzite ZnO, which has C_{6v} symmetry, are A1 + 2E2 + E1. In general, polar phonons exhibit transverse optical (TO) and longitudinal optical (LO) phonons that have different frequencies because of their A1 and E1 symmetry. E2 is a nonpolar mode that consists of two different frequencies. The E2 (high) mode corresponds to oxygen atoms, whereas the E2 (low) mode corresponds to Zn vibrations. From multiple research reports [8], the phonon modes of ZnO have the following locations in Raman spectra (Figure 5.4a): the A1 (TO) mode is at 380 cm^{-1}, the E1 (TO) mode is at 408 cm^{-1}, the E1 (LO) mode is at 583 cm^{-1}, the E2 (low) mode is at 331 cm^{-1}, and the E2 (high) mode is at 437 cm^{-1}. The peak at 576 cm^{-1} is the A1 (LO) mode of the ZnO nanocrystal, and this is similar to that of various ions implanted in a ZnO single crystal; it has been proven that this is correlated to Vo and interstitial Zn. A higher intensity of the A1 (LO) mode indicates a higher defect density in ZnO films.

5.3.2.2 X-ray Photoelectron (XPS) Spectroscopy

X-ray photoelectron spectroscopy (XPS) is another spectroscopic measurement that can be used to detect defects in catalysts. In such a case, XPS results may show additional peaks, or the peaks may shift slightly because of defects that change the bonding energies, or new peaks may appear. The low binding energy peak located at

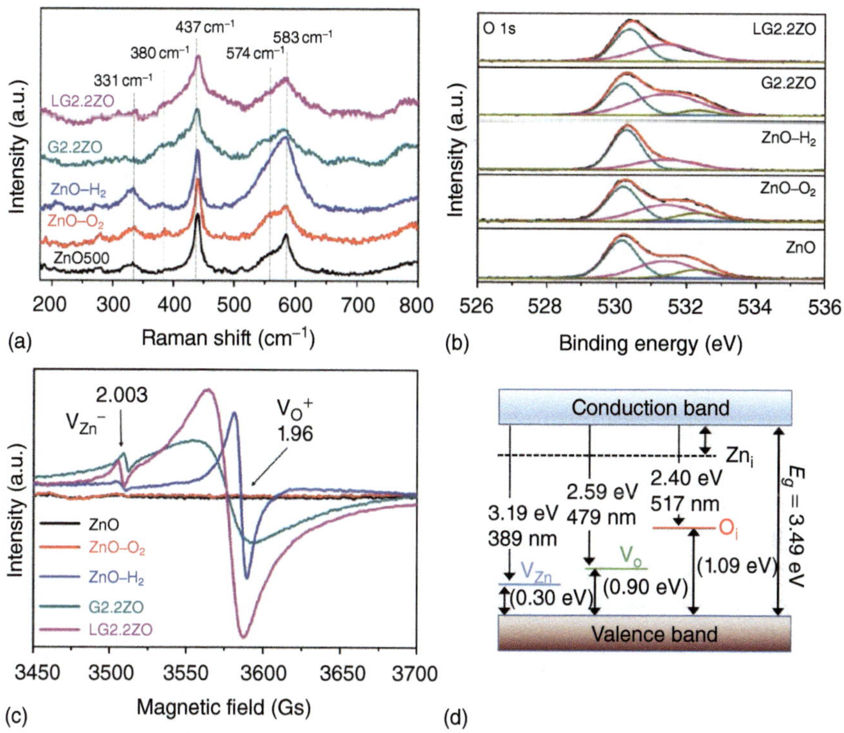

Figure 5.4 (a) Raman scattering spectra of ZnO; (b) O1s XPS spectra of ZnO; (c) EPR spectrum of ZnO under UV irradiation; (d) schematic diagram depicting various defects and ascribed transitions. Source: Reproduced with permission from Wang et al. [8]/Elsevier.

~529 eV can be ascribed to lattice oxygen in the ZnO structure. The medium binding energy peak at ~531 eV is typically attributed to oxygen deficiencies/vacancies in ZnO or OH groups adsorbed on the surface (Figure 5.4b) [8]. Moreover, the atomic ratio of lattice oxygen to Zn at the surface of each sample can be obtained using the peak area and the atomic sensitivity factor (ASF) of Zn2p and O1s ($O1s_{ASF} = 0.63$, $Zn2p_{3/2ASF} = 5.3$) [17]. When ZnO does not have vacancies, the atomic ratio of lattice O to Zn at the surface of the sample is about 1; when ZnO has Vo, the atomic ratio is less than 1, and when ZnO has V_{Zn}, the atomic ratio is greater than 1.

5.3.2.3 Electron Paramagnetic Resonance (EPR) Spectroscopy

Electron paramagnetic resonance (EPR) spectroscopy is a method that can be used to study materials that have unpaired electrons. Materials with various vacancies may have unpaired electrons, and therefore, EPR signals of such materials may be quite different from those of materials without vacancies. This technology can be used to help demonstrate the defects that exist in materials, especially those in metallic oxides. The V_O^0 type of defect captures two electrons and is neutral in the lattice; as long as the spins of two captured electrons compensate, the center is diamagnetic and cannot be detected using EPR. However, a vacancy can be observed as a broad signal if the spins of the two captured electrons do not compensate for each other. The V_O^+ vacancy has an unpaired electron, and thus, the signal can be detected using EPR spectroscopy. The V_O^{2+} vacancy does not have any unpaired electrons, and hence, it is diamagnetic and cannot be detected. The sharp resonance signals at g = 1.960 and g = 2.005 are assigned to oxygen vacancies (V_O^+) and zinc vacancies (V_{Zn}^-), respectively (Figure 5.4c) [8]. Moreover, vacancy defects may further react with photogenerated electrons or holes under light irradiation; this changes their signal position; in this case, there is a new resonance signal at g = 2.019, which results when a zinc vacancy with a photogenerated hole is irradiated by UV light. It is also possible that multiple defects interact with each other to form a new signal response; in this case, the signal at g = ~2.0025 is attributed to V_O–V_{Zn} divacancy [18]. Hence, considering that many vacancy defects are present in the metastable state, the EPR spectra can be measured under photoexcitation irradiation or at low temperatures.

5.3.2.4 Photoluminescence (PL) Spectra

Researchers matched the defect energy level with the fluorescence emission spectrum and calculated the emission wavelength range of various defects. Based on calculations performed by Figure 5.1, the defect levels of V_{Zn}, V_O, and O_i are 0.30, 0.9, and 1.09 eV above the VB, respectively, the level of Zn_i, which is 0.22 eV below the CB (Figure 5.1). Hence, the transition from CB → V_{Zn} may lead to radiative emission at 3.19 eV (389 nm, Figure 5.4d); that from CB → V_O may lead to radiative emission at 2.59 eV (479 nm); that from CB → O_i results in radiative emission of 2.40 eV (517 nm). Therefore, the type and relative concentration of defects in a sample can be determined according to the position and height of the emission peaks.

In addition to the above methods, defects can be analyzed more intuitively with the continuous innovations that have been made in characterization technologies in recent years. Among these developments, positron annihilation lifetime spectroscopy (PALS) provides direct insight into the types and concentrations of defects [19]; positrons are very sensitive to defects, and thus, different vacancies can be identified by measuring the lifetime of the positron. This can make up for shortcomings of other experimental technologies, and PALS can be used for almost all forms of substances (e.g. metals, ionic compounds, polycrystalline, and amorphous). PALS has become a powerful technique for investigating defects in materials. Scanning transmission electron microscopy (STEM) using a spherical aberration-corrected method can be used to observe materials at the atomic level. This technology can be used to directly inspect defects in a material, and the defects can even be counted [20]. These advanced methods can help researchers obtain in-depth information about defects, even at the atomic level.

5.4 Mechanism of Surface Defects on Photocatalytic Reaction Behavior

5.4.1 Roles of Defects in Gas Adsorption

V_O and V_{Zn} generally result in isolated energy levels, and these significantly enhance optical absorption, charge generation, and separation. Moreover, vacancy defects can regulate the Fermi level of ZnO itself and act as adsorption sites that are involved in the surface adsorption and reaction processes. First, according to boundary adsorption theory, electron transfer during gas adsorption depends on the difference between the gas adsorption level and the Fermi level of ZnO itself. If a defect is introduced into the lattice structure, the bandgap width becomes narrower, and the Fermi level changes to match the difference between the levels and to regulate electron transfer.

Wang et al. [17] measured the Mott–Schottky plots for samples under UV irradiation and compared the Fermi levels of different ZnO samples. The Fermi level of ZnO with V_O or V_{Zn} defects is lower than that of perfect ZnO materials, and the influence of V_{Zn} on the Fermi level is more obvious because the level of V_{Zn} is closer to the VB. The experiment further proves that the adsorption electron transfer of H_2 gas in different ZnO samples changed through the gas sensing test system. In addition, V_O and V_{Zn} enrich photogenerated electrons and holes when samples are irradiated by light, and this induces the adsorption of gas molecules, which results in different electron transfers and adsorption modes. For example, H_2 (ads) accepted the electrons from V_O in a nondissociative mode, and H_2 (ads) donated electrons to V_{Zn} in a dissociative mode (Figure 5.5) [21]. It is worth noting that oxygen vacancies often form double-adsorption sites with surrounding metal atoms and can be used for gas adsorption [22]. Reactants tend

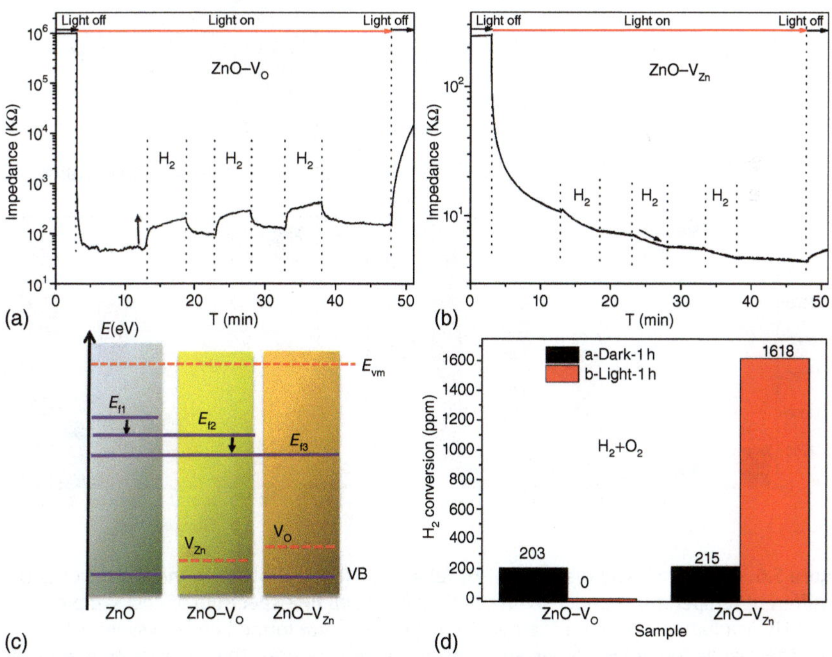

Figure 5.5 Gas sensitivity response of H_2 adsorption on the surface of ZnO−V_O (a) and ZnO−V_{Zn} (b) in a N_2 atmosphere under UV irradiation; (c) schematic diagram of changes in the position of the Fermi energy level under UV irradiation for different ZnO samples; (d) H_2 conversions in catalytic H_2 oxidation over ZnO. Source: Adapted from Wang et al. [17].

to result in chemisorption and obtain electrons from oxygen vacancies, and this is more conducive to electron transfer during gas adsorption. Here, we discuss the influence that vacancy defects have on gas adsorption behavior because the adsorption of gas is an essential part of heterogeneous photocatalytic reactions, and this further affects the reaction direction and activity of photocatalytic reactions. For example, experiments show that photothermal synergy dramatically changed electron transfer of CO adsorption and H_2 via regulating the Fermi level and

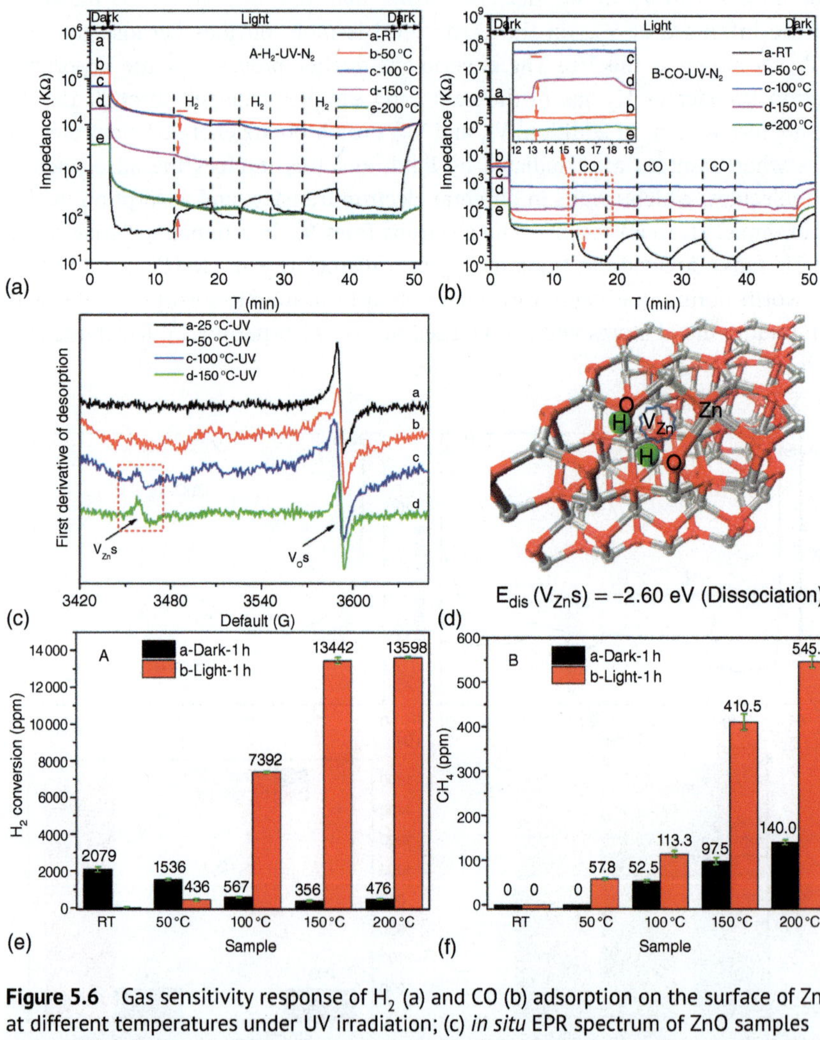

Figure 5.6 Gas sensitivity response of H_2 (a) and CO (b) adsorption on the surface of ZnO at different temperatures under UV irradiation; (c) *in situ* EPR spectrum of ZnO samples under UV irradiation at different temperatures; V_{Zn} is *in situ* formed under a synergistic photothermal action; (d) H_2 dissociative adsorption at V_{Zn}s sites; the dissociation energy of H_2 is −2.60 eV; H_2 conversions in the photothermal catalytic H_2 oxidation (e) and CH_4 yields in the CO reduction (f) over ZnO catalysts under different photothermal conditions. Source: Reproduced with permission from Wang et al. [21]/Elsevier.

providing V_{Zn} adsorption sites, respectively. CO adsorbed on a catalyst can accept electrons; this process is known as pre-reduction, and it tends to undergo reduction reactions. In contrast, CO and H_2 that are adsorbed can donate electrons to the catalyst; this is regarded as a pre-oxidation process, and it is more likely to undergo oxidation reactions (Figure 5.6). CO/H_2 molecules accept/donate electrons and also improve the reduction of CO [21, 23]. In other words, defects change the electron transfer of gas adsorption and adsorption modes by adjusting the Fermi level and providing adsorption sites, and they further affect the selectivity of photocatalytic reactions.

5.4.2 Defects Function as a Double-Edged Sword in Regulating Photocatalytic Performance

It is generally believed that surface defects (V_O and V_{Zn}) are beneficial for improving photocatalytic performance because defects extend the range of light absorption and promote visible light harvesting by causing the band-gap to be narrower and by lowering the Fermi level of ZnO. This facilitates the separation and transport of photogenerated charge carriers (Figure 5.7) [24]. Moreover, defects are used as active

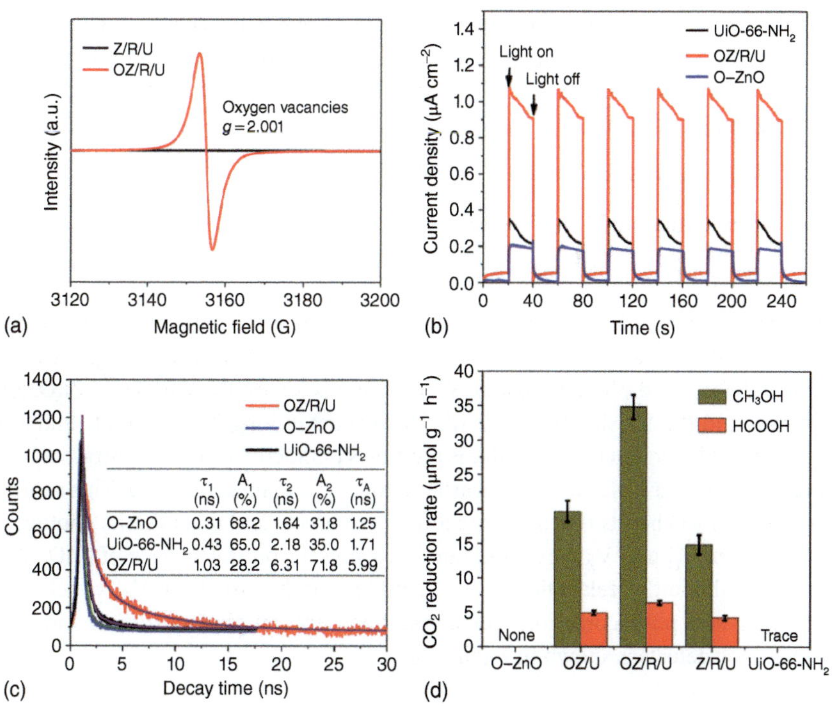

Figure 5.7 (a) EPR spectra of Z/R/U and OZ/R/U (with Vo); (b) time-resolved transient PL decay; (c) photocurrent responses of UiO-66-NH_2, O-ZnO, and OZ/R/U; (d) photocatalytic CO_2 reduction rate of the as-prepared samples. Source: Reproduced with permission from Meng et al. [24]/American Chemical Society.

sites; for example, V_O decreases the Gibbs free energy for the gas (N_2, O_2, and CH_4) activation on the surface of the catalyst by improving the surface charge density [25].

Although defects have the above positive effects, if the concentration of defects is too high, they may have a negative effect on photocatalytic performance. For piezoelectric materials, such as $ZnSnO_3$, the ferroelectric property of $ZnSnO_3$ nanowires can be destroyed by an excessively high concentration of V_O. This leads to poor piezocatalytic performance [26]. Also, Zou and coworkers [27] showed that oxygen vacancies reduce the lifetime of holes to some extent, and this adversely affects charge separation and injection. Furthermore, excess defects may act as charge recombination centers for photogenerated carriers, which can have a detrimental effect on catalytic activity.

Experimental and theoretical results provide a comprehensive understanding that defects (especially the V_O) function as a double-edged sword in regulating photocatalytic performance. As a result, a superior photocatalytic ability is achieved at a moderate concentration of defects.

5.4.3 Defect Engineering Regulates Photocorrosion of ZnO

5.4.3.1 Relationship Between Defects and Photocorrosion

A photocorrosion process causes holes to be transported to the surfaces of ZnO, and these can react with lattice oxygen. The Zn—O bond is attacked, allowing oxygen to escape from the surface. Equations (5.1)–(5.3) summarize this photoinduced process of a ZnO semiconductor:

$$\text{Photocatalysis: } ZnO + h\nu \rightarrow ZnO + h^+ + e^- \tag{5.1}$$

$$\text{Recombination: } h^+ + e^- \rightarrow h\nu \tag{5.2}$$

$$\text{Photocorrosion: } ZnO + 2h^+ \rightarrow Zn^{2+} + 1/2O_2 \tag{5.3}$$

Surface vacancy defects cause electron density to be localized, and this results in lattice relaxation. Wang et al. [10] believe that there is a certain relationship between vacancy defects and photocorrosion. When there is V_O on the surface of ZnO, Zn atoms around the V_O site relax inward. This results in long Zn—O bonds that are more susceptible to fracturing, and this promotes photocorrosion. In contrast, when there is V_{Zn} on the surface, O atoms around the V_{Zn} site relax outward. This results in shorter Zn—O bonds that are more difficult to break, and this inhibits photocorrosion. When V_O and V_{Zn} are simultaneously present on the surface of ZnO, V_{Zn} suppresses the lattice relaxation of V_O by inhibiting the growth of V_O, which further suppresses photocorrosion. The correlation between photocorrosion and surface vacancy defects is shown in Figure 5.8.

5.4.3.2 Constructing an Electron Channel Through Electron Transfer upon the Adsorption of Molecules and Its Role in Inhibiting Photocorrosion of ZnO

V_O also functions as a double-edged sword for ZnO. On one hand, V_O captures photogenerated electrons; this inhibits the recombination of photogenerated carriers and

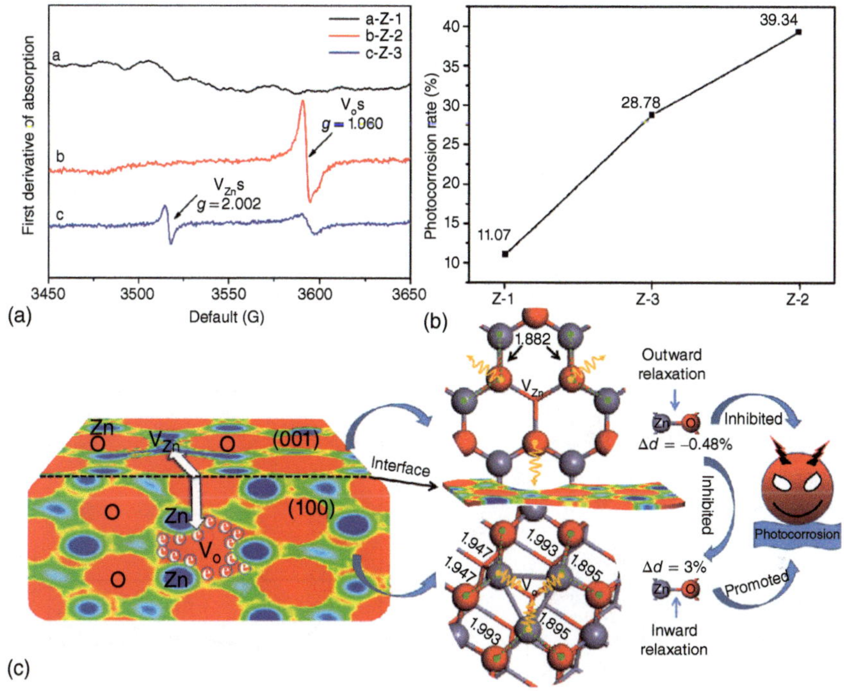

Figure 5.8 (a) EPR spectra and (b) photocorrosion rates of Z-1 and Z-2 with Vo and of Z-3 with V_{Zn} samples; (c) proposed mechanism of the relationship between surface vacancy defects and photocorrosion. Source: Reproduced with permission from Wang et al. [10]/American Chemical Society.

extends the lifetime of photogenerated electrons. On the other hand, excess electrons become enriched around V_O and cause Zn atoms to relax inward; this intensifies photocorrosion. Therefore, we hope to retain the positive effect of oxygen vacancies and reduce their negative effect.

Additionally, an effective electron transfer channel is created via gas adsorption (such as O_2 (curve a in Figure 5.9a), H_2 (curve b in Figure 5.9a) and H_2O (curve c in Figure 5.9a)) in which the adsorbed gas accepts electrons from V_O sites. This results in an excess of electrons around V_O being transferred. Simultaneously, electrons transferred from V_O activate O atoms in O_2 and H_2O molecules. In this way, activated O atoms compensate for O_{lat} that is lost during photocorrosion or also compensate for V_O. As a result, the lattice relaxation of ZnO could further weaken, and this inhibits photocorrosion (Figure 5.9). To verify the above conclusion, we added a trace amount of H_2O to the CO oxidation system, and the stability of ZnO was obviously improved (Figure 5.9d). This chapter explains the relationship between vacancy defects and zinc oxide photocorrosion and provides a possible approach that can be used to solve semiconductor photocorrosion.

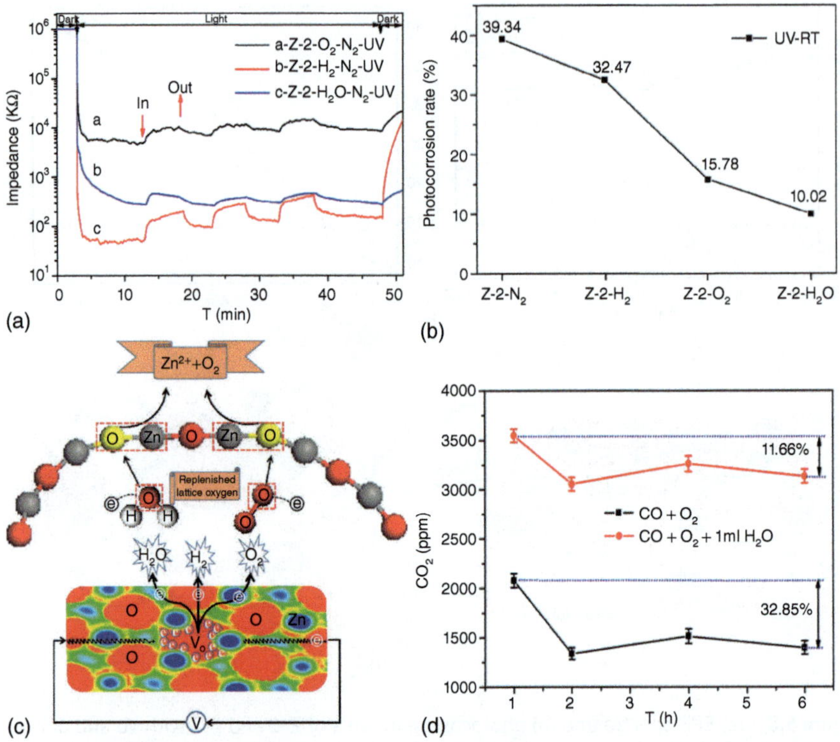

Figure 5.9 (a) Changes in the impedance of a Z-2 (with V_O) sample to (a) O_2, (b) H_2, and (c) H_2O and their corresponding photocorrosion rates (b) when the sample was irradiated by UV light at room temperature in a N_2 atmosphere; (c) schematic illustration showing the mechanism for constructing electron transfer channels that inhibit the photocorrosion of ZnO; (d) formation of CO_2 in the photocatalytic oxidation of CO with O_2 in the absence and presence of H_2O on a Z-2 catalyst when it is irradiated by UV light at room temperature. Source: Reproduced with permission of Wang et al. [10]/American Chemical Society.

5.5 Conclusions and Prospects

This chapter summarizes a strategy for constructing ZnO with defects and the relationship between defects and gas adsorption, surface reactions, and photocorrosion. First, we presented a brief understanding of the four main defects (V_O, V_{Zn}, O_i, and Zn_i) on the surface of zinc oxide and its electronic properties. Then, the synthesis methods for defects are summarized, such as annealing at different conditions, metal and nonmetal doping, electron energies, and light irradiation. Furthermore, the mechanisms of constructing defects using each method are also analyzed. A thorough discussion of their characterization methods and basic mechanisms for defects is presented because the detection of defects is necessary to study them. Finally, defects and defect levels regulate the bandgap and Fermi energy level of ZnO and act as adsorption or reaction sites; these function as a double-edged sword in photocatalytic reactions. As a result, it is suggested that preparing zinc oxide with a medium concentration of defects can obtain excellent photocatalytic performance.

Many researchers have focused on studying the defects of ZnO, but there is still a large field to explore and many barriers to overcome. For example, the existing methods provide us with more accurate information, but researchers still face challenges in obtaining more information about defects. To further understand the role that defects play in photocatalytic reactions, it is necessary to develop additional useful characterization methods, especially *in situ* technologies. Also, the method to construct defects needs to be more flexible and simple. Preparing ZnO or other photocatalysts that have stable surface defects can expand the idea of engineering defects to industrial applications.

References

1 Han, B., Wei, W., Li, M.J. et al. (2019). A thermo-photo hybrid process for steam reforming of methane: highly efficient visible light photocatalysis. *Chem. Commun.* 55: 7816–7819.

2 Zhang, C.J., Fei, W.H., Wang, H.Q. et al. (2020). pn heterojunction of BiOI/ZnO nanorod arrays for piezo-photocatalytic degradation of bisphenol A in water. *J. Hazard. Mater.* 399: 123109.

3 Goktas, S. and Goktas, A. (2021). A comparative study on recent progress in efficient ZnO based nanocomposite and heterojunction photocatalysts: a review. *J. Alloys Compd.* 863: 158734.

4 Zhang, S.B., Wei, S.H., and Zunger, A. (2001). Shallow donor muonium states in II-VI semiconductor compounds. *Phys. Rev. B* 63: 075205.

5 Wei, J.C., Jiang, L.L., Huang, M.L. et al. (2021). Intrinsic defect limit to the growth of orthorhombic HfO_2 and $(Hf,Zr)O_2$ with strong ferroelectricity: first-principles insights. *Adv. Funct. Mater.* 31: 2104913.

6 Yuan, H.Y., Aljneibi, S.A.A.A., Yuan, J.R. et al. (2019). ZnO nanosheets abundant in oxygen vacancies derived from metal-organic frameworks for ppb-level gas sensing. *Adv. Mater.* 31: 1807161.

7 Kim, H., Lee, Y., Lee, Y. et al. (2020). Realization of excitation wavelength independent blue emission of ZnO quantum dots with intrinsic defects. *ACS Photonics* 7 (3): 723–734.

8 Wang, A.Q., Zhang, L., Rahimi, M.G. et al. (2020). Defect engineering of ZnO for electron transfer in O_3 catalytic decomposition. *Appl. Catal., B* 277: 119223.

9 Deng, B., Rosa, A.L., Frauenheim, T. et al. (2014). Oxygen vacancy diffusion in bare ZnO nanowires. *Nanoscale* 6: 11882–11886.

10 Wang, Z.M., Wang, H., Wang, X.X. et al. (2021). Correlation between photocorrosion of ZnO and lattice relaxation induced by its surface vacancies. *J. Phys. Chem. C* 125: 3242–3255.

11 Wu, G.D., Zhao, G.Q., Sun, J.H. et al. (2019). The effect of oxygen vacancies in ZnO at an Au/ZnO interface on its catalytic selective oxidation of glycerol. *J. Catal.* 377: 271–282.

12 Knutsen, K.E., Galeckas, A., Zubiaga, A. et al. (2012). Zinc vacancy and oxygen interstitial in ZnO revealed by sequential annealing and electron irradiation. *Phys. Rev. B* 86: 121203(R).

13 Wang, B., Wang, X.H., Lu, L. et al. (2018). Oxygen-vacancy-activated CO_2 splitting over amorphous oxide semiconductor photocatalyst. *ACS Catal.* 8 (1): 516–525.

14 Yang, J., Lee, J., Lee, J., and Yi, W. (2019). Oxygen annealing of the ZnO nanoparticle layer for the high-performance PbS colloidal quantum-dot photovoltaics. *J. Power Sources* 421: 124–131.

15 Chang, F.M., Brahma, S., Huang, J.H. et al. (2019). Strong correlation between optical properties and mechanism in defciency of normalized self-assembly ZnO nanorods. *Sci. Rep.* 9: 905.

16 Ansari, S.A., Manjunatha, C., Parveen, N. et al. (2021). Mechanistic insights into defect chemistry and tailored photoluminescence and photocatalytic properties of aliovalent cation substituted $Zn_{0.94}M_{0.06-x}Li_xO$ (M: Fe^{3+}, Al^{3+}, Cr^{3+}) nanoparticles. *Dalton Trans.* 50: 14891.

17 Wang, Z.M., Wang, K., Wang, H. et al. (2018). The correlation between surface defects and the behavior of hydrogen adsorption over ZnO under UV light irradiation. *Catal. Sci. Technol.* 8: 3260.

18 Bian, Z., Tachikawa, T., Zhang, P. et al. (2014). A nanocomposite superstructure of metal oxides with effective charge transfer interfaces. *Nat. Commun.* 5: 3038.

19 Geng, Z.G., Kong, X.D., Chen, W.W. et al. (2018). Oxygen vacanciesin ZnO nanosheets enhance CO_2 electrochemical reduction to CO. *Angew. Chem. Int. Ed.* 130: 6162–6167.

20 Wang, Z.M., Wang, H., Xiao, M.Q. et al. (2022). Constructing a channel to regulate the electron transfer behavior of CO adsorption and light-driven CO reduction by H_2 over CuO-ZnO. *ACS Appl. Mater. Interfaces* 14 (19): 22531–22543.

21 Wang, Z.M., Wang, H., Wang, X.X. et al. (2021). Thermo-driven photocatalytic CO reduction and H_2 oxidation over ZnO via regulating the electron transfer behavior of reactant gas adsorption. *Chin. J. Catal.* 42: 1538–1552.

22 Wang, Z.M., Xiao, M.Q., Wang, X.X. et al. (2022). Thermo-driven photocatalytic CO_2 hydrogenation over NiO_x/Nb_2O_5 via regulating the electron transfer behavior of reactant gas adsorption. *Appl. Surf. Sci.* 592: 153246.

23 Wang, Z.M., Wang, X.X., Wang, H. et al. (2020). The role of electron transfer behavior induced by CO chemisorption on visible-light-driven CO conversion over WO_3 and $CuWO_4/WO_3$. *Appl. Catal., B* 265: 118588.

24 Meng, J.C., Chen, Q., Lu, J.Q., and Liu, H. (2019). Z-scheme photocatalytic CO_2 reduction on a heterostructure of oxygen-defective ZnO/reduced graphene oxide/UiO-66-NH_2 under visible light. *ACS Appl. Mater. Interfaces* 11 (1): 550–562.

25 Wang, J., Hu, C.Y., Xia, Y., and Zhang, B. (2021). Mesoporous ZnO nanosheets with rich surface oxygen vacancies for UV-activated methane gas sensing at room temperature. *Sens. Actuators, B* 333: 129547.

26 Wang, Y.C. and Wu, J.M. (2019). Effect of controlled oxygen vacancy on H_2-production through the piezocatalysis and piezophototronics of ferroelectric R3C $ZnSnO_3$ nanowires. *Adv. Funct. Mater.* 30: 1907619.
27 Zhang, Y.C., Afzal, N., Pan, L. et al. (2019). Structure-activity relationship of defective metal-based photocatalysts for water splitting: experimental and theoretical perspectives. *Adv. Sci.* 6: 1900053.

Part II

Photocatalytic Splitting of Water to Produce Hydrogen

6

Strategies for Promoting Overall Water Splitting with Particulate Photocatalysts via Single-Step Visible-Light Photoexcitation

Jiadong Xiao[1], Xiaoping Tao[1], and Kazunari Domen[1,2]

[1]Shinshu University, Research Initiative for Supra-Materials, Interdisciplinary Cluster for Cutting Edge Research, 4-17-1 Wakasato, Nagano-shi, Nagano 380-8553, Japan
[2]The University of Tokyo, Office of University Professors, 2-11-16 Yayoi, Bunkyo-ku, Tokyo 113-8656, Japan

6.1 Introduction

The Sun provides clean, free energy in abundance. In fact, approximately 1.3×10^5 TW of solar power reaches the Earth's surface, which is more than three orders of magnitude greater than the current global energy consumption [1]. Among the various approaches to solar energy conversion, solar-driven water splitting is one of the most promising means of converting solar energy into hydrogen as a storable fuel as well as a feedstock for various industrial processes [1]. Direct water splitting using sunlight and particulate semiconductor photocatalysts via single-step photoexcitation is particularly simple and inexpensive [2] and has been shown to be potentially scalable [3]. The water splitting reaction is an energetically uphill endothermic process with a Gibbs free energy change of 237 kJ mol^{-1} under standard conditions [4]. The single-step photoexcitation mechanism comprises the three steps summarized in Figure 6.1. Initially, a semiconductor absorbs photons having energies larger than its bandgap such that electrons in the valence band (VB) of the semiconductor are excited to the conduction band (CB), leaving positively charged holes in the VB. The VB of the photocatalyst must be situated at an energy level that is more positive than the redox potential associated with water oxidation (1.23 V vs. a reversible hydrogen electrode [RHE]), while the CB should be more negative than the redox potential for proton reduction (0 V vs. RHE). Each photogenerated electron in the CB and hole in the VB can produce an exciton as a result of the Coulomb force, which needs to be overcome to produce free charges. During this process, some excitons and free charges can undergo annihilation via radiative or nonradiative recombination. The surviving photogenerated electrons and holes subsequently migrate to the semiconductor surface through diffusion or as a result of the driving force provided by a built-in electric field. Finally, the photogenerated electrons and holes will react with adsorbed species on the surface of the photocatalyst to evolve hydrogen or oxygen, respectively. It is typically

UV-Visible Photocatalysis for Clean Energy Production and Pollution Remediation: Materials, Reaction Mechanisms, and Applications, First Edition. Edited by Xinchen Wang, Masakazu Anpo, and Xianzhi Fu.
© 2023 WILEY-VCH GmbH. Published 2023 by WILEY-VCH GmbH.

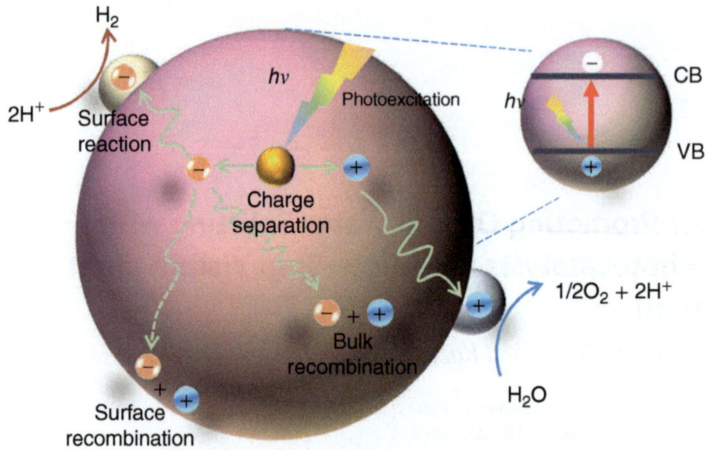

Figure 6.1 Schematic illustration of the main processes occurring during photocatalytic overall water splitting (OWS) via single-step photoexcitation.

necessary to load cocatalysts on the semiconductor surface to accelerate these respective reactions.

The total solar energy conversion efficiency (η) of a photocatalytic material can be defined as the product of the efficiencies associated with light absorption (η_{LA}), charge separation (η_{CS}), and surface conversion (η_{SC}). The η_{LA} value for a photocatalyst determines the upper limit of the theoretical solar-to-chemical conversion efficiency for a given photocatalyst. A solar-to-hydrogen (STH) energy conversion efficiency in the range of 5–10% is required for photocatalytic overall water splitting (OWS) to be economically viable in conjunction with an external quantum efficiency (EQE) of 30–60% based on the adsorption of radiation at wavelengths up to 600 nm [2]. Note that, in the case of a semiconductor photocatalyst with an EQE of unity, a bandgap of 2.7 eV (equivalent to an absorption edge of approximately 460 nm) corresponds to an STH value of 5% [2]. It is therefore vital to achieve high EQE values over photocatalysts responsive up to at least 500 nm. To this end, effective approaches to promoting charge separation and surface conversion reactions for the photocatalyst/cocatalyst systems are required.

The present chapter primarily focuses on strategies that have been applied to promote OWS based on visible-light-responsive particulate photocatalysts via single-step photoexcitation. The $SrTiO_3$:Al/Rh/Cr_2O_3/CoOOH, an almost perfect ultraviolet (UV)-activated photocatalyst/cocatalyst system, is first discussed along with useful means of improving the EQE associated with OWS. Following this, the main techniques that have been used to develop high-performance visible-light-responsive particulate photocatalysts for OWS are examined in detail. These techniques include the control of defects in semiconductors based on the design of new precursors and aliovalent doping, dual-cocatalyst loading, and coating with surface nanolayers. Based on a comparison between the optimized UV-activated system and existing less than ideal visible-light-activated OWS systems, approaches to designing improved long-wavelength-responsive photocatalyst/cocatalyst systems are assessed.

6.2 $SrTiO_3$:Al/Rh/Cr_2O_3/CoOOH: A Model Particulate OWS Photocatalyst

Strontium titanate ($SrTiO_3$) was one of the first materials found to promote photocatalytic water splitting via single-step excitation [5]. Since that time, as a result of persistent effort over four decades, the author's group has demonstrated photocatalytic OWS with a quantum efficiency of almost unity. This was accomplished using a modified aluminum-doped strontium titanate ($SrTiO_3$:Al) photocatalyst together with facet-selective loading of Rh/Cr_2O_3 and CoOOH as cocatalysts [3]. This material has been characterized by selected-area electron diffraction (SAED, Figure 6.2a), transmission electron microscopy (TEM, Figure 6.2b), and crystal orientation analysis (Figure 6.2c). A combination of photocatalyst design strategies was employed to eliminate the efficiency losses associated with charge recombination in this material. Specifically, a molten salt-assisted synthesis method was used to improve the crystallinity of the photocatalyst particles and to ensure a regular crystal morphology, both of which affected the concentration of lattice defects such as grain boundaries between particles and the movement of photogenerated carriers to the particle surfaces. In addition, Al-doping was used to inhibit the formation of Ti^{3+} defects in $SrTiO_3$ crystals that otherwise acted as carrier recombination centers,

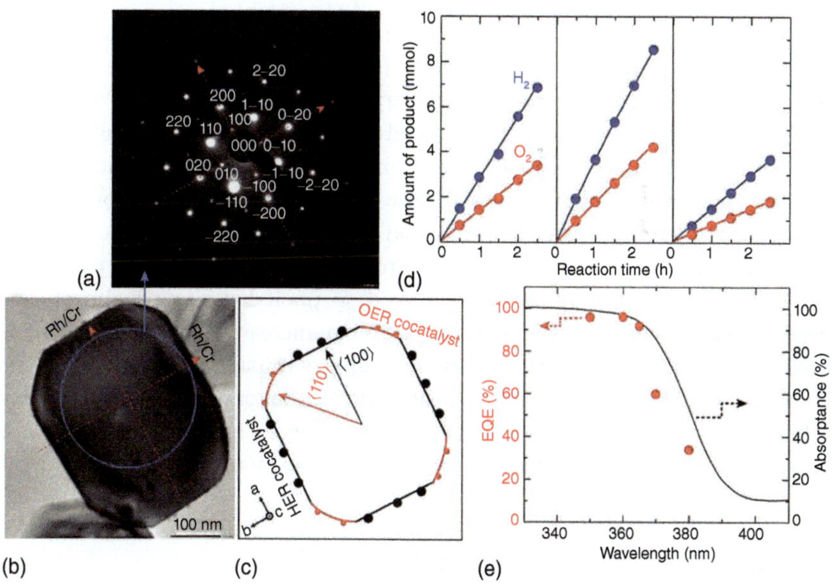

Figure 6.2 (a) SAED pattern, (b) corresponding TEM image, and (c) crystal orientation diagram for a $SrTiO_3$:Al/Rh (0.1 wt%)/Cr_2O_3 (0.05 wt%)/CoOOH (0.05 wt%) particle. (d) Time courses of H_2 and O_2 evolution over $SrTiO_3$:Al loaded with Rh (0.1 wt%)/Cr_2O_3 (0.05 wt%) via two-step photodeposition (left), Rh (0.1 wt%)/Cr_2O_3 (0.05 wt%)/CoOOH (0.05 wt%) via three-step photodeposition (middle), and Rh (0.1 wt%)-Cr (0.1 wt%) via co-impregnation (right) in response to irradiation with a Xe lamp (300 W, full arc). (e) Ultraviolet-visible diffuse reflectance spectrum of bare $SrTiO_3$:Al (black line) and wavelength dependence of EQE during OWS over $SrTiO_3$:Al/Rh/Cr_2O_3/CoOOH (red symbols). Source: Takata et al. [3]/reproduced with permission from Springer Nature.

thus promoting internal charge separation. Finally, and most importantly, exposed nonequivalent {110} and {100} crystal facets have different work functions on the $SrTiO_3$:Al particles generated an internal electric field, resulting in anisotropic charge separation on the different facets.

Anisotropic facet engineering is a precondition for further improvements in the selective loading of Rh/Cr_2O_3 and CoOOH cocatalysts on the electron-collecting {100} and hole-collecting {110} planes, respectively. This loading has been found to provide better results than the loading of only Rh/Cr_2O_3 by two-step photodeposition or co-impregnation methods and has given the highest H_2 and O_2 generation rates of 3.8 and 1.9 mmol h^{-1}, respectively, under UV light (Figure 6.2d). It should be noted that Rh/Cr_2O_3 forms a Rh@Cr_2O_3 core–shell nanostructure upon loading and that the resulting Rh nanoparticles promote both the hydrogen evolution reaction and the backward oxygen reduction reaction. In contrast, Cr_2O_3 prevents the migration of molecular O_2 to the Rh surfaces and so inhibits the backward reactions that would otherwise generate water. As a result, EQEs of up to 96% were achieved over $SrTiO_3$:Al/Rh/Cr_2O_3/CoOOH during OWS under light irradiation in the wavelength range of 350–360 nm (Figure 6.2e). These results correspond to the upper limit of quantum efficiency for OWS over $SrTiO_3$. As a consequence of the limitations imposed by a low η_{LA}, the observed STH energy conversion efficiency for this material was only 0.65%, although this was the highest value reported for a one-step OWS system at that point in time.

$SrTiO_3$:Al/Rh/Cr_2O_3/CoOOH has an optimal cocatalyst/photocatalyst structure and so provides efficient OWS free from charge recombination losses. As such, the strategies adopted to develop this material could also be useful in the design of high-performance, visible-light-activated water splitting photocatalysts. It is important to inhibit or eliminate the formation of detrimental structural and chemical defects in a photocatalyst. These defects can generate recombination and/or deep trapping centers and so can lead to performance losses associated with bulk/surface charge recombination. Anisotropic facet engineering can potentially result in anisotropic charge separation, and so photodepositing hydrogen and oxygen evolution cocatalysts on selected facets can increase charge separation and accelerate surface reactions. In addition, co-loading of suitable hydrogen and oxygen evolution cocatalysts on a site-selective basis provides better performance compared with the use of a hydrogen evolution cocatalyst alone. The OWS performance is also sensitive to the loading method, sequence, composition, and content.

6.3 Current Strategies Promoting OWS with Visible-Light-Activated Particulate Photocatalysts

Although UV-activated photocatalytic OWS has been successfully demonstrated, there are, to the best of the author's knowledge, few visible-light-activated photocatalysts capable of splitting water into H_2 and O_2 via one-step photoexcitation. A majority of these photocatalytic materials were developed by the author's group, including $(Ga_{1-x}Zn_x)(N_{1-x}O_x)$ [6, 7], TaON [8, 9], Ta_3N_5 [10],

$LaMg_xTa_{1-x}O_{1+3x}N_{2-3x}$ [11, 12], and $Y_2Ti_2O_5S_2$ [13]. These compounds exhibit absorption edges of approximately 475, 500, 600, 600, and 650 nm, respectively. This section summarizes the main strategies that enabled one-step OWS and/or notably improved the OWS performance using these visible-light-responsive systems in our recent work.

6.3.1 Defect Control of the Semiconductor Material

6.3.1.1 New Precursor Designs

The precursor material used in the synthetic process greatly affects the properties of the resulting semiconductor, and so the design and use of a suitable precursor is important when employing photocatalytic particles to promote OWS. Ta_3N_5 nanorods epitaxially grown on lattice-matched cubic $KTaO_3$ particles [10] and Zr-doped $TaON/Ta_3N_5$ nanoparticles [9] are two typical examples, both of which showed increased photocatalytic OWS performance primarily as a consequence of the use of a particular precursor.

Ta_3N_5 is an n-type semiconductor with a narrow bandgap (2.1 eV) and suitable CB and VB positions that straddle the water redox potentials. As such, this material is one of the most promising photocatalysts for producing sustainable hydrogen via solar photocatalytic water splitting [14]. However, the conventional synthesis of Ta_3N_5 involves prolonged nitridation of crystalline Ta_2O_5, during which polycrystals with grain boundaries, reduced Ta^V, and anion defects tend to form. These defects subsequently act as recombination centers for charge carriers, thus preventing the migration of carriers and limiting OWS efficiency. As a result of these issues, one-step excitation OWS with Ta_3N_5 was not realized until cubic $KTaO_3$ particles were used as the precursor material in 2018 [10]. A very brief nitridation step produced Ta_3N_5 nanorod single crystals rapidly and selectively on the edges of lattice-matched cubic-like $KTaO_3$ particles via the volatilization of K species, without the formation of grain boundaries (Figure 6.3a–f). When combined with Rh/Cr_2O_3 as a cocatalyst, this material was found to promote OWS in response to visible light, representing the first-ever Ta_3N_5-based photocatalyst for OWS via single-step photoexcitation [10]. Rh/Cr_2O_3-modified $Ta_3N_5/KTaO_3$ exhibited stable, stoichiometric evolution of hydrogen and oxygen under simulated sunlight (AM 1.5G) with an estimated STH efficiency of 0.014% (Figure 6.3g). The EQE values at 320 nm (due to excitation of the $KTaO_3$), 420 nm, and 500 nm were determined to be 2.2%, 0.22%, and 0.024%, respectively, as shown in Figure 6.3h. This limited efficiency can be ascribed to the low incident light utilization and relatively low proportion of Ta_3N_5 single crystals in the $Ta_3N_5/KTaO_3$. Even so, these results confirmed that the fabrication of highly crystalline semiconductor photocatalysts containing fewer defects along with the utilization of an appropriate precursor can play a critical role in achieving efficient solar water splitting.

Another example highlighting the importance of the synthetic precursor is the recently developed OWS-active $Zr\text{-}TaON/Ta_3N_5$ photocatalyst [9]. In this work, amorphous hydrated Ta_2O_5 ($Ta_2O_5{\cdot}3.3H_2O$) with an average particle size of 15 nm (Figure 6.4a) was fabricated as a completely new precursor for tantalum-based

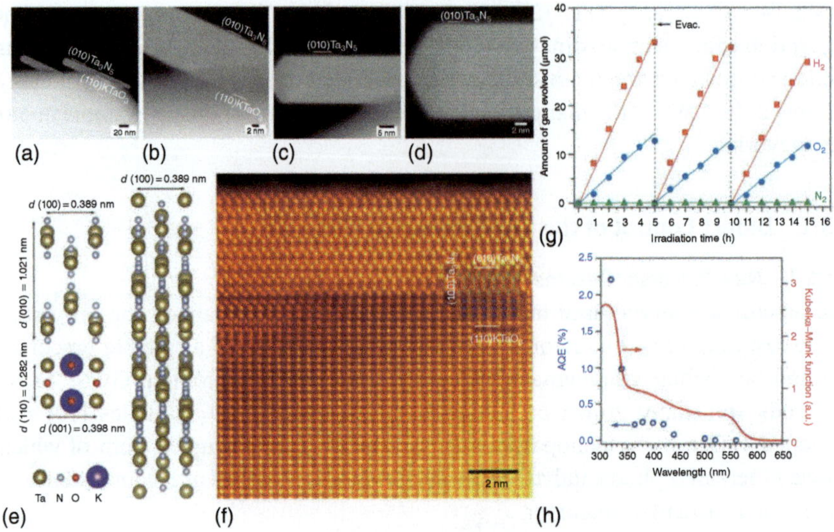

Figure 6.3 (a–d) Annular dark field (ADF)–scanning transmission electron microscopy (STEM) images of $Ta_3N_5/KTaO_3$ viewed along the Ta_3N_5 [001] direction. (e) Crystal structures of Ta_3N_5 and $KTaO_3$ projected from different directions (*d* is the interplanar distance). (f) Colorized and magnified annular dark-field STEM image of $Ta_3N_5/KTaO_3$ synthesized with a nitridation time of 0.25 hour, viewed along the Ta_3N_5 [001] direction. (g) Time courses of gas evolution on $Ta_3N_5/KTaO_3$ loaded with Rh/Cr_2O_3 in response to simulated sunlight (AM 1.5G). (h) Apparent quantum efficiency (AQE) during OWS over $Ta_3N_5/KTaO_3$ loaded with Rh/Cr_2O_3 (blue symbols) as a function of the irradiation wavelength. Source: Wang et al. [10]/reproduced with permission from Springer Nature.

(oxy)nitride materials. The Zr-TaON/Ta_3N_5 (3 h) photocatalyst obtained from the nitridation of Zr-impregnated $Ta_2O_5{\cdot}3.3H_2O$ comprised segregated Zr-TaON and Zr-Ta_3N_5 nanoparticles having particle sizes of approximately 63 nm that were free from grain boundaries (Figure 6.4b). In contrast, a Zr-TaON/Ta_3N_5 (3 h) photocatalyst synthesized from standard crystalline Ta_2O_5 particles approximately 445 nm in size (Figure 6.4d) comprised micron-sized polycrystalline particles with numerous grain boundaries (Figure 6.4e) that proved to be detrimental with regard to OWS. As a result of the elimination of grain boundaries and the significantly reduced particle size stemming from the utilization of a new precursor, the Zr-TaON/Ta_3N_5 (3 h) photocatalyst (Figure 6.4c) provided an OWS efficiency that was improved by an order of magnitude compared with that of the latter material (Figure 6.4f).

6.3.1.2 Aliovalent Doping

The compositional modification of semiconductor material by doping with foreign ions is a simple means of improving photocatalytic performance via chemical defect control [14]. Even so, this technique has rarely been used successfully in visible-light-driven photocatalytic OWS processes. In the case of the Zr-TaON/Ta_3N_5 photocatalyst described above [9], doping with Zr^{IV} improved the OWS efficiency of the material by a factor of approximately five (Figure 6.5a). An optimized Zr-TaON/Ta_3N_5 (3 h) sample with a Zr/Ta molar ratio of 7.2%, loaded with a

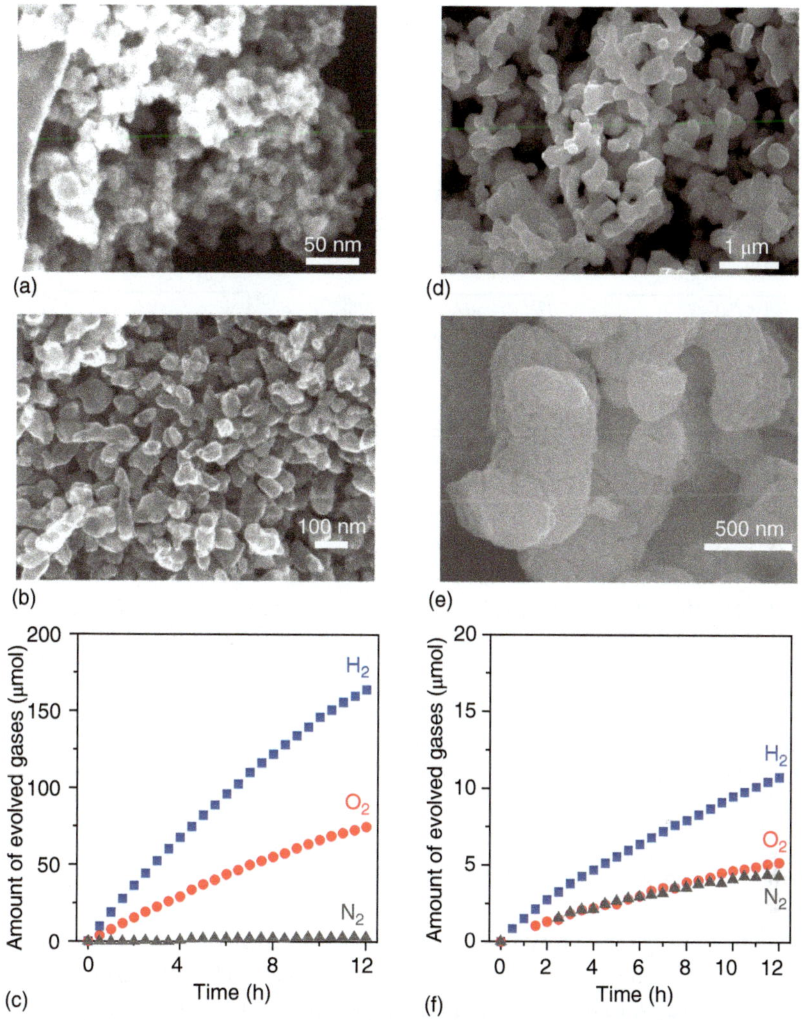

Figure 6.4 Scanning electron microscopy (SEM) images of (a) the amorphous $Ta_2O_5 \cdot 3.3H_2O$ precursor and (b) the resulting Zr-TaON/Ta_3N_5 (3 h) photocatalyst. The term "3 h" indicates the nitridation period. (c) Time courses of gas evolution during OWS over Ru/Cr_2O_3/IrO_2-loaded Zr-TaON/Ta_3N_5 (3 h) (i.e. the material in panel b) under visible light ($\lambda > 380$ nm). SEM images of (d) crystalline Ta_2O_5 precursor and (e) resulting Zr-TaON/Ta_3N_5 (3 h) photocatalyst. (f) Time courses of gas evolution during OWS over Ru/Cr_2O_3/IrO_2-loaded Zr-TaON/Ta_3N_5 (3 h) (i.e. the material in panel e) under visible light ($\lambda > 380$ nm). Source: Xiao et al. [9]/reproduced with permission from John Wiley & Sons.

Ru/Cr_2O_3/IrO_2 cocatalyst system exhibited EQE values of 0.81%, 0.73%, 0.66%, 0.27%, and 0% at 365, 380, 420, 460, and 500 nm (Figure 6.5b), respectively, along with an STH energy conversion efficiency of 0.009% (Figure 6.5c). Note that only the Ru/Cr_2O_3/IrO_2-loaded Zr-TaON particles actively promoted OWS via one-step excitation, while the Zr-Ta_3N_5 in Zr-TaON/Ta_3N_5 (3 h) was most likely an inert byproduct that consumed photons but did not promote OWS [9]. A Ta $4f_{7/2}$ component

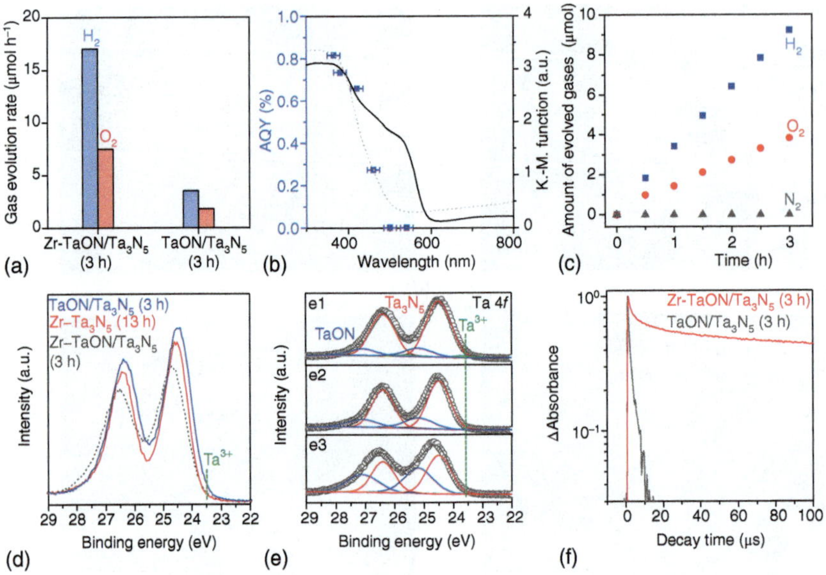

Figure 6.5 (a) Effect of Zr-doping on photocatalytic OWS performance. (b) EQE values (symbols) as a function of incident light wavelength during OWS over Ru/Cr_2O_3/IrO_2-loaded Zr-TaON/Ta_3N_5 (3 h). The solid and dashed lines are the diffuse reflectance spectroscopy (DRS) data obtained from Zr-TaON/Ta_3N_5 (3 h) and Zr-TaON (0.5 h), respectively. (c) Time courses of gas evolution during OWS over Ru/Cr_2O_3/IrO_2-loaded Zr-TaON/Ta_3N_5 (3 h) under simulated sunlight (0.8 sun, 30 cm^2 irradiation area). (d) Ta 4*f* XPS spectra obtained from pristine TaON/Ta_3N_5 (3 h), Zr-Ta_3N_5 (13 h) and Zr-TaON/Ta_3N_5 (3 h). (e) Deconvoluted Ta 4*f* XPS spectra of (b1) TaON/Ta_3N_5 (3 h), (b2) Zr-Ta_3N_5 (13 h) and (b3) Zr-TaON/Ta_3N_5 (3 h). (f) Transient decay profiles of the conduction band and/or shallowly trapped electrons (monitored at 5000 nm) for Zr-TaON/Ta_3N_5 (3 h) and TaON/Ta_3N_5 (3 h). Source: Reproduced with permission from Xiao et al. [9]; © 2022 John Wiley & Sons, Inc.

with a binding energy of 23.6 eV was evident in the spectrum obtained from Zr-free TaON/Ta_3N_5 (3 h) using X-ray photoelectron spectroscopy (XPS) but was not present in the Zr-TaON/Ta_3N_5 (3 h) or Zr-Ta_3N_5 (13 h) spectra (Figure 6.5d,e). This peak was assigned to Ta^{3+} cations based on a previous report [15] and the fact that Ta^{4+} ($I = 7/2$ (100%)), which generates an electron paramagnetic resonance signal [16], was not detected. The elimination of Ta^{3+} as a consequence of Zr doping resulted in an increase in the charge carrier population in the photocatalyst and extended the lifetimes of these carriers (Figure 6.5f). This effect is believed to be the primary reason for the increased OWS activity following Zr^{IV} doping. Thus, these data demonstrate that aliovalent doping can decrease the concentration of chemical defects and so promote the solar energy conversion efficiencies of OWS photocatalysts.

6.3.2 Dual-Cocatalyst Loading

During single-step OWS, the hydrogen and oxygen evolution reactions need to occur on the same photocatalyst particle. In this scenario, it is essential to load cocatalysts on the photocatalyst particle surfaces to decrease the overpotential and so accelerate surface reactions. These cocatalysts can also enhance photogenerated charge separation by modulating the degree of band bending and changing the built-in electric field in the spatial charge region [4, 17]. The cocatalysts designed for $SrTiO_3$:Al demonstrate that, while OWS can proceed using a semiconductor photocatalyst modified solely with a hydrogen evolution cocatalyst, co-loading of dual cocatalysts (i.e. hydrogen and oxygen evolution cocatalysts) enhances the reaction kinetics of existing OWS systems. Note that M/Cr_2O_3 (M = Rh, Ru, Pt, Pd, Ir, etc.) having an $M@Cr(III)O_{1.5-m}(OH)_{2m} \cdot xH_2O$ core–shell nanostructure is a typical hydrogen evolution cocatalyst. In this material, the metal cores act as electron collectors and as the hydrogen evolution cocatalyst, while the Cr shells serve as molecular sieves, preventing reverse reactions involving oxygen molecules [2, 18].

A GaN:ZnO solid solution is a representative photocatalyst capable of splitting water via single-step visible-light excitation. As shown in Figure 6.6a,b, the co-loading of Rh/Cr_2O_3 serving as a water reduction cocatalyst and Mn_3O_4 as a water oxidation cocatalyst on the surface of this material can enhance the OWS activity remarkably [19]. The modification of TaON [8, 9] and $Y_2Ti_2O_5S_2$ [13] with

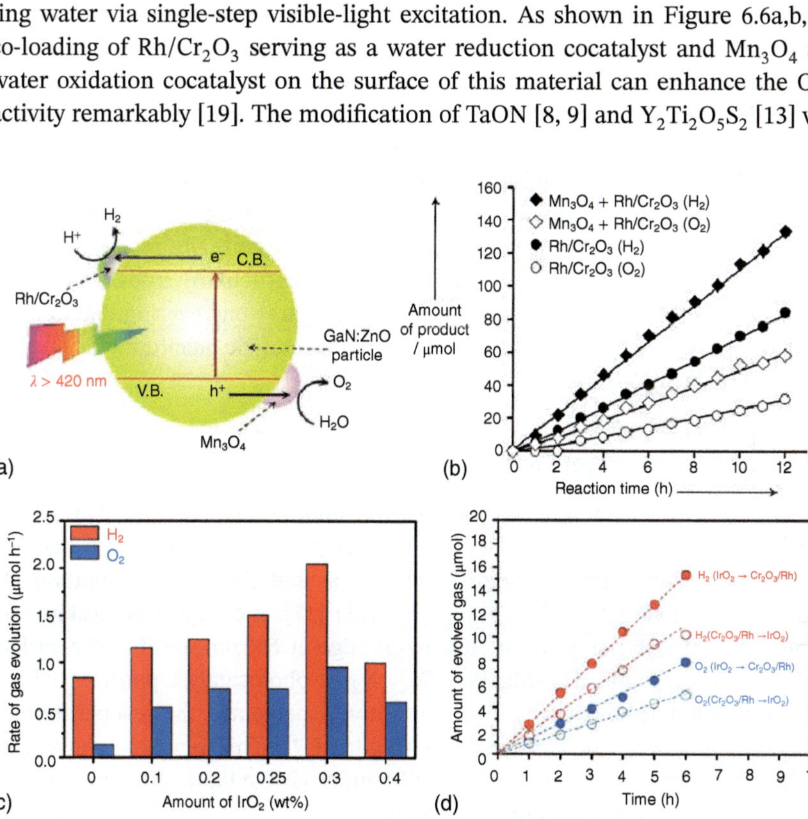

Figure 6.6 (a) Diagram showing water splitting on $Rh/Cr_2O_3/Mn_3O_4$-decorated GaN:ZnO particle. (b) Time courses of H_2 and O_2 evolution during OWS using GaN:ZnO catalysts loaded with $Rh/Cr_2O_3/Mn_3O_4$ or Rh/Cr_2O_3 under visible light. Source: Reproduced with permission from Maeda et al. [19]; © 2010 John Wiley & Sons, Inc. (c) Dependence of gas evolution rates over $Cr_2O_3/Rh/IrO_2$-modified $Y_2Ti_2O_5S_2$ powder on the amount of IrO_2. The Rh and Cr loadings were fixed at 1.5 and 1.5 wt%, respectively. (d) Effect of dual-cocatalyst loading sequence on photocatalytic activity of $Y_2Ti_2O_5S_2$ during OWS. Source: Adapted from Wang et al. [13].

dual cocatalysts was also found to be indispensable to achieving stoichiometric water splitting via single-step excitation. As demonstrated in Figure 6.6c, Cr_2O_3/Rh-modified $Y_2Ti_2O_5S_2$ produced H_2 and O_2 under visible light, but the H_2/O_2 ratio greatly deviated from the expected stoichiometric value of two, most likely due to the ongoing oxidation of S^{2-} ions in the $Y_2Ti_2O_5S_2$. Co-loading with IrO_2 was, therefore, necessary in order to extract holes and reduce the water oxidation energy barrier, such that oxygen evolution was promoted and the stoichiometric evolution of H_2 and O_2 was achieved. Notably, $Y_2Ti_2O_5S_2$ loaded first with Cr_2O_3/Rh and then with IrO_2 split water 30% less efficiently than that loaded with IrO_2 and then Cr_2O_3/Rh (Figure 6.6d). This effect was attributed to the nonselective deposition of IrO_2, such that part of the Cr_2O_3/Rh cocatalyst was masked.

In these systems, the majority of cocatalysts are deposited on the photocatalyst surface non-selectively and at relatively high concentrations (such as 2–4 wt% of a noble metal relative to the mass of the photocatalyst) using impregnation or adsorption methods. Although both the cocatalyst loading method and density were optimized to allow the photocatalytic system to take full advantage of randomly dispersed active sites, there will inevitably be a mismatch between the deposition sites and the cocatalysts. That is, some of the reduction cocatalysts will be deposited at hole-accumulating sites, and/or some of the oxidation cocatalysts will be located at electron-accumulating sites. This will lead to severe charge recombination and greatly limit the OWSs performance. This scenario essentially results from the nonoptimal design of the semiconductor material, particularly from the absence of spatial charge separation based on anisotropic facet engineering. Hence, this engineering is a vital aspect of improving the photocatalytic quantum efficiency, as demonstrated by the design of $SrTiO_3$:Al [3].

6.3.3 Surface Nanolayer Coating

In addition to the loading of cocatalysts, the photocatalyst surface can also be modified to modulate the complex ion transfer processes taking place between the interface of the semiconductor or cocatalyst and the aqueous solution [4]. In the case of $LaMg_xTa_{1-x}O_{1+3x}N_{2-3x}$ ($x \geq 1/3$) [11], a complex perovskite-type oxynitride photocatalyst with an absorption edge at 600 nm, coating the surface of the $RhCrO_y$-decorated $LaMg_xTa_{1-x}O_{1+3x}N_{2-3x}$ photocatalyst particles with a nanolayer of amorphous oxyhydroxide effectively prevented the reverse reaction and self-oxidation of the photocatalyst (Figure 6.7). This rendered the system capable of splitting water photocatalytically under visible light, although the EQE value at c. 440 nm was only approximately 0.03%.

A coating of amorphous oxides of Ti, Ta, Nb, or Si allows the permeation of oxygen from the interior to the exterior of the system, likely due to the flexible lattice of the amorphous oxide layer and the large oxygen partial pressure gradient across the layer. However, permeation in the opposite direction is unlikely (Figure 6.7a). As such, oxygen molecules will be unable to pass through to the reduction sites, and so the reverse reaction is inhibited, and OWS activity is increased. Double coating with both SiO_2 and MO_x (M = Ti, Zr, Nb, or Ta) amorphous layers using a modified

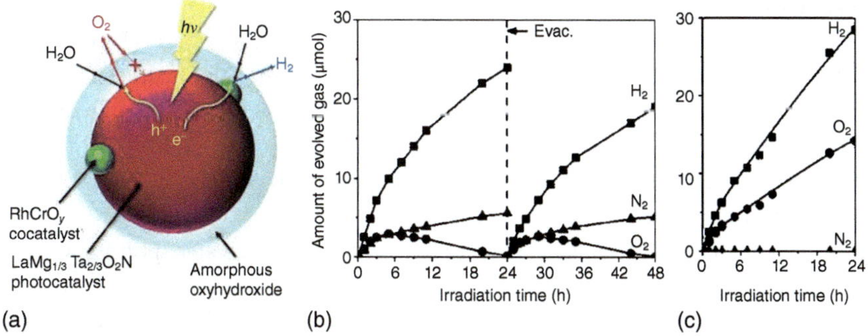

Figure 6.7 (a) Diagram of water splitting reaction mechanism on surface-coated photocatalyst. Gas evolution during water splitting on (b) $RhCrO_y/LaMg_{1/3}Ta_{2/3}O_2N$ and (c) $TiOXH/RhCrO_y/LaMg_{1/3}Ta_{2/3}O_2N$. Source: Reproduced with permission from Pan et al. [11]; © 2015 John Wiley & Sons, Inc.

photodeposition method was found to increase water splitting activity several times as a result of the more effective suppression of the oxygen reduction reaction [12]. This technique improved the EQE value of the system to approximately 0.18% at c. 440 nm [12], when combined with a slight refinement of the photocatalyst quality.

6.4 Concluding Remarks

To date, no particle-based photocatalytic OWS system has simultaneously exhibited a high EQE and long-wavelength responsiveness. Thus, the development of such materials remains a critical challenge. Although several photocatalytic systems are able to split water in response to wavelengths as high as 600–650 nm via single-step photoexcitation based on the use of new precursors, aliovalent doping, dual-cocatalyst loading, and surface nanolayer coating, the EQE values at 420 nm remain limited to approximately 0.4%. As indicated by the design of the $SrTiO_3$:Al/Rh/Cr_2O_3/CoOOH system, which shows an EQE of almost unity, spatial charge separation via anisotropic facet engineering of the semiconductor photocatalyst is an essential aspect of achieving a high EQE. However, this technique has not yet been applied to existing visible-light-activated OWS photocatalysts, leading to severe performance losses as a consequence of charge recombination. Thus, it will be vital to fabricate single-crystalline fine particles of visible-light-driven photocatalysts with different exposed crystal facets that can be modified site-selectively with suitable cocatalysts. Such work is expected to provide sufficient STH energy conversion efficiencies in one-step OWS systems.

Acknowledgments

This work was financially supported by the Artificial Photosynthesis Project of the New Energy and Industrial Technology Development Organization (NEDO).

References

1 Hisatomi, T. and Domen, K. (2019). Reaction systems for solar hydrogen production via water splitting with particulate semiconductor photocatalysts. *Nat. Catal.* 2: 387–399.

2 Chen, S., Takata, T., and Domen, K. (2017). Particulate photocatalysts for overall water splitting. *Nat. Rev. Mater.* 2: 1–17.

3 Takata, T., Jiang, J., Sakata, Y. et al. (2020). Photocatalytic water splitting with a quantum efficiency of almost unity. *Nature* 581: 411–414.

4 Tao, X., Zhao, Y., Wang, S. et al. (2022). Recent advances and perspectives for solar-driven water splitting using particulate photocatalysts. *Chem. Soc. Rev.* 51: 3561–3608.

5 Domen, K., Naito, S., Soma, M. et al. (1980). Photocatalytic decomposition of water vapour on an NiO-$SrTiO_3$ catalyst. *J. Chem. Soc., Chem. Commun.* 543–544.

6 Maeda, K., Takata, T., Hara, M. et al. (2005). GaN:ZnO solid solution as a photocatalyst for visible-light-driven overall water splitting. *J. Am. Chem. Soc.* 127: 8286–8287.

7 Ohno, T., Bai, L., Hisatomi, T. et al. (2012). Photocatalytic water splitting using modified GaN:ZnO solid solution under visible light: long-time operation and regeneration of activity. *J. Am. Chem. Soc.* 134: 8254–8259.

8 Maeda, K., Lu, D., and Domen, K. (2013). Direct water splitting into hydrogen and oxygen under visible light by using modified TaON photocatalysts with d^0 electronic configuration. *Chem. Eur. J.* 19: 4986–4991.

9 Xiao, J., Nishimae, S., Vequizo, J.J.M. et al. (2022). Enhanced overall water splitting by a zirconium-doped TaON-based photocatalyst. *Angew. Chem. Int. Ed.* 61: e202116573.

10 Wang, Z., Inoue, Y., Hisatomi, T. et al. (2018). Overall water splitting by Ta_3N_5 nanorod single crystals grown on the edges of $KTaO_3$ particles. *Nat. Catal.* 1: 756–763.

11 Pan, C., Takata, T., Nakabayashi, M. et al. (2015). A complex perovskite-type oxynitride: the first photocatalyst for water splitting operable at up to 600 nm. *Angew. Chem. Int. Ed.* 54: 2955–2959.

12 Pan, C., Takata, T., and Domen, K. (2016). Overall water splitting on the transition-metal oxynitride photocatalyst $LaMg_{1/3}Ta_{2/3}O_2N$ over a large portion of the visible-light spectrum. *Chem. Eur. J.* 22: 1854–1862.

13 Wang, Q., Nakabayashi, M., Hisatomi, T. et al. (2019). Oxysulfide photocatalyst for visible-light-driven overall water splitting. *Nat. Mater.* 18: 827–832.

14 Xiao, J., Vequizo, J.J.M., Hisatomi, T. et al. (2021). Simultaneously tuning the defects and surface properties of Ta_3N_5 nanoparticles by Mg-Zr codoping for significantly accelerated photocatalytic H_2 evolution. *J. Am. Chem. Soc.* 143: 10059–10064.

15 Xiao, Y., Feng, C., Fu, J. et al. (2020). Band structure engineering and defect control of Ta_3N_5 for efficient photoelectrochemical water oxidation. *Nat. Catal.* 3: 932–940.

16 Hunter, J.A., Lindsell, W.E., McCullough, K.J. et al. (1990). Bis (η^5-cyclopentadienyl) complexes of niobium(IV) and tantalum(IV); electron spin resonance and electrochemical studies and the molecular structure of [Ta(SCOPh)$_2$(η^5-C_5H_5)$_2$]. *J. Chem. Soc., Dalton Trans.* 2145–2153.

17 Karkas, M.D., Verho, O., Johnston, E.V., and Akermark, B. (2014). Artificial photosynthesis: molecular systems for catalytic water oxidation. *Chem. Rev.* 114: 11863–12001.

18 Yoshida, M., Takanabe, K., Maeda, K. et al. (2009). Role and function of noble-metal/Cr-layer core/shell structure cocatalysts for photocatalytic overall water splitting studied by model electrodes. *J. Phys. Chem. C* 113: 10151–10157.

19 Maeda, K., Xiong, A., Yoshinaga, T. et al. (2010). Photocatalytic overall water splitting promoted by two different cocatalysts for hydrogen and oxygen evolution under visible light. *Angew. Chem. Int. Ed.* 49: 4096–4099.

7

Integration of Redox Cocatalysts for Photocatalytic Hydrogen Evolution

Muhammad Tayyab[1], Yujie Liu[1], Zehong Xu[1], Summan Aman[2], Wenhui Yue[1], Rana M. Irfan[3], Liang Zhou[1], and Jinlong Zhang[1]

[1] *East China University of Science & Technology, School of Chemistry and Molecular Engineering, 130 Meilong Road, Shanghai 200237, P.R. China*
[2] *University of Gujrat, Department of Chemistry, Gujrat, 50700, Pakistan*
[3] *Sogang University, Department of Physics, Seoul, 04107, Korea*

7.1 Introduction

Most of the world's energy demand is currently fulfilled by CO_2-emitting technologies, which have well-known effects on climate change [1]. Therefore, the world's issues concerning the energy crisis and environmental pollution necessitate the world's future in hydrogen production by solar energy, which has zero impact on the environment [2]. Photocatalysis is a method that can harness and store solar radiation due to its renewable capability and incomparable abundance on Earth [3]. Solar energy is already being converted to Earth's biomass by a process of natural photosynthesis; however, such solar-to-biofuel conversion efficacy is comparatively low, which is unable to fulfill the world's future requirements and the fossilization process [2]. In contrast, artificial photosynthesis empowers the conversion of solar radiation into more valuable chemicals and fuels [3]. Interestingly, the solar-to-biofuel conversion efficiency can be improved via the modulation and rational design of the photocatalysis system, which surpasses that of natural photosynthesis [4]. As a result, semiconductor photocatalysis is viewed as a promising strategy for resolving the current energy, environmental, and global economic crises.

However, photocatalytic water splitting has relatively low efficiency for solar-to-hydrogen transformation, which is far from the needed standard for practical applications. The low efficiency of hydrogen production by photocatalysis is recognized as the following issues in these three vital steps of a bare photocatalyst: (i) inefficient consumption of the solar radiation and excitation of the photogenerated charge carriers; (ii) incompetent separation and transfer of photogenerated electron–hole pairs; and (iii) sluggish kinetics of the surface

UV-Visible Photocatalysis for Clean Energy Production and Pollution Remediation: Materials, Reaction Mechanisms, and Applications, First Edition. Edited by Xinchen Wang, Masakazu Anpo, and Xianzhi Fu.
© 2023 WILEY-VCH GmbH. Published 2023 by WILEY-VCH GmbH.

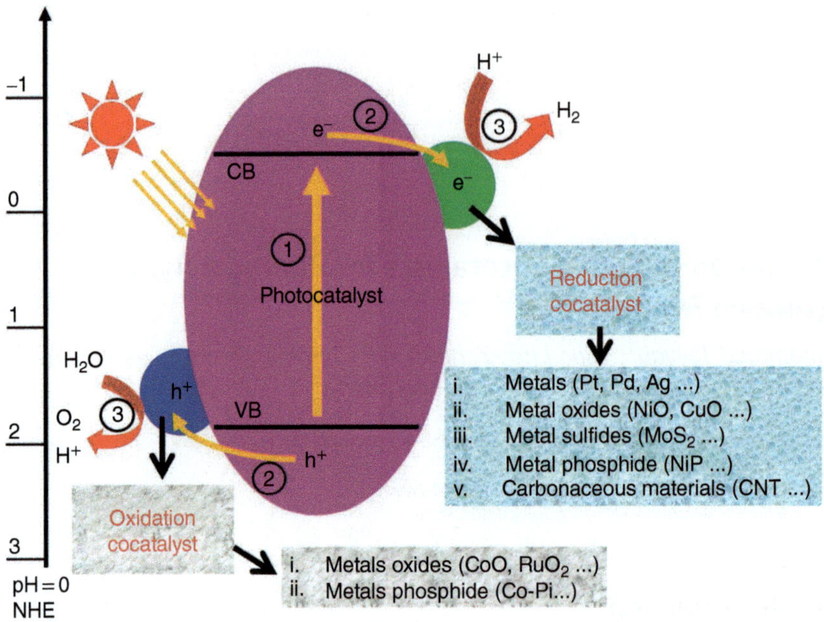

Figure 7.1 A schematic illustration of the dual redox cocatalysts and their respective reactions on the surface of the photocatalyst.

redox reactions (Figure 7.1) [3, 5]. All these consecutive steps overall affect the efficiency of hydrogen production by photocatalysis [6]. Furthermore, it is well acknowledged that the recombination rate of photogenerated electron–hole pairs on the surface of the photocatalysts is relatively very rapid (10^{-12}–10^{-11} seconds), as compared to their transfer and trapping rates (10^{-9}–10^{-7} seconds) [4]. As such, transfer, charge trapping, and redox reactions on the surface of a photocatalyst are generally regarded to be the rate-determining steps. Hence, to overcome the above problems, different strategies were adopted to design and fabricate efficient photocatalysts by applying heterojunction, doping, vacancies, defects, loading of cocatalysts, etc. [7, 8].

A cocatalyst deposition, such as an oxidation or reduction cocatalyst, has been demonstrated to be a dependable strategy for accelerating charge transfer and suppressing surface charge recombination on photocatalysts (Figure 7.1) [9, 10]. The simultaneous loading of dual cocatalysts on a photocatalyst appears to be a more advantageous tactic to boost the surface catalytic competence with respect to the anchoring of a solo cocatalyst [4]. Furthermore, it should be prominent that the synergetic effects of the dual redox cocatalysts act as a vital role in boosting the efficiencies of redox reactions on the surface of photocatalysts. Nowadays, this strategy has attracted attention in the field of photocatalytic hydrogen evolution reaction (HER) as impressive improvements have been attained in boosting the catalytic reaction, selectivity, and stability.

7.2 Fundamentals of Dual Cocatalysts

7.2.1 Classification of Cocatalysts on the Basis of the Functional Mechanism

The term "dual redox cocatalysts" discussed below refers to oxidation and reduction cocatalysts, which are classified on the basis of their functions and the trapped charge carriers during the photocatalytic HER. Generally, metal-based molecular materials, metal phosphates (Ni–Pi and Co–Pi), and metal oxides (MnO_x and IrO_2) play a vital role in obtaining holes from photocatalysts, thus accelerating the photocatalytic oxygen evolution reaction [9]. Figure 7.2a describes the loading of oxidation cocatalysts on photocatalysts, which causes an actually reduced activation energy of the water molecules. In particular, the major challenge for water oxidation on solo photocatalysts is the photocatalyst's comparatively negative valence band hybridization, which results in poor oxidation capability and hence fails to achieve the greater energy barrier for O—O bond formation and O—H bond cleavage [11]. In comparison, oxidation cocatalysts, particularly spinel oxides, are capable of offering additional active centers for water oxidation due to their customized electronic structures, such as the metal-oxygen bond, electrons of e_g orbital, and the number of d electrons [11].

In contrast, non-noble metals, non-noble metal sulfides, non-noble metal phosphides, non-noble metal nitrides, non-noble metal carbides, metal-free elemental materials (tellurene, graphene, phosphorene, graphdiyne, and carbon nanotubes), precious metals, and their derivatives have been determined to act as reduction cocatalysts to contribute to the HER [10]. Most reduction cocatalysts possess a higher work function (low Fermi level) and metallic properties result in a Schottky barrier between the photocatalysts and the reduction cocatalysts, which promotes electron transfer from the photocatalysts to the cocatalysts (Figure 7.2b) [9]. The Fermi level difference between semiconducting photocatalysts and metallic reduction cocatalysts drives the charge flow until the Fermi level equilibrium is achieved, and then band bending and Schottky barriers are formed.

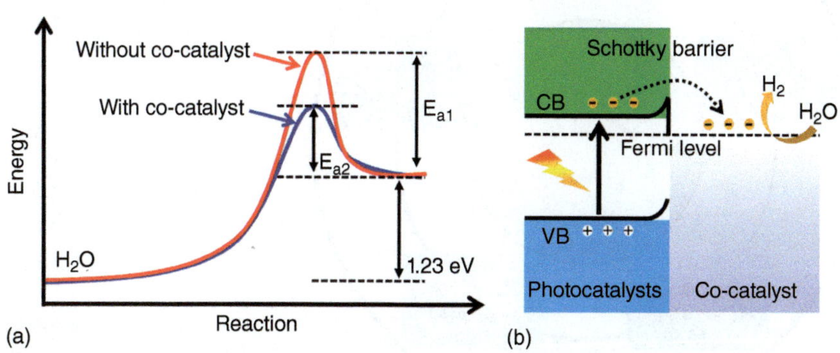

Figure 7.2 Functional mechanisms of (a) an oxidation cocatalyst and (b) a reduction cocatalyst.

7.2.2 The Advantages of the Design of Dual Cocatalysts

In general, the efficiency of photocatalytic hydrogen evolution is closely related to both the oxidation and reduction half-reactions; thus, handling both of them is equally important. The introduction of dual redox cocatalysts allows charge carriers to be separated from photocatalyst surfaces, which considerably extends their lifetime and prevents the recombination of electron–hole pairs. In particular, the dual redox cocatalyst plays five crucial roles in tuning the durability, activity, and selectivity of the photocatalysis system (Figure 7.3) [11]. First, dual redox cocatalysts can get their associated photoinduced charge carriers from the photocatalysts for spatial charge separation, which is a crucial step for reaching an excessive solar-to-chemical conversion efficiency [12]. Second, dual cocatalysts with abundant active sites and excellent electronic conductivity improve hydrogen evolution activity by enabling charge carriers to move faster to their catalytic sites [13]. As a third benefit, dual cocatalysts provide a redox pathway that has relatively lower overpotentials than pure photocatalysts. Fourth, the deposition of dual redox cocatalysts optimizes the adsorption and the desorption properties of reactants and products, respectively. Finally, self-oxidative degradation behaviors, particularly in oxynitride and sulfide photocatalysts, can be prevented by eliminating photoexcited electrons or holes from the photocatalyst to the redox cocatalyst sites, respectively [14]. As a result of the advantages triggered by dual redox cocatalysts, integrated photocatalytic devices are broadly used to efficiently catalyze HER.

7.2.3 The Effect of Redox Cocatalyst Parameters on Photocatalysis

It has been demonstrated that multiple factors influence the performance of a dual cocatalyst, including element composition, size, deposition position, deposition

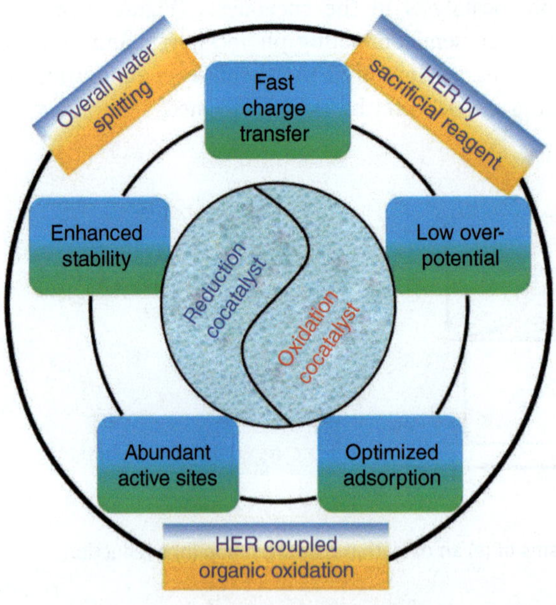

Figure 7.3 The hydrogen evolution reactions and advantages of the dual redox cocatalysts.

method, dispersion, and loading amount. When these parameters have been optimized, the efficiency of solar-to-chemical conversion will be maximized. In the case of a deficiency in the redox cocatalyst loading, surface catalysis is compromised, whereas excess loading shields the photocatalysts from solar radiation, which hinders solar energy harvesting. Therefore, the deposition amount of redox cocatalysts must be determined to achieve optimal results. The deposition method also plays a major role in increasing the catalytic activity of the photocatalysts in addition to the loading amount. In the photodeposition method, for example, redox cocatalysts such as MnO_x, Pt, and Au can be attached directly to the photocatalysts by undergoing the photooxidation/reduction of the metal precursors [11]. In comparison with other methods of deposition, photodeposition provides a rapid and shorter charge transfer distance from photocatalysts to cocatalysts.

It has been shown that photocatalytic activity and selectivity are crucially affected by the elemental composition of cocatalysts. As an example, metallic phosphide cocatalysts have been extensively used as electron sinks to enhance photocatalytic activity in semiconductors due to the formation of the Schottky barrier, whereas the enhancement range is always determined by the chemical composition of the phosphide cocatalyst. For example, when coupled with the CdS semiconductor, the cobalt phosphide Co_2P cocatalyst had significantly stronger hydrogen production ability than CoP, which was attributed to the higher Fermi level of Co_2P [15]. Hence, it is suggested that adjusting the chemical composition of the cocatalyst helps to attain greater charge migration from the photocatalyst to the cocatalyst by getting a higher Fermi level of the cocatalyst.

7.2.4 Design Principles of Dual Cocatalysts

Basically, dual redox cocatalyst design should be based on a charge transfer reaction that is widely recognized as being the rate-determining step. Taking the cocatalysts-electrolyte interface as an instance, before the deposition of the dual redox cocatalyst on the photocatalysts, the distinctive properties of the cocatalysts, especially the product desorption and reactant adsorption abilities, are crucially included in the redox cocatalysts, which frankly favor the photocatalytic performance. For the optimized adsorption of hydrogen atoms, Yu et al. developed a reduction cocatalyst based on sulfur-rich transition metal chalcogenide (core–shell Au@NiS_{1+x}) [10]. The authors stated that the Au@NiS_{1+x} cocatalyst sites simultaneously increase the active site number, conductivity, and efficiency to enhance the HER.

Furthermore, the coupling of photocatalyst and cocatalyst through a strong interfacial interaction has shown its progressive effect on rapid interfacial charge transfer from the photocatalyst to the cocatalyst. Therefore, a deep understanding of the kinetic and dynamics details of the interfacial hole and electron transfer is of utmost importance for the construction and rational design of an incorporated photocatalytic system comprising photocatalyst-oxidation-cocatalyst and photocatalyst-reduction-cocatalyst interfaces. Usually, the coupling of electronic interaction between cocatalysts and photocatalysts can be regulated by performing a thermal treatment, creating a chemical bond, and establishing a p–p interaction. As an example, because of the similar graphitic crystallinity between carbon nitride (C_3N_4) and N-doped carbon (N–C), the N–C has been used as a bridge to enhance the interface interaction between Pt and carbon nitride, which led to an outstanding improvement in HER, as recently described by Zhang's group [16].

7.3 Recent Advances in the Configuration of Dual Redox Cocatalysts/Photocatalyst

A selection of emerging configurations for dual redox cocatalysts on photocatalysts is displayed in Figure 7.4. Dual redox cocatalyst with distinctive functionalities that are randomly dispersed on the photocatalyst can be easily recognized. In contrast to the design of unregulated distribution and loading of dual cocatalysts, the spatially separated preparation of dual redox cocatalysts is usually synthesized by a defined step-by-step technique, which is rather challenging. The configurations of spatially separated dual cocatalysts can be further categorized into four groups based on their architecture: for example, tip/side distribution, York-shell distribution, facet-dependent distribution, and center/edge distribution.

7.3.1 Random Distribution

The unregulated distribution and deposition states of dual cocatalysts make them easier to attach to the surface of a photocatalyst, where they are usually utilized to promote both oxidation and reduction reactions simultaneously. In 2019, Wang and coworkers designed polymeric carbon nitride (PCN) nanosheets decorated with RhO_x and Rh as oxidative and reductive cocatalysts, respectively (Figure 7.5a).

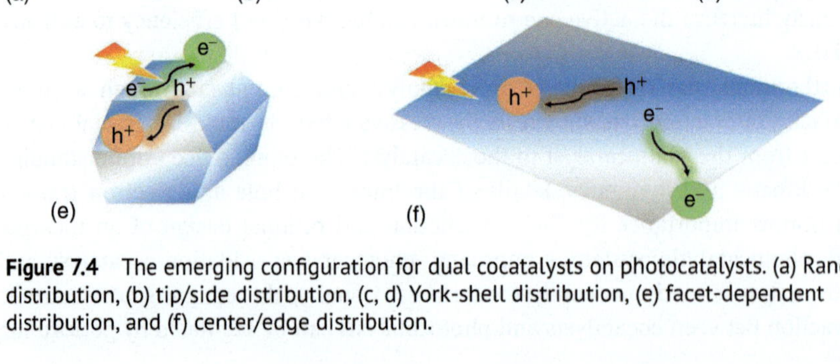

Figure 7.4 The emerging configuration for dual cocatalysts on photocatalysts. (a) Random distribution, (b) tip/side distribution, (c, d) York-shell distribution, (e) facet-dependent distribution, and (f) center/edge distribution.

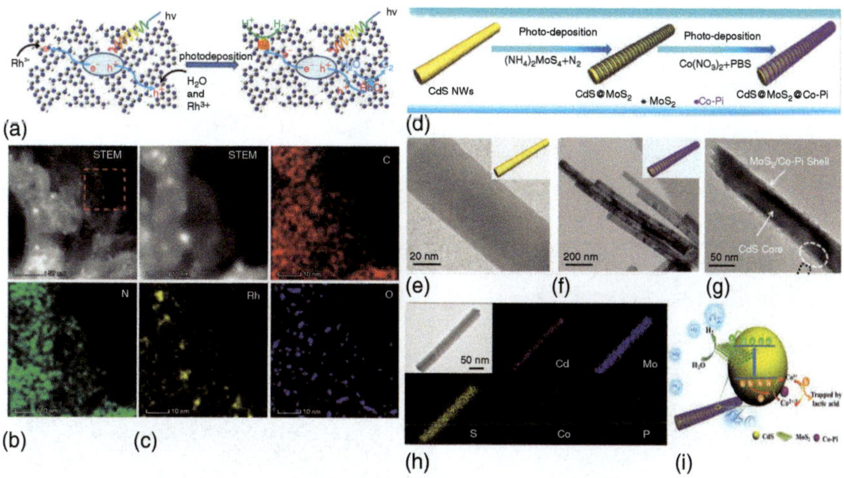

Figure 7.5 (a) The PCN nanosheets decorated with RhO_x and Rh as dual redox cocatalyst. (b, c) The STEM and EDX elemental mapping images. Source: Pan et al. [17]/reproduced with permission from Elsevier. (d) CdS NWs loaded with MoS_2 and CoP. (e–h) TEM and elemental mapping images, and (i) proposed mechanism of HER. Source: Lu et al. [18]/reproduced with permission from American Chemical Society.

Scanning transmission electron microscopy (STEM) certified the successful loading of dual redox cocatalysts with good dispersion on 2D PCN sheets (Figure 7.5b,c) [17]. The resultant Rh–RhO_x/PCN photocatalyst showed higher water splitting activity than Rh/PCN and RhO_x/PCN, thus demonstrating the functionality of each cocatalyst for photocatalytic redox reactions. Furthermore, taking 1D CdS nanowires (NWs) as an example, MoS_2 and CoP could be attached to the surface of NWs by the two-step photodeposition method (Figure 7.5d), as evidenced by the TEM analysis (Figures 7.6e–h) [18]. In this dual redox cocatalyst system, MoS_2 is a site to accelerate the migration of electrons from CdS NWs, whereas CoP acts as a hole's reservoir as well as an active site for the consumption of lactic acid as a sacrificial reagent (Figure 7.6i).

The integration of such randomly distributed dual redox cocatalysts on photocatalysts provides a high degree of flexibility, as it is free from the restrictions of the categories of photocatalysts and cocatalysts. However, the unregulated distribution and accumulation of redox cocatalysts will result in overlapping between reduction and oxidation catalysts, which may result in a failure of the charge separation. Due to the ability to separate the photoexcited charge carriers from the photocatalytic systems, it is essential to design dual redox cocatalysts with spatial separation.

7.3.2 Spatially Separated Distribution

7.3.2.1 Tip/Side Distribution

Tip/side configurations of dual redox cocatalysts tend to be found mostly in 1D photocatalysts. Normally, oxidation and reduction cocatalysts are attached to the side and tip of the 1D photocatalytic system, respectively. For instance, Li et al.

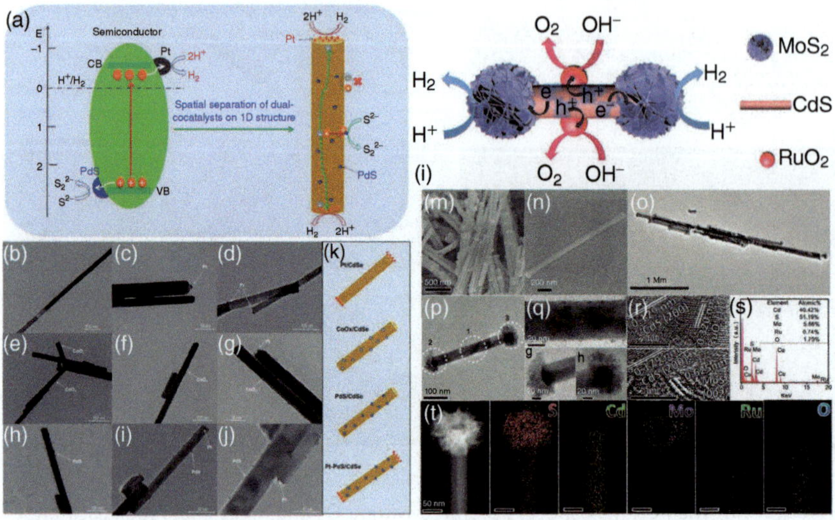

Figure 7.6 (a) The HER mechanism over CdSe nanorods decorated with PdS and Pt, (b–k) confirmation of configuration by TEM images. Source: Reproduced with permission from Khan et al. [19]/Royal Society of Chemistry. (l) The migration of photoexcited charges over $RuO_2/CdS/MoS_2$ composite. (m–t) The TEM, HRTEM, EDX, and elemental mapping of the $RuO_2/CdS/MoS_2$ composite. Source: Qiu et al. [20]/reproduced with permission from John Wiley & Sons.

fabricated CdSe nanorods decorated with the oxidation cocatalyst (PdS) and the reduction cocatalyst (Pt) attached at the side and tip of the CdSe nanorod, respectively (Figure 7.6a) [19]. In their research, it was claimed that the photoexcited holes and electrons migrate to the side and tip of the catalyst, thereby causing the selective deposition of dual redox cocatalysts. The configuration of the CdSe nanorods decorated with dual redox cocatalysts is confirmed by TEM images (Figure 7.6b–k). Recent work by Qiu's group shows another example of separated dual redox cocatalysts with a tip/side structure (Figure 7.6l) [20]. This case involved a site-selective ethylenediamine (EDA) modification to induce anisotropic growth of MoS_2 NPs as a reduction catalyst on CdS nanorods. After that, the EDA molecules at the side of the CdS nanorods assisted the selective growth of RuO_2 as an oxidation cocatalyst via a strong coordination effect between the $-NH_2$ and the Ru^{3+}groups, as evidenced by the TEM, HRTEM, EDX, and elemental mapping of the $RuO_2/CdS/MoS_2$ composite (Figure 7.6m–t). The HER occurs at MoS_2 NPs and the sacrificial reagent provided the electrons by taking up the holes from the oxidation cocatalyst.

To date, the only reports that exist on the tip/side configuration have mainly been based on the 1D cadmium chalcogenide photocatalytic systems, which are prone to severe toxicity and photocorrosion. Furthermore, a reduction cocatalyst can only be attached to the tip/side structure of a 1D photocatalyst and has narrow operating ranges (MoS_2, Pt, and Pd). As a result, this integrated photocatalytic system, equipped with dual redox cocatalysts, allows a low degree of flexibility in using cocatalysts and photocatalysts.

7.3.2.2 York-Shell Distribution

A York-shell structure is a hybrid nanomaterial comprising of an inner core and a shell, which helps to harvest solar energy in the hollow area while also providing a large surface area for catalysis. The dual redox cocatalysts are loaded in a spatially separated arrangement based on the York-shell structure, where the shell indicates semiconductor photocatalysts and the oxidation and reduction cocatalysts alternate along with the shell. Recently, innovative work from Yong's group for the separated arrangement of the heterojunction York-shell structure of the dual redox cocatalysts should be mentioned [21]. In this work, a sacrificial template method was used to synthesize the Z-scheme heterojunction Pt/g-C_3N_4/TiO_2/IrO_x with a York-shell structure, in which IrO_x was allocated inside of the heterojunction shell as a hole collector, and the Pt as an electron sink was attached to the outer shell of the heterojunction (Figure 7.7a). The HRTEM, TEM, and elemental mapping confirmed the successful preparation of the Z-Scheme heterojunction Pt/g-C_3N_4/TiO_2/IrO_x (Figure 7.7b–g). In another example, Qin et al. employed carbon nano coils (CNCs) as templates and atomic layer deposition (ALD) for fabricating a porous tubular Pt/TiO_2/CoO_x photocatalyst system with spatially separated dual redox cocatalysts (Figure 7.7h–j) [22]. The HRTEM was employed to explore the distribution state of CoO_x and Pt NPs on TiO_2 nanotubes (Figure 7.7k–m). Light irradiation caused the TiO_2 charge carriers to move inward and outward, where they took part in redox processes.

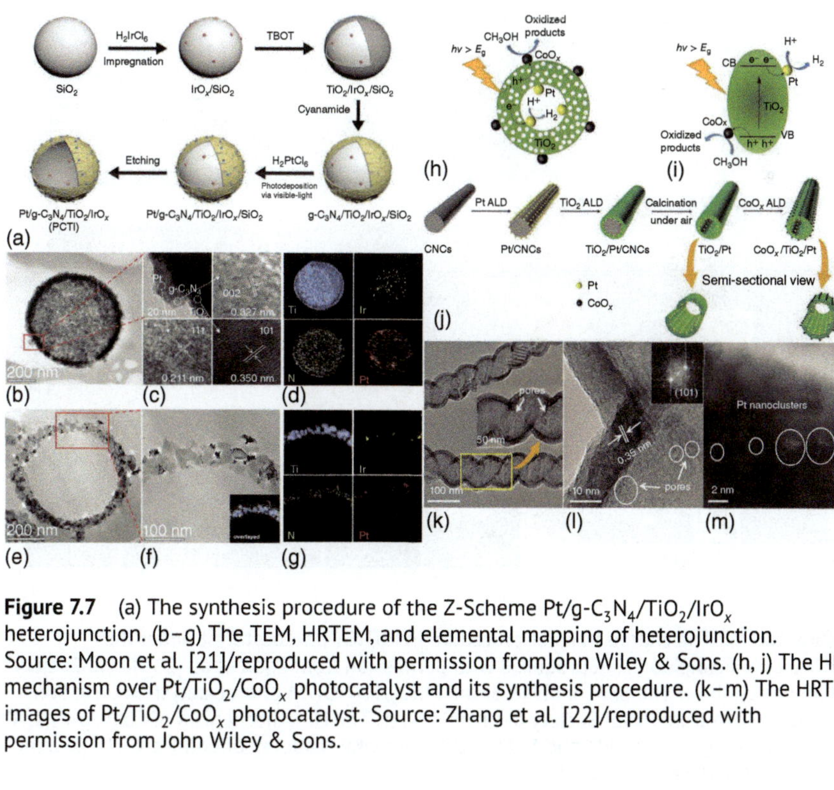

Figure 7.7 (a) The synthesis procedure of the Z-Scheme Pt/g-C_3N_4/TiO_2/IrO_x heterojunction. (b–g) The TEM, HRTEM, and elemental mapping of heterojunction. Source: Moon et al. [21]/reproduced with permission fromJohn Wiley & Sons. (h, j) The HER mechanism over Pt/TiO_2/CoO_x photocatalyst and its synthesis procedure. (k–m) The HRTEM images of Pt/TiO_2/CoO_x photocatalyst. Source: Zhang et al. [22]/reproduced with permission from John Wiley & Sons.

No doubt, the York-shell distribution of dual redox cocatalysts overcomes the fundamental issues in photocatalytic water splitting, such as charge recombination, inefficient light absorption, and sluggish reaction kinetics, and it has relatively high scalability. However, there is a weakness that goes unnoticed in this York-shell structure. It is essential that the shell has an appropriate porous structure so that the reactants can pass through it quickly; otherwise, the cocatalysts can starve reactants due to limited transport.

7.3.2.3 Facet-Dependent Distribution

Recent research on photocatalyst facet engineering revealed that photoinduced charges can be pushed into a variety of facets. Therefore, the reduction and oxidation reactions are favored by different facets in a crystallized semiconductor. In the presence of an embedded ferroelectric field, the photoexcited electrons and holes in a single domain and single crystals of ferroelectric materials can migrate in opposite directions (Figure 7.8a) [23]. Liu and coworkers used the ferroelectric field to selectively deposit the oxidation cocatalyst (MnO_x) and reduction cocatalyst (Pt)

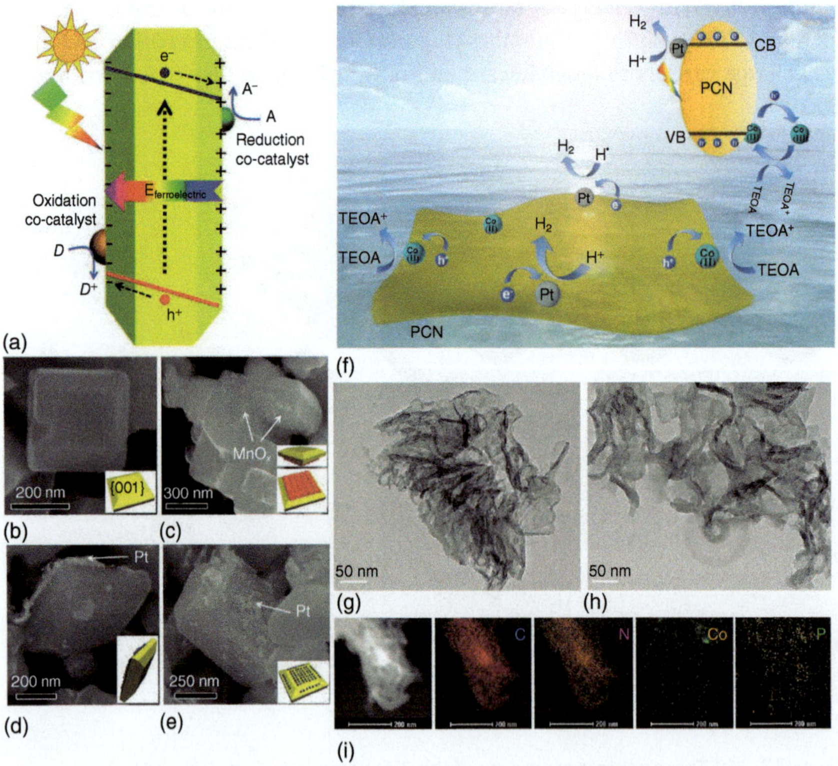

Figure 7.8 (a) The charge transfer procedure in $PbTiO_3$ photocatalyst under the ferroelectric field. Source: Reproduced with permission from Zhen et al. [23]/Royal Society of Chemistry. (b–e) The SEM analysis of the $MnO_x/PbTiO_3/Pt$ photocatalyst. Source: Sun et al. [24]/2014 from Royal Society of Chemistry. (f) The HER mechanism of Co(II)/g-C_3N_4/Pt photocatalyst, (g–i) The TEM and STEM-EDX elemental mapping of Co(II)/g-C_3N_4. Source: Sun et al. [24]/2014 from Elsevier.

on two opposing surfaces of the $PbTiO_3$ photocatalytic system for H_2 evolution, as confirmed by SEM analysis (Figure 7.8b–e) [23]. A unique advantage of constructing these separate dual redox cocatalysts on different facets is the ability to optimize the reactant atmosphere within each of the separate sections, in contrast to the York-shell arrangement. The second most important point is that the separate arrangement of the dual redox cocatalysts can also yield a well-established interface between the photocatalysts and cocatalysts in the photocatalytic system. It also had some drawbacks, like long-term stability, a limited selection of photocatalysts and cocatalysts, etc.

7.3.2.4 Center/Edge Distribution

Separated dual redox cocatalysts with the center/edge structure are mostly constructed with 2D photocatalysts that have voids in the basal plane. For example, Lei et al. smartly constructed the 2D C_3N_4 nanosheets to host the Pt reduction cocatalyst at the edge of the nanosheets and Co(II) as an oxidation cocatalyst mostly in the center of the nanosheet as demonstrated graphically (Figure 7.8f) and further confirmed the dual redox cocatalyst loading by TEM and STEM-EDX elemental mapping (Figures 7.8g–i) [24]. There was an optimized content of dual redox cocatalysts at some distance from each other that overcomes the hurdle of charge recombination and boosts the hydrogen evolution. Therefore, the dual redox cocatalysts are highly functional due to the various chemical and physical properties of 2D photocatalysts, especially ultrathin 2D materials. The result of such a structure is the ability to integrate multiple functionalities, such as the ability to shorten bulk diffusion distances, have high surface separation efficiency, and have superior electron conductivity. However, in center/edge distribution, there are some difficulties in obtaining the optimized contents of dual redox cocatalysts, which is the main bottleneck to achieving this distribution.

7.4 Major Types of Photocatalytic Water Splitting

The photocatalytic HER by water splitting can be categorized into three major types as displayed (Figure 7.3). The first category is the overall water split. In this system, H_2 evolves from the reduction site and oxygen is produced from oxidation sites [11]. H_2 is highly required in the transportation, agricultural, and other industries as a carbon-free fuel. However, oxygen is technically and economically less valuable when it is derived from the hydrolysis of water. Furthermore, oxygen production causes durability and kinetics limitations in the reaction system; it is a major bottleneck of this process. So, to eliminate the risk of simultaneous production of oxygen and H_2, sacrificial reagents (like $Na_2S{\cdot}9H_2O/Na_2SO_3$, ascorbic acid, TEOA, CH_3OH, formic acid, lactic acid, etc.) are used to consume the holes of photocatalyst, and this is called the HER by sacrificial reagent [21]. The sacrificial reagents make the H_2 evolution uneconomical, and on the other side, it is also a waste of energy. So nowadays, researchers are focusing on third category coupling reactions. It is also called HER coupled with organic oxidation. In this category, the electrons from reduction sites are employed for H_2 evolutions whereas the holes from the oxidation site are effectively utilized for the oxidation of the organic moiety [6]. Therefore, in the future,

Table 7.1 Recently reported photocatalytic systems integrated with dual redox cocatalyst for hydrogen evolution.

Photocatalyst	OC[a)]	Loading method	RC[b)]	Loading method	Light	Sacrificial reagent	H_2 evolution	References
CdS	NiS	Hydrothermal	Ag_2S	Photodeposition	420 nm	Lactic acid	48.28 mmol g^{-1} h^{-1}	[12]
CdS	Co	Physical mixing	CoO	Heating in tube furnace	420 nm	Na_2S/Na_2SO_3	19.97 mmol g^{-1} h^{-1}	[2]
CdS	Co-Pi	Photo-deposition	MoS_2	Photodeposition	420 nm	Lactic acid	40.50 mmol g^{-1} h^{-1}	[18]
CdS	PdS	Photo-deposition	Pt	Photodeposition	420 nm	Na_2S/Na_2SO_3	8.77 mmol g^{-1} h^{-1}	[25]
CdS	Ti(IV)	Impregnation	Ni(II)	Impregnation	420 nm	Na_2S/Na_2SO_3	3.44 mmol g^{-1} h^{-1}	[26]
$Mn_{0.5}Cd_{0.5}S$	$Cu_{2-x}S$	Chemical-deposition	MoS_2	Photodeposition	420 nm	Na_2S/Na_2SO_3	13.75 mmol g^{-1} h^{-1}	[27]
ZnCdS	PdS	Chemical-deposition	GQDs	hydrothermal deposition	420 nm	Na_2S/Na_2SO_3	10.34 mmol g^{-1} h^{-1}	[14]
TiO_2	IrO_x	Adsorption	Pd	Photodeposition	Xe-lamp (300 W)	Methanol	7.74 mmol g^{-1} h^{-1}	[28]
$CaIn_2S_4$	MnO_x	Photo-deposition	AuCu	Photodeposition	420 nm	Na_2S/Na_2SO_3	95.75 mmol g^{-1} h^{-1}	[13]
$g\text{-}C_3N_4/TiO_2$	IrO_x	Impregnation	Pt	Photodeposition	Xe-lamp (300 W)	Methanol	8.15 mmol g^{-1} h^{-1}	[21]

a) Oxidation cocatalyst.
b) Reduction cocatalyst.

dual redox cocatalyst decorated photocatalysts or heterojunctions like S-Scheme heterojunction are a hot research topic for HER coupled with organic oxidation. Some recently reported photocatalysts loaded with dual redox cocatalysts for HER are shown in Table 7.1.

7.5 Conclusions

The water splitting by photocatalytic systems for HER has some critical problems due to the bare photocatalysts such as charge recombination, insufficient light absorption, and sluggish reaction kinetics. Currently, the simultaneous embedding of the dual redox cocatalysts onto a photocatalytic system to tackle these limitations of artificial photosynthesis is a hot research topic. In the present chapter, we have tried to shade the importance and need of dual redox cocatalysts for artificial photosynthesis, especially for HER. Furthermore, we have reviewed the essential design principles and emerging configurations of the dual redox cocatalysts and provided a side-by-side comparison to reveal their strengths and deficiencies. In parallel, we have discussed how to choose a suitable pair of redox cocatalysts for a photocatalytic system of HER, and how some key lessons that emerged out of the relevant studies can be applied to further investigations of photocatalytic H_2 evolution.

References

1 Rahman, M.Z., Edvinsson, T., and Gascon, J. (2022). Hole utilization in solar hydrogen production. *Nat. Rev. Chem.* 6: 243–258.

2 Ren, X., Wei, S., Wang, Q. et al. (2021). Rational construction of dual cobalt active species encapsulated by ultrathin carbon matrix from MOF for boosting photocatalytic H_2 generation. *Appl. Catal., B* 286: 119924.

3 Meng, A., Zhang, L., Cheng, B., and Yu, J. (2019). Dual cocatalysts in TiO_2 photocatalysis. *Adv. Mater.* 31: 1807660.

4 Qiu, B., Du, M., Ma, Y. et al. (2021). Integration of redox cocatalysts for artificial photosynthesis. *Energy Environ. Sci.* 14: 5260–5288.

5 Wolff, C.M., Frischmann, P.D., Schulze, M. et al. (2018). All-in-one visible-light-driven water splitting by combining nanoparticulate and molecular co-catalysts on CdS nanorods. *Nat. Energy* 3: 862–869.

6 Tayyab, M., Liu, Y., Min, S. et al. (2022). Simultaneous hydrogen production with the selective oxidation of benzyl alcohol to benzaldehyde by a noble-metal-free photocatalyst VC/CdS nanowires. *Chin. J. Catal.* 43: 1165–1175.

7 Takata, T., Jiang, J., Sakata, Y. et al. (2020). Photocatalytic water splitting with a quantum efficiency of almost unity. *Nature* 581: 411–414.

8 Cheng, C., He, B., Fan, J. et al. (2021). An inorganic/organic S-scheme heterojunction H_2-production photocatalyst and its charge transfer mechanism. *Adv. Mater.* 33: 2100317.

9 Li, Y.H., Peng, C., Yang, S. et al. (2015). Critical roles of co-catalysts for molecular hydrogen formation in photocatalysis. *J. Catal.* 330: 120–128.

10 Gao, D., Xu, J., Wang, L. et al. (2022). Optimizing atomic hydrogen desorption of sulfur-rich NiS_{1+x} cocatalyst for boosting photocatalytic H_2 evolution. *Adv. Mater.* 34: 2108475.

11 Zhang, J., Bai, T., Huang, H. et al. (2020). Metal–organic-framework-based photocatalysts optimized by spatially separated cocatalysts for overall water splitting. *Adv. Mater.* 32: 2004747.

12 He, B., Bie, C., Fei, X. et al. (2021). Enhancement in the photocatalytic H_2 production activity of CdS NRs by Ag_2S and NiS dual cocatalysts. *Appl. Catal., B* 288: 119994.

13 Ding, J., Li, X., Chen, L. et al. (2019). Site-selective deposition of reductive and oxidative dual cocatalysts to improve the photocatalytic hydrogen production activity of $CaIn_2S_4$ with a surface nanostep structure. *ACS Appl. Mater. Interfaces* 11: 835–845.

14 Wang, F., Su, Y., Min, S. et al. (2018). Synergistically enhanced photocatalytic hydrogen evolution performance of ZnCdS by co-loading graphene quantum dots and PdS dual cocatalysts under visible light. *J. Solid State Chem.* 260: 23–30.

15 Bi, W., Zhang, L., Sun, Z. et al. (2016). Insight into electrocatalysts as Co-catalysts in efficient photocatalytic hydrogen evolution. *ACS Catal.* 6: 4253–4257.

16 Wang, J., Zhou, Q., Shen, Y. et al. (2019). Carbon nitride Co-catalyst activation using N-doped carbon with enhanced photocatalytic H_2 evolution. *Langmuir* 35: 12366–12373.

17 Pan, Z., Wang, S., Niu, P. et al. (2019). Photocatalytic overall water splitting by spatially-separated Rh and RhO_x cocatalysts on polymeric carbon nitride nanosheets. *J. Catal.* 379: 129–137.

18 Lu, K.-Q., Qi, M.-Y., Tang, Z.-R., and Xu, Y.-J. (2019). Earth-abundant MoS_2 and cobalt phosphate dual cocatalysts on 1D CdS nanowires for boosting photocatalytic hydrogen production. *Langmuir* 35: 11056–11065.

19 Khan, K., Tao, X., Zhao, Y. et al. (2019). Spatial separation of dual-cocatalysts on one-dimensional semiconductors for photocatalytic hydrogen production. *J. Mater. Chem. A* 7: 15607–15614.

20 Qiu, B., Cai, L., Zhang, N. et al. (2020). A ternary dumbbell structure with spatially separated catalytic sites for photocatalytic overall water splitting. *Adv. Sci.* 7: 1903568.

21 Moon, H.S., Hsiao, K.-C., Wu, M.-C. et al. (2022). Spatial separation of cocatalysts on Z-scheme organic/inorganic heterostructure hollow spheres for enhanced photocatalytic H_2 evolution and in-depth analysis of the charge-transfer mechanism. *Adv. Mater.* 2200172.

22 Zhang, J., Yu, Z., Gao, Z. et al. (2017). Porous TiO_2 nanotubes with spatially separated platinum and CoO_x cocatalysts produced by atomic layer deposition for photocatalytic hydrogen production. *Angew. Chem. Int. Ed.* 56: 816–820.

23 Zhen, C., Yu, J.C., Liu, G., and Cheng, H.-M. (2014). Selective deposition of redox co-catalyst(s) to improve the photocatalytic activity of single-domain ferroelectric $PbTiO_3$ nanoplates. *Chem. Commun.* 50: 10416–10419.

24 Sun, K., Shen, J., Liu, Q. et al. (2020). Synergistic effect of Co(II) hole and Pt-electron cocatalysts for enhanced photocatalytic hydrogen evolution performance of P-doped g-C_3N_4. *Chin. J. Catal.* 41: 72–81.

25 Yan, H., Yang, J., Ma, G. et al. (2009). Visible-light-driven hydrogen production with extremely high quantum efficiency on Pt–PdS/CdS photocatalyst. *J. Catal.* 266: 165–168.

26 Yu, H., Huang, X., Wang, P., and Yu, J. (2016). Enhanced photoinduced-stability and photocatalytic activity of CdS by dual amorphous cocatalysts: synergistic effect of Ti(IV)-hole cocatalyst and Ni(II)-electron cocatalyst. *J. Phys. Chem. C* 120: 3722–3730.

27 Liu, X., Liu, Q., Wang, P. et al. (2018). Efficient photocatalytic H_2 production via rational design of synergistic spatially-separated dual cocatalysts modified $Mn_{0.5}Cd_{0.5}S$ photocatalyst under visible light irradiation. *Chem. Eng. J.* 337: 480–487.

28 Ma, Y., Chong, R., Zhang, F. et al. (2014). Synergetic effect of dual cocatalysts in photocatalytic H_2 production on Pd–IrO_x/TiO_2: a new insight into dual cocatalyst location. *Phys. Chem. Chem. Phys.* 16: 17734–17742.

8

Polymeric Carbon Nitride-based Materials in Aqueous Suspensions for Water Photo-splitting and Photo-reforming of Biomass Aqueous Solutions to Generate H_2

E.I. García-López[1], G. Marcì[2], and L. Palmisano[2]

[1] *University of Palermo, Chemical and Pharmaceutical Sciences and Technologies (STEBICEF), Department of Biological, Viale delle Scienze, 90128, Palermo, Italy*
[2] *University of Palermo, "Schiavello-Grillone" Photocatalysis Group, Department of Engineering, Viale delle Scienze, 90128, Palermo, Italy*

8.1 Introduction

Fast population growth and economic development have led to an increase in the use of fossil fuels, so the consumption of energy resources has given rise to negative environmental implications. Heterogeneous photocatalysis can play an essential role in the development of sustainable energy and is considered very useful to turn solar energy into chemical energy. Furthermore, it is a low-cost ecological technology that has also shown its importance from the point of view of environmental remediation [1, 2]. Hydrogen is a storable, clean, and ecological fuel, and its combustion involves the emission of water without other gases or particulates; it also has an excellent energy content per mass, which corresponds to 122 kJ g^{-1} [3].

Fossil fuels, at present, represent the main hydrogen source (c. 95%), but many technologies are under development to produce green hydrogen from renewable resources, such as water and biomass [4]. In this context, the photocatalytic processes based on the use of semiconductors irradiated under UV or visible light have received particular attention for their production. Green hydrogen can be obtained through the photocatalytic splitting of water or the photo-reforming of an organic or inorganic substrate with the aim of replacing a part of traditional fossil fuels, thus alleviating the emerging energy crisis.

Photocatalytic reactions are initiated by the excitation of a semiconductor, which is irradiated with suitable wavelengths, equal to or greater than its bandgap energy, producing the promotion of electrons (e^-) from the valence band (VB) to the conduction band (CB) leaving a corresponding number of holes (h^+) in the VB. The generated electron–hole pairs can be recombined by releasing energy or migrating to the surface where they are trapped on active sites of the semiconductor.

As schematized in Figure 8.1, in anaerobic conditions, the water cleavage may produce hydrogen and oxygen in a process named water photo-splitting. This occurs when the energy of the electrons is sufficient to reduce the protons to H_2, i.e. when

UV-Visible Photocatalysis for Clean Energy Production and Pollution Remediation: Materials, Reaction Mechanisms, and Applications, First Edition. Edited by Xinchen Wang, Masakazu Anpo, and Xianzhi Fu.
© 2023 WILEY-VCH GmbH. Published 2023 by WILEY-VCH GmbH.

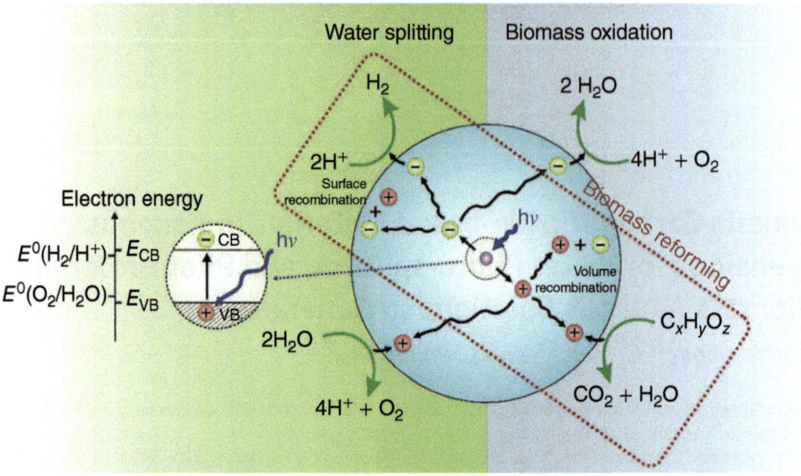

Figure 8.1 Reaction pathways of water splitting and biomass photo-reforming under anaerobic conditions, along with biomass oxidation in the presence of oxygen. Source: Reproduced with permission from Huang et al. [5]/Elsevier.

$E_{CB} < E\ (H_2/H^+)$, and that of the holes to oxidize the oxygen of the water to O_2, i.e. when $E_{VB} > E\ (O_2/H_2O)$. As an alternative to H_2O, an organic molecule can play the role of a trap for photogenerated holes that act directly or indirectly through the formation of hydroxyl radicals. The latter can lead to partial oxidation of the organic molecule and ultimately to CO_2 and H_2O. This process, which uses organic molecules to trap the holes, is called photo-reforming. The organic species can also be a biomass derivative, indicated in Figure 8.1 as $C_xH_yO_z$. The reduction of H^+ occurs under anaerobic conditions, always obtaining the production of H_2, as shown in Figure 8.1. The rate of H_2 formation is much higher in the photo-reforming process than in water splitting because the oxidation reaction of $C_xH_yO_z$ with photogenerated holes is irreversible and also because it is a favorite with respect to the reaction of water oxidation. Conversely, H_2 and O_2 derived from water splitting can easily reform H_2O, decreasing the efficiency of photo-conversion. Suppression of electron–hole recombination is also more difficult during water photo-splitting.

The overall photocatalytic process to generate H_2 can be considered for both water splitting and biomass photo-reforming. The water splitting process simultaneously involves the oxygen evolution reaction (OER) shown in Eq. (8.1) and H_2 evolution reaction (HER) as reported in Eqs. (8.2) and (8.3):

$$2\,H_2O + 4\,h^+ \rightarrow O_2 + 4\,H^+ \tag{8.1}$$

$$2\,H^+ + 2\,e^- \rightarrow H_2 \tag{8.2}$$

or

$$2\,H_2O + 2\,e^- \rightarrow H_2 + 2\,OH^- \tag{8.3}$$

Therefore, the photo-reforming of aqueous solutions containing oxygenated organic compounds combines photocatalytic water splitting and oxidation of organic substrates in a single process. The photo-reforming reaction is nonselective and organic or inorganic species, including biomass compounds, may be used for the production of hydrogen. Photo-reforming has not received the same attention as water photo-splitting despite its interest and reported high efficiencies. A wide range of biomass-derived substances with a broad variety of oxygen-containing functionalities would be used as hole scavengers in the photo-reforming reaction, as schematized in Figure 8.2. Monosaccharides, such as pentoses (ribose, arabinose) and hexoses (glucose, galactose, fructose, mannose), alcohols (methanol, ethanol, propanol, butanol, glycerol), amines such as triethanolamine (TEOA) and organic acids (acetic acid, formic acid) have been widely used as a hole scavengers.

For example, in the presence of di- or poly-saccharides, the hydrogen production rate is one or two orders of magnitude higher than that obtained from pure water [6]. It is worth noting that the selectivity of photo-reforming processes exceeds

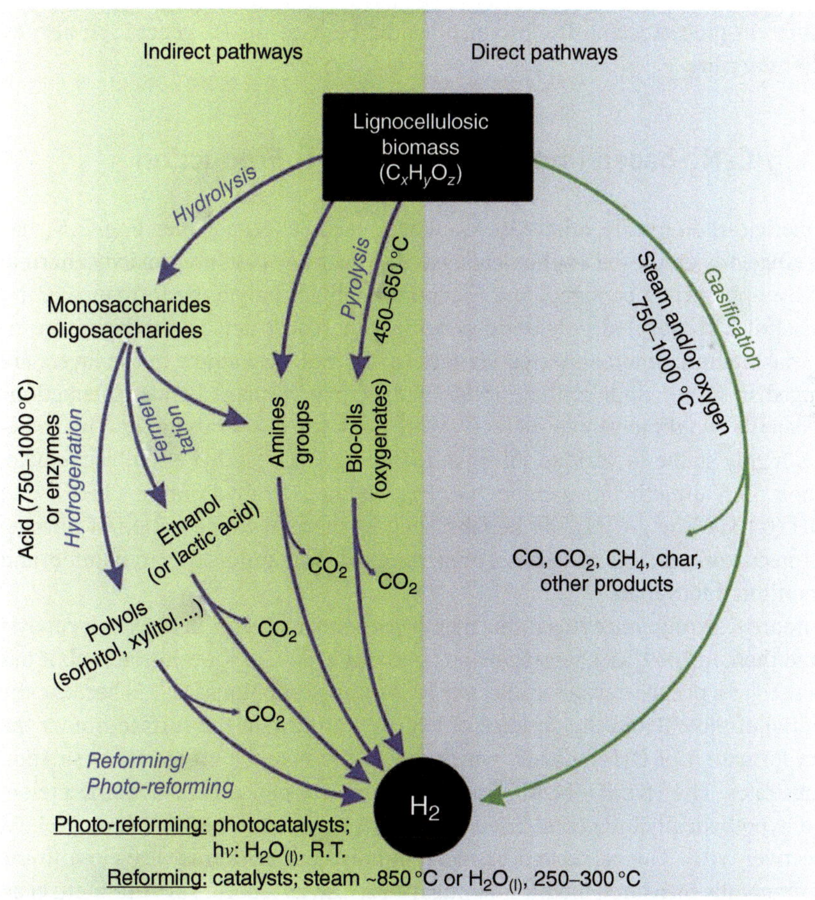

Figure 8.2 Scheme of the reactions capable of obtaining H_2 from lignocellulosic biomass, including photo-reforming. Source: Reproduced with permission from Huang et al. [5]/Elsevier.

that of thermocatalytic analogs, due to the milder (ambient) conditions under which they are performed. The use of more complicated raw materials, such as wood, rice husk, sawdust, and seaweed, has been also explored. When the organic molecules used are partially oxidized rather than mineralized, the photo-reforming process, which ultimately always provides H_2, can be more widely considered as a synthesis process, for example in the case of alcohols selectively oxidized to aldehydes [7].

The most important biomass feedstocks are lignocellulose, starch, sugar-based crops, and vegetable oil crops, which are the lignocellulosic substrates present in the biosphere in greater proportion. Considering the lignocellulosic derivatives, bio-reforming can be divided into direct or indirect processes [8], as shown in Figure 8.2. The direct gasification is carried out at high temperatures and pressures and gives rise to H_2 and CO, CO_2, CH_4, etc., and the indirect transformation of $C_xH_yO_z$ is performed by several routes, as schematized in Figure 8.2. In the latter process, monomeric (or oligomeric) oxygenated substances are generated based on different processes and the products would be used for H_2 generation also by photo-reforming.

8.2 *g*-C_3N_4-based Photocatalysts for H_2 Production

Polymeric carbon nitride, otherwise known as melon, C_3N_4 polymer, or g-C_3N_4 has been efficiently employed in photocatalysis due to its low cost, non-toxicity, thermal stability, appropriate bandgap, and exceptional photocatalytic performance. C_3N_4 consists of a conjugated polymeric system, being constructed from s-triazine or tri-s-triazine units interconnected via tertiary amines. The atoms in the layers are arranged in honeycomb configurations with strong covalent bonds. Interactions between the two-dimensional sheets (2D) are weak van der Waals forces. This material is highly stable in various solvents, including H_2O, diethyl ether, acetic acid, alcohols, *N,N*-dimethylformamide (DMF), toluene, tetrahydrofuran (THF), and NaOH (0.1 M). The g-C_3N_4 can be fabricated by thermal condensation of nitrogenous precursors, such as melamine, urea, thiourea, cyanamide, dicyandiamide, and ammonium thiocyanate.

Antonietti's group introduced this metal-free semiconductor in 2006 as a catalyst [9] and then, in 2009, as a heterogeneous photocatalyst for H_2 evolution [10]. It has been used as a photocatalyst for selective redox transformations [11, 12] because the potential of the VB and the absence of hydroxyl groups on the surface hinder the direct formation of OH· radicals, which are responsible for unselective oxidation of substrates. The HOMO–LUMO gaps of the melem molecule, polymeric melon, and a hypothetically fully condensed g-C_3N_4 are reported to be 3.5, 2.6, and 2.1 eV, respectively [10]. The calculated bandgap of polymeric melon is very close to the experimentally measured medium-bandgap of 2.7 Ev, as reported by Antonietti et al. with the edges of the CB and VB lying at −1.13 and + 1.57 V (vs. NHE at pH = 7) [13].

The photocatalytic activity of C_3N_4 has been hampered by several important drawbacks, which are challenges to be overcome, as highlighted in Figure 8.3. In particular, the low mobility and high recombination rate of the charge carriers considerably limit their practical use.

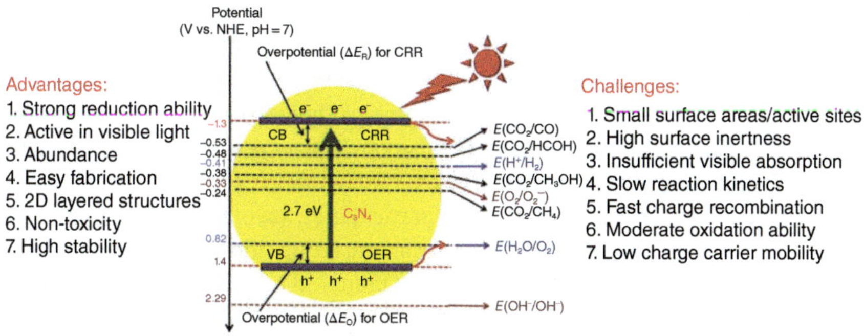

Figure 8.3 Advantages and disadvantages of *g*-C_3N_4 in photocatalysis along with the estimated position of the *g*-C_3N_4 band edges at pH = 7 and reduction potentials of the relevant reactions related to water splitting and CO_2 reduction. Source: Reproduced with permission from Wen et al. [14]/Elsevier.

Different approaches have been explored to improve/modify/optimize the structure of C_3N_4 using top-down strategies such as acid treatment, exfoliation, or etching, as well as bottom-up approaches [15]. Some synthetic strategies, such as the design of nanostructures, alteration of the electronic structure through the incorporation of doping species, generation of point defects through vacancies, chemical modifications with substituents and functional groups, supramolecular pre-organization, noble metal deposition, were examined to improve the electron–hole separation, increase the specific surface area, and modify the bandgap value [14, 15].

Graphitic C_3N_4 has been used as a photocatalyst for water photo-splitting for the first time to obtain H_2 and O_2 under visible light irradiation by Antonietti's group [10]. The production of H_2 after 72 hours of reaction went from 0.1–4 in the presence of bare g-C_3N_4 to 11 μmol/h when 3% by weight of Pt nanoparticles was added to the semiconductor under irradiation with a wavelength longer than 420 nm. On the other hand, under UV irradiation at $\lambda > 300$ nm the H_2 production was c. 240 μmol h^{-1} in the presence of Pt/g-C_3N_4 photocatalyst. The authors also measured the evolution of O_2 from the surface of g-C_3N_4 loaded with RuO_2 [10]. Very few papers report the evolution of H_2 from H_2O without the use of a hole scavenger. Ong et al. [16] report only one example in the presence of a composite of g-C_3N_4/polypyrrole. Polypyrrole injects electrons into the g-C_3N_4 CB while the holes in the VB of g-C_3N_4 react with water, giving rise to H_2O_2. The authors do not propose a trap for polypyrrole VB holes, which could hardly be transferred to solution species, hence a self-oxidation of the polymer (sacrificial agent) would likely occur [17]. Alternatively, Liu et al. report O_2 and H_2 evolution in the absence of scavengers using metal-free carbon nanodot/g-C_3N_4 nanocomposite as a photocatalyst. The quantum efficiencies were 16, 6.3, and 4.4% at $\lambda = 420$, 580 and 600 nm, respectively, determining an overall solar energy conversion efficiency of about 2.0%. The oxidation mechanism of water is a process with two electrons steps producing H_2O_2, which subsequently decomposes to give O_2 and H_2O. The rate increased with the loading of the carbon nanodots because they catalyzed the rate-limiting step, i.e. the decomposition of H_2O_2 [18]. The results reported by Liu et al. are surprisingly very good.

As reviewed by Naseri et al. [19], a hole scavenger is almost always used to minimize electron recombination, often with good quantum yields; however, the reaction cannot be considered a photo-splitting of the water, but rather a photo-reforming of the scavenger used. In the aqueous system, the oxidation reaction determines the rate of the overall reaction, therefore the use of electron donors (hole scavengers) can significantly improve the activity [20].

Noble metals such as Rh, Au, Ag, and in particular Pt, are often used as cocatalysts because they trap the photogenerated electrons and reduce the overpotential in order to facilitate multi-electron transfer reactions. Often, RuO_2 is also present to sink the holes, thus avoiding the recombination of the photoproduced pairs (h^+/e^-) [21].

The wide work function and low activation energy of Pt allow this metal to be the most effective cocatalyst for H_2 evolution [22]. The optimal amount of Pt follows a bell shape because the photocatalytic activity increases as the amount of metal increases. However, a further increase could lead to a decrease in the number of active sites on the semiconductor photocatalyst and even a shielding effect on incident light, and in this case, the metal would act as an electron/hole recombination center. Furthermore, the presence of noble metal nanoparticles on the semiconductor surface can give rise to the so-called surface plasmon resonance (SPR), which would make the photocatalytic process active under visible irradiation.

Transition metals such as Fe, Co, and Ni or inorganic compounds as $Cu(OH)_2$, MoS_2, WS_2, NiS, NiO, $Ni(OH)_2$, or CoP have been also used as cocatalysts for the photo-reforming. The analogous layered structures of inorganic semiconductors, such as MoS_2 and g-C_3N_4 gave rise to a composite that remarkably increases the photocatalytic H_2 evolution. This performance has been attributed to the similar layered geometries of the solid photocatalysts that allow the composite to enhance the charge carrier mobility at the interfaces and so their lifetime [23]. It has been reported that the mobility of electrons becomes higher and the binding energies of the bound electron/hole pairs are reduced by increasing the dimension of conjugated polymers from 0D to 1D, 2D, and 3D. It has been observed a decrease in the electron/hole recombination rate in the presence of graphene nanosheets/g-C_3N_4 composite for the photocatalytic H_2 evolution. Xian et al. observed that an amount of 1% wt of graphene on g-C_3N_4 was the optimum, obtaining 3 times more H_2 than by using the pristine g-C_3N_4. The heterojunction between g-C_3N_4 and reduced graphene oxide (rGO) increased electrical conductivity, improved carrier separation, and was able to store and transport electrons at the reaction sites [24].

Enhanced performance for photocatalytic H_2 production was attributed by Sun et al. to the synergic effect of rGO nanosheets and Pt nanoparticles [25]. Indeed, the rGO nanosheet, acting as an electron transfer mediator, trapped the electrons photogenerated by the g-C_3N_4 and then transferred them to the Pt co-catalyst, while the Pt nanoparticle acted as an active reduction site to promote the H_2 evolution reaction.

Yan et al. report that in g-C_3N_4/rGO composites, the carbon nitride plays the role of a photocatalyst while the rGO can store and transport electrons to reaction sites, thus improving the activity as an electron transfer medium [26].

Organic polymers like poly(3-hexylthiophene) can exhibit semiconducting properties (bandgap c. 2 eV), and its composite with g-C_3N_4, using Pt as a cocatalyst, has been used for H_2 production from aqueous ascorbic acid under visible light [27]. The poly(3-hexylthiophene)/g-C_3N_4 composite enabled outstanding activities (>300 mmol g_{cat}^{-1} h^{-1}). These results are outstandingly high for irradiation in the range $\lambda > 500$ nm. While significant deactivation of 30% reduced H_2 production rates after several days of operation, these results encourage further investigation of this inexpensive carbon-based material.

Interestingly, solar light can also be used as an irradiation source to obtain H_2 from water by TEOA photo-reforming using rGO nanosheets/C_3N_4 composites [28]. Spinel ferrite-g-C_3N_4 systems also take advantage of solar light for H_2 generation through photo-reforming [29]. For instance, the evolved H_2 rate by a ferrite-g-C_3N_4 composite gave rise to 10 times more H_2 than that obtained with the bare g-C_3N_4. The composite's efficiency was justified by claiming its optimized light absorption capacity. Composites of Nb_2O_5/g-C_3N_4 exhibited high visible light absorption, resulting in a remarkable photocatalytic activity under simulated solar irradiation using as hole scavengers TEOA or methanol. An amount of 110 mmol $g_{cat}^{-1}\cdot h^{-1}$of H_2 was produced in the presence of a niobium oxide photocatalyst with 10% by weight of g-C_3N_4, a value more than double that obtained with the two bare components [30]. The enhanced photocatalytic activity has been attributed to the fast photogeneration of electron–hole pairs at the Nb_2O_5/g-C_3N_4 interface through a direct Z-scheme, as suggested in Figure 8.4.

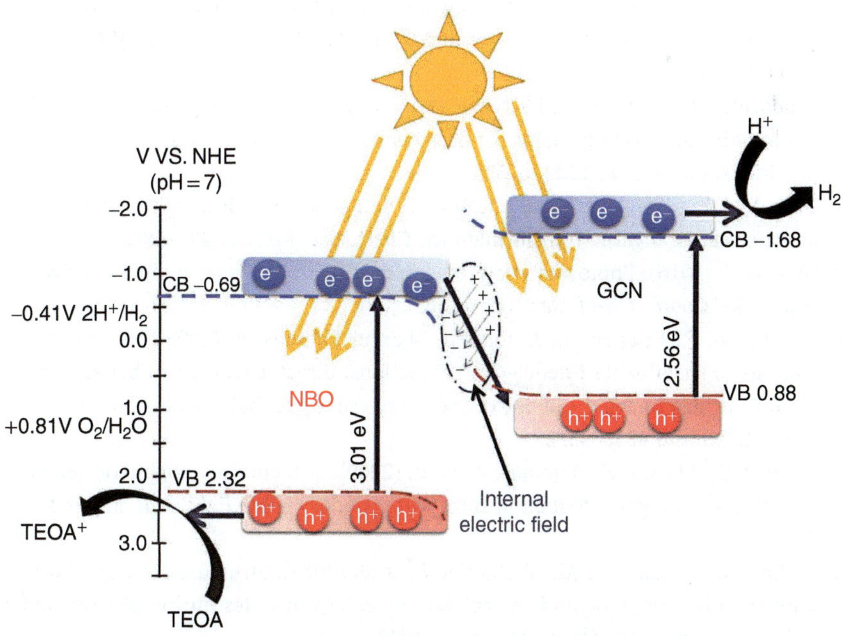

Figure 8.4 Scheme of the possible photocatalytic mechanism in the presence of Nb_2O_5/g-C_3N_4 photocatalysts. Source: Idrees et al. [30]/MDPI/CC BY4.0.

8.3 Conclusions

Hydrogen production by heterogeneous photocatalysis in the presence of renewable solar energy is one of the most versatile and environmentally benign paths for research to pursue. In this manner, H_2 can be produced at room temperature and atmospheric pressure by a simple, efficient, low-cost, and sustainable process, with the use of a photocatalyst, biomass, solar light, and water. This chapter briefly summarizes the latest progress of C_3N_4-based semiconductors for photo-assisted H_2 production in water suspensions by both photocatalytic water splitting and light-induced oxidation of biomass compounds.

References

1 Wang, X., Anpo, M., and Fu, X. (ed.) (2019). *Current Developments in Photocatalysis and Photocatalytic Materials: New Horizons in Photocatalysis*. Elsevier.

2 García-López, E.I. and Palmisano, G. (ed.) (2021). *Materials Science in Photocatalysis*. Elsevier.

3 Ng, C.H., Teo, S.H., Mansir, N. et al. (2021). Recent advancements and opportunities of decorated graphitic carbon nitride toward solar fuel production and beyond. *Sustainable Energy Fuels* 5: 4457–4511.

4 Turner, J.A. (2004). Sustainable hydrogen production. *Science* 305: 972–974.

5 Huang, C.W., Nguyen, B.S., Wu, J.C.S., and Nguyen, V.H. (2020). A current perspective for photocatalysis towards the hydrogen production from biomass-derived organic substances and water. *Int. J. Hydrogen Energy* 45: 18144–18159.

6 Kondarides, D.I., Vasileia, M., Daskalaki, V.M. et al. (2008). Hydrogen production by photo-induced reforming of biomass components and derivatives at ambient conditions. *Catal. Lett.* 122: 26–32.

7 Lang, X., Chen, X., and Zhao, J. (2014). Heterogeneous visible light photocatalysis for selective organic transformations. *Chem. Soc. Rev.* 43: 473–486.

8 Puga, A.V. (2016). Photocatalytic production of hydrogen from biomass-derived feedstocks. *Coord. Chem. Rev.* 315: 1–66.

9 Goettmann, F., Fischer, A., Antonietti, M., and Thomas, A. (2006). Metal-free catalysis of sustainable Friedel–Crafts reactions: direct activation of benzene by carbon nitrides to avoid the use of metal chlorides and halogenated compounds. *Chem. Commun.* 4530–4532.

10 Wang, X.C., Maeda, K., Thomas, A. et al. (2009). A metal-free polymeric photocatalyst for hydrogen production from water under visible light. *Nat. Mater.* 8: 76–80.

11 Hollmann, D., Karnahl, M., Tschierlei, S. et al. (2014). Structure-activity relationships in bulk polymeric and sol–gel-derived carbon nitrides during photocatalytic hydrogen production. *Chem. Mater.* 26: 1727–1733.

12 Marcì, G., García-López, E.I., and Palmisano, L. (2018). Polymeric carbon nitride (C_3N_4) as heterogeneous photocatalyst for selective oxidation of alcohols to aldehydes. *Catal. Today* 315: 126–137.

13 Cui, Y.J., Ding, Z.X., Liu, P. et al. (2012). Metal-free activation of H_2O_2 by g-C_3N_4 under visible light irradiation for the degradation of organic pollutants. *Phys. Chem. Chem. Phys.* 14: 1455–1462.

14 Wen, J., Xie, J., Chen, X., and Li, X. (2017). A review on g-C_3N_4-based photocatalysts. *Appl. Surf. Sci.* 391: 72–123.

15 Niu, P., Zhang, L., Liu, G., and Cheng, H. (2012). Graphene-like carbon nitride nanosheets for improved photocatalytic activities. *Adv. Funct. Mater.* 22: 4763–4770.

16 Ong, W.J., Tan, L.L., Ng, Y.H. et al. (2016). Graphitic carbon nitride (g-C_3N_4)-based photocatalysts for artificial photosynthesis and environmental remediation: are we a step closer to achieving sustainability? *Chem. Rev.* 116: 7159–7329.

17 Sui, Y., Liu, J., Zhang, Y. et al. (2013). Dispersed conductive polymer nanoparticles on graphitic carbon nitride for enhanced solar-driven hydrogen evolution from pure water. *Nanoscale* 5: 9150–9155.

18 Liu, J., Liu, Y., Liu, N. et al. (2015). Metal-free efficient photocatalyst for stable visible water splitting via a two-electron pathway. *Science* 347: 970–974.

19 Naseri, A., Samadi, M., Pourjavadi, A. et al. (2017). Graphitic carbon nitride (g-C_3N_4)-based photocatalysts for solar hydrogen generation: recent advances and future development directions. *J. Mater. Chem. A* 5: 23406–23433.

20 Cao, S., Low, J., Yu, J., and Jaroniec, M. (2015). Polymeric photocatalysts based on graphitic carbon nitride. *Adv. Mater.* 27: 2150–2176.

21 Yang, J., Wang, D., Han, H., and Li, C. (2013). Roles of cocatalysts in photocatalysis and photoelectrocatalysis. *Acc. Chem. Res.* 46: 1900–1909.

22 Maeda, K., Wang, X., Nishihara, Y. et al. (2009). Photocatalytic activities of graphitic carbon nitride powder for water reduction and oxidation under visible light. *J. Phys. Chem.C* 113: 4940–4947.

23 Hou, Y., Laursen, A.B., Zhang, J. et al. (2013). Layered nanojunctions for hydrogen-evolution catalysis. *Angew. Chem. Int. Ed.* 52: 3621–3625.

24 Lightcap, I.V., Kosel, T.H., and Kamat, P.V. (2010). Anchoring semiconductor and metal nanoparticles on a two-dimensional catalyst mat. storing and shuttling electrons with reduced graphene oxide. *Nano Lett.* 10: 577–583.

25 Sun, Q., Wang, P., Yu, H.G., and Wang, X.F. (2016). In situ hydrothermal synthesis and enhanced photocatalytic H_2-evolution performance of suspended rGO/g-C_3N_4 photocatalysts. *J. Mol. Catal. A: Chem.* 424: 369–376.

26 Yan, J.Q., Peng, W., Zhang, S.S. et al. (2020). Ternary Ni_2P/reduced graphene oxide/g-C_3N_4 nanotubes for visible light-driven photocatalytic H_2 production. *Int. J. Hydrogen Energy* 45: 16094–16104.

27 Zhang, X.H., Peng, B.S., Zhang, S., and Peng, T.Y. (2015). Robust wide visible-light-responsive photoactivity for H_2 production over a polymer/polymer heterojunction photocatalyst: the significance of sacrificial reagent. *ACS Sustain. Chem. Eng.* 3: 1501–1509.

28 García-López, E.I., Meo, P.L, B. Megna, L. Palmisano, and Marcì, G. (2022). C_3N_4/reduced graphene oxide based photocatalysts for H_2 evolution from aqueous solutions of oxygenated organic molecules. *Catal. Today* 2022, https://doi.org/10.1016/j.cattod.2022.11.026.

29 Acharya, R., Pati, S., and Parida, K. (2022). A review on visible light driven spinel ferrite-g-C_3N_4 photocatalytic systems with enhanced solar light utilization. *J. Mol. Liq.* 357: 119105.

30 Idrees, F., Dillert, R., Bahnemann, D. et al. (2019). In-situ synthesis of Nb_2O_5/gC_3N_4 heterostructures as highly efficient photocatalysts for molecular H_2 evolution under solar illumination. *Catalysts* 9: 169.

9

Organic Supramolecular Materials for Photocatalytic Splitting of Water to Produce Hydrogen

Xianjie Chen[1] *and Yongfa Zhu*[1,2]

[1] *Southwest University of Science and Technology, State Key Laboratory of Environment-friendly Energy Materials, School of Materials and Chemistry, Qinglong Road, Mianyang 621010, P.R. China*
[2] *Tsinghua University, Department of Chemistry, Shuangqing Road, Beijing 100084, P.R. China*

9.1 Introduction

Hydrogen gas can be seen as a future renewable energy carrier/fuel because it gives water as a combustion product without evolving CO_2. Hydrogen is considered as the cleanest and storable energy carrier of the future if it can be produced from a renewable energy source via a CO_2-neutral and efficient route. Solar water splitting is a renewable and sustainable hydrogen production method because it can utilize sunlight, the most abundant energy source on Earth, and water, the most abundant natural resource available on Earth. The key to achieving solar hydrogen production is to develop stable, efficient, and inexpensive photocatalysts, which should be active in visible light and especially should split water in natural sunlight. But still, many material-related issues hinder its widespread use to develop a suitable photocatalyst. According to thermodynamics, there are three main features involved in the water splitting reaction. One, it is a highly endothermic process. Second, the change in Gibb's free energy is positive ($\Delta G = 238\ \text{kJ mol}^{-1}$). Third, it is a four-electron process that requires 1.23 V potential. The fundamental equation that shows the required energy to produce hydrogen and oxygen from water is as follows:

$$2H_2O \rightarrow 4H^+ + O_2 \quad 1.23\ \text{V vs. NHE}$$

$$4H^+ \rightarrow 2H_2 \quad 0\ \text{V vs. NHE}$$

$$2H_2O \rightarrow 2H_2 + O_2 \quad 1.23\ \text{V vs. NHE}$$

Therefore, according to these equations, the basic requirements for a semiconductor to split water are suitable bandgap (more than 1.23 V) and the conduction band

UV-Visible Photocatalysis for Clean Energy Production and Pollution Remediation: Materials, Reaction Mechanisms, and Applications, First Edition. Edited by Xinchen Wang, Masakazu Anpo, and Xianzhi Fu.
© 2023 WILEY-VCH GmbH. Published 2023 by WILEY-VCH GmbH.

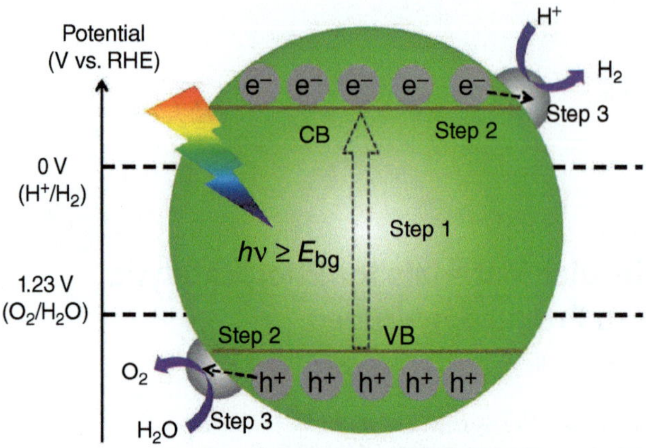

Figure 9.1 The basic process of semiconductor-based photocatalytic water splitting for hydrogen production.

(CB) minima should be more negative than the reduction potential of H^+/H_2 (0 V vs. normal hydrogen electrode [NHE]). The valence band (VB) maxima should be more positive than the reduction potential of O_2/H_2O (1.23 V vs. NHE). Figure 9.1 shows the basic process of semiconductor-based photocatalytic water splitting for hydrogen production. It is particularly difficult to obtain a simple, cost-effective, and highly active semiconductor material that satisfies all the crucial requirements:

(1) The ideal bandgap should be 1.5–3.2 eV for effective utilization of the solar spectrum.
(2) The CB minimum should be more negative than the water reduction (H^+/H_2) level and the VB maximum should be more positive than the water oxidation (O_2/H_2O) level concerning the potential of NHE.
(3) Stability in both acidic and basic mediums.
(4) Photocorrosion against resistance.
(5) The surface area, porosity, or reactive facets should be larger for higher active sites.

In addition to all these requirements, a photocatalyst has to suppress the recombination of photoinduced charge carriers generated during the reaction and enhance the photocatalytic process. In 1972, Fujishima and Honda ignited the overall photocatalytic water splitting using TiO_2 electrodes under ultraviolet (UV) light, which has been considered a landmark event in the photocatalytic field. However, the main drawback of TiO_2 photocatalyst is that it possesses a wider bandgap (3.2 eV) and faster recombination rates of photoinduced charge carriers. It should be noted that still more efforts are being made to achieve water splitting in visible light using modified TiO_2 as a photocatalyst. The design of an effective photocatalyst is important to achieve higher efficiency for hydrogen generation and to transfer the process to large-scale hydrogen generation realization.

In this chapter, we briefly overviewed the recent research in visible-light-driven photocatalytic water splitting, mainly through organic supramolecular semiconductor materials. First, the necessity of a photocatalyst for achieving water splitting with drawbacks from metal-based systems is mentioned, followed by a brief description of the organic semiconductor for photocatalytic water splitting. Second, the design concept of organic supramolecular photocatalysts for visible-light-driven water splitting is reviewed. Finally, recent advances in visible-light-driven organic semiconductors, including perylene imide (PDI) and porphyrin supramolecular photocatalysts as well as other new arrivals, and perspectives are discussed in detail.

9.2 Organic Supramolecular Photocatalysts for Water Splitting

During the past three decades, various UV and visible-light-driven photocatalysts have been explored, including metal oxides, sulfides, nitrides, phosphides, metal (oxy) nitrides, and graphene-based materials [1]. Notably, metal oxides with d^0 and d^{10} electronic configurations, such as $SrTiO_3$, Ta_2O_5, Zn_2GeO_4, Bi_2WO_6, $K_4Nb_6O_{17}$, $Sr_{0.25}H_{1.5}Ta_2O_6$, and $Sr_{0.4}H_{1.2}Nb_2O_6 \cdot H_2O$ materials have been employed as catalysts for photocatalytic water splitting in the visible light. Although, most of these photocatalysts are limited in practical applications due to their poor visible-light absorbance, photocorrosion in sulfide compounds, and fast recombination of charge carriers. Out of this, recently, oxynitrides and oxysulfides emerged as visible light active photocatalysts that utilize visible light with promising performance with high quantum yields for overall water splitting [2]. The abovementioned materials are basically inorganic semiconductors, whereas they have some disadvantages, such as their earth abundance, crystallinity, the toxicity of heavy metals, and cumbersome synthesis process. Therefore, it is of utmost urgency to develop cost-effective and earth-abundant materials as new visible-light-driven photocatalysts for water splitting, to increase the efficiency of the process.

Metal-free photocatalysts are more advantageous because of their earth abundance, lightweight, cost-effective, easy fabrication, and good mechanical flexibility, which make them promising catalysts in photocatalytic water splitting. The first organic semiconductor used for photoreduction of water to hydrogen is poly(p-phenylene), which is only active in UV light (1–1.7 $\mu mol\,h^{-1}$ at $\lambda > 290$ nm) [3]. Conjugated linear poly(phenylene) can catalyze hydrogen evolution in conjunction with methyl viologen (1 $\mu mol\,h^{-1}$ in $\lambda > 420$ nm), but they are only modestly active under UV irradiation and their performance under visible light is very poor. In addition, poly(azomethine), a conjugated polymer system, generates around 7 $\mu mol\,h^{-1}$ in $\lambda > 300$ nm. Since 2009, polymeric graphitic carbon nitride (g-C_3N_4) has been discovered as a polymeric semiconductor, metal-free photocatalyst, which fulfills the basic requirements for a water splitting catalyst and revolutionizes the field after its first phenomenal report by Wang et al. [4]. Generally, g-C_3N_4 has

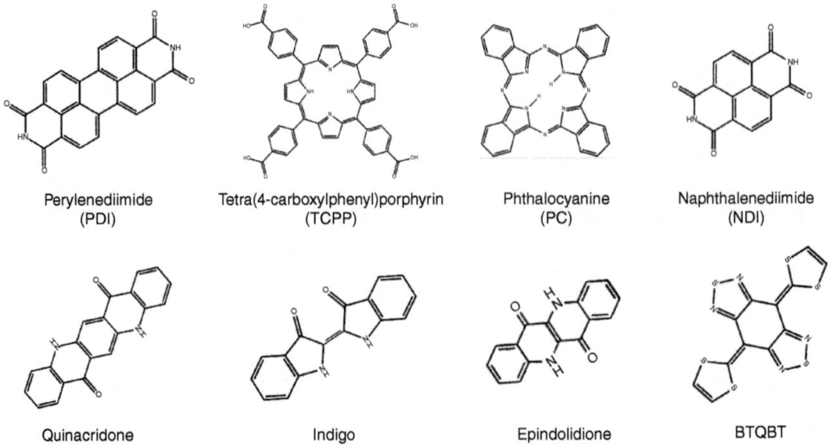

Figure 9.2 Molecular structures of the pigments.

been considered the most stable allotrope among various carbon nitrides under mild conditions. The main advantages of using these polymers are abundance, stability, and visible-light response in the presence of a sacrificial donor or recent reports on overall water splitting in the presence of suitably located cocatalysts [5]. However, the photocatalytic efficiency of g-C_3N_4 (0.1%) is still low due to its nonporous nature, faster recombination, and lower electrical conductivity.

Surprisingly, some counterparts form bulk materials via non-covalent interactions and possess crystalline features, which are usually renamed pigments. These pigments present particular optical/electronica/chemical properties different from monomeric dyes, owing to various self-assembled structures [6]. The dye–dye interactions can strongly affect the structures, morphologies, and coloristic properties of bulk materials, for instance, forming organic semiconductors. As shown in Figure 9.2, this type of organic semiconductor is usually composed of many semi-planar molecules with large conjugated π-bond and rigid structures, such as porphyrin, phthalocyanine (PC), 1,8:4,5-Naphthalenetetracarboxdiimide (NDI), bis(1,2,5-thiadiazolo)-*p*-quinobis(1,3-dithiole) (BTQBT), Quinacridone, Indigo, perylene monoimide (PMI), epindolidione, indanthrene, perylene-3,4,9,10-tetracarboxylic acid diimide (PDI).

Herein, we present the overview of the electronic structural modulation to photocatalytic applications of PDI and porphyrin self-assembly supramolecular photocatalysts. Firstly, the basic characteristics of PDI and porphyrin molecules and their self-assembly are introduced. Next, the possible electronic modulation approaches are discussed, including modifying perylene areas, tuning π–π stacking via side-chain substituents, and constructing PDI and porphyrin self-assembly supramolecular composites. Subsequently, the applications of photocatalytic water splitting into H_2/O_2 are exemplified to highlight the significance of the organic self-assembled supramolecular materials, with high efficiency of solar-light utilization stemming from particular π–π stacking structures. Finally, the outlooks and perspectives on the further development of PDI and porphyrin self-assembly supramolecular photocatalysts are envisioned.

9.2.1 PDI-based Supramolecular Photocatalysts for Hydrogen Production

Perylene diimide derivatives act as one common type of these pigments (abbreviated as PI, PDI, PBI, or PTCDI from perylene-3,4,9,10-tetracarboxylic acid diimide). The molecule unit consists of a PDI core and side chains. Interestingly, due to the great π-extended structure of the perylene ring of an individual molecule, the intrinsic π–π stacking interaction of each building block occurs along the one-dimensional direction if suitable solvents are given [7]. Together with hydrogen bonds or other noncovalent interactions, it eventually affords supramolecular self-assembly with a highly ordered structure. Chemical modification of the perylene ring and side chains can easily tune self-assembled modes and supramolecular constructions [6]. Consequently, a variety of morphologies, e.g. hollow nanosphere, wire, and so on [8], can be constructed in specific dispersion media (Figure 9.3). The process of forming supramolecular architectures demonstrates an essential dynamic equilibrium between gel and sol; therefore, environmental factors (such as solvent, temperature, and concentration) are of utmost importance toward pathway-controlled self-assembly of well-organized structures [9]. As a result, modulating microscopic molecular stacking arrangements between PDI molecules is a sticky challenge because the balance state is strongly sensitive to environmental conditions. Different π–π stacking modes could result in distinctly different electron-transfer pathways and thus change photocatalytic performances [10].

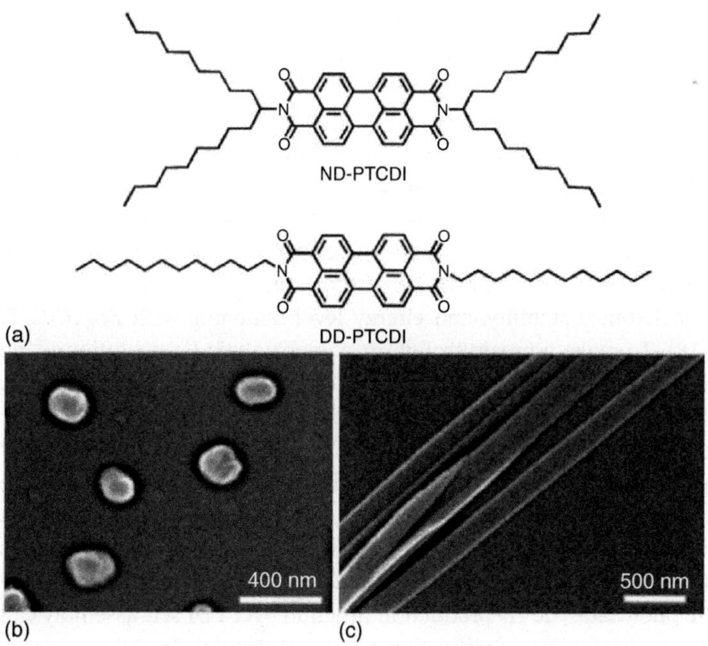

Figure 9.3 (a) Molecular Structures of ND- and DD-PTCDI, (b) SEM image of ND-PTCDI aggregates, (c) SEM image of DD-PTCDI nanobelts. Source: Balakrishnan et al. [8]/reproduced with permission from American Chemical Society.

The aggregation endows it with a broader light-absorption range, even extending to the near-infrared region via changing the molecular stacking columns (such as J- and H-type aggregation) as has been evidenced by experiments and DFT calculations [11]. Importantly, electron transfer along the π–π stacking direction is a salient characteristic in photocatalysis [12], which can effectively prevent the combination of photogenerated electrons and holes.

Furthermore, as shown in Figure 9.4, the supramolecular can form a band-like electronic structure with CB and VB levels, and enable photoinduced charge delocalization along the π–π stacking direction [14]. Hence, as a metal-free photocatalyst, PDI self-assembly possesses an electronic structure similar to common inorganic semiconductors, delivering three basic photocatalytic processes including light harvesting, charge separation, and charge transport, and importantly, exhibits some impressive performances largely attributable to the π–π stacking structure and consequential electron transfer along the π–π stacking direction. Given these properties, PDI self-assembly can effectively utilize the wider solar-light spectrum and inhibit the combination of photogenerated electrons and holes; therefore, PDI self-assembly-based photocatalysts and their composites have been employed in the photocatalytic degradation of organic contaminants [13], photocatalytic water splitting [14], disease therapy [15], and photocatalytic organic synthesis [16]. Although some exceptional reviews on PDI self-assembly have been published in the fields of supramolecular, optoelectronics, synthesis, and biology [17], to the best of our knowledge, there are currently no comprehensive reviews on photocatalysis of PDI self-assembly. The modulation of electronic structures of photocatalysts plays a dominant role in improving their photocatalytic performance. Therefore, it is necessary to systematize the relationship between the structure, electronic structure, and photocatalytic performance of PDI self-assembly to provide a reference for future photocatalytic research on self-assembly analogs.

Generally, the potential of the CB of PDI self-assembly is lower than the reduction potential of H^+/H_2; therefore, the kinetic process of its photocatalytic H_2 production is unfavorable in thermodynamics. In terms of photocatalytic H_2 production, PDIs were initially considered as dye sensitizers or cocatalysts to assist photocatalytic hydrolysis of host photocatalysts, mainly benefiting from their strong visible-light absorption, photochemical stability, and energy level matching with $Zn_{0.5}Cd_{0.5}S$, TiO_2 or C_3N_4 [18]. To make a breakthrough in photocatalytic H_2 evolution using PDI self-assembly, it needs to be combined with other photocatalysts bearing H_2 production ability. Chen and coworkers succeeded in achieving the photocatalytic H_2 production performance via constructing a C_3N_4/PDI composite catalyst for the enhanced performance, mainly because the photogenerated holes in C_3N_4 are assumed to rapidly transfer to the PDI molecules, which effectively prevented the recombination of photogenerated electrons and holes in C_3N_4 [19]. All these studies regarded PDI as a kind of auxiliary catalyst but not host photocatalyst.

Afterward, the photocatalytic H_2 production function over PDI self-assembly can be achieved via an appropriate modification to PDI molecules. In 2016, Li et al. revealed that PDI with incorporation of bipyridyl moieties into the network led to the enhancement of photocatalytic H_2 production performance [20]. Recently, our group synthesized perylenetetracarboxylic acid (PTA) supramolecular nanosheets with a monolayer thickness of ~1.5 nm (Figure 9.5), which showed highly efficient

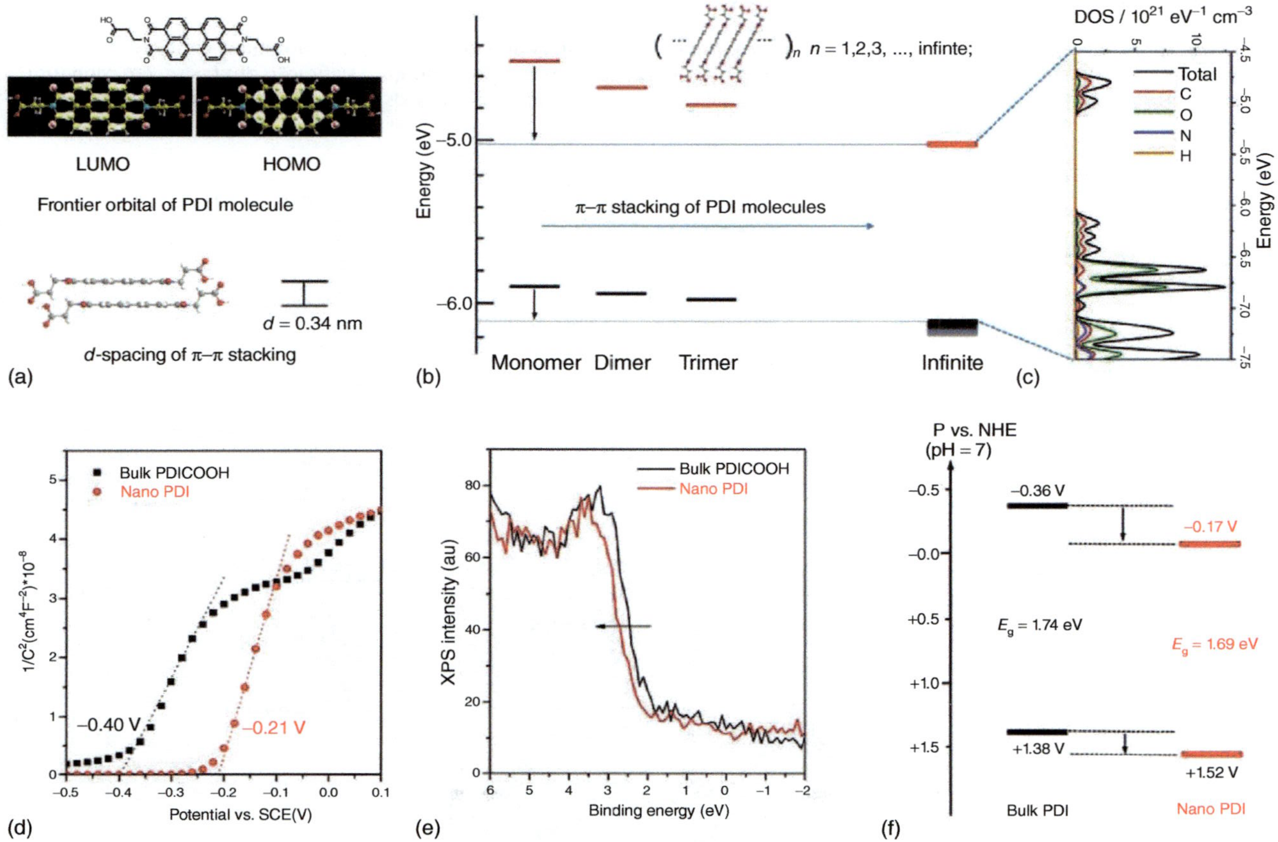

Figure 9.4 (a) Top: the frontier molecular orbitals of PDI molecule; Bottom: d-spacing of $\pi-\pi$ stacking. (b) Energy diagram representing the theoretically evaluated HOMO and LUMO levels of $\pi-\pi$ stacked PDI aggregates. (c) DOS of PDI aggregates with infinite degree of $\pi-\pi$ stacking. (d) Mott–Schottky plots of bulk PDI and nano PDI samples. (e) XPS valence band spectrum of nano PDI. (f) Diagram representing experimentally evaluated HOMO (VB) and LUMO (CB) levels of bulk PDI and nano PDI. Source: Reproduced with permission from Wang et al. [13]/Elsevier.

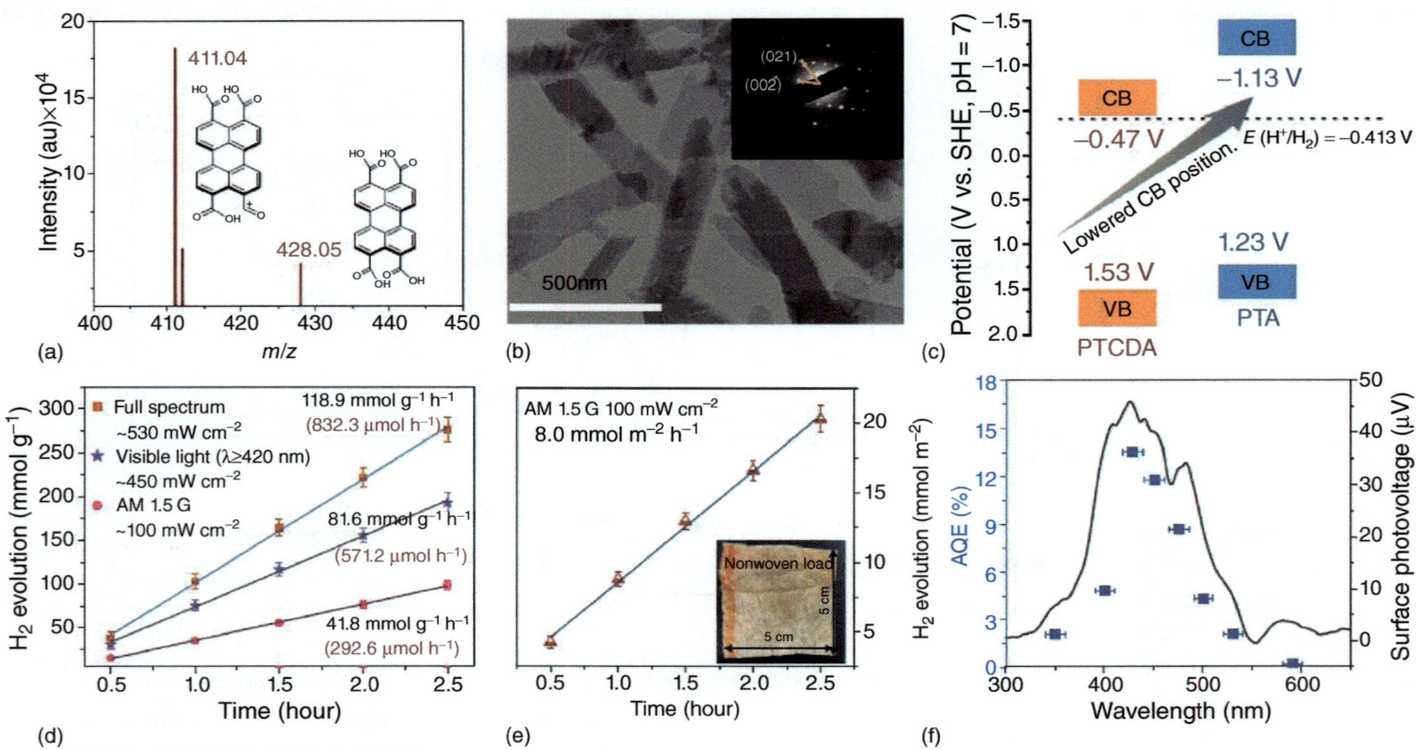

Figure 9.5 (a) Time-of-flight secondary ion mass spectrometry, (b) TEM image and (c) the band structures of PTA nanosheets. (d) Photocatalytic hydrogen evolution performance of PTA, (e) the hydrogen evolution performance under AM 1.5 G of PTA loaded on nonwoven fabrics, (f) the wavelength-dependent AQE for photocatalytic hydrogen evolution over PTA. Source: Guo et al. [21]/reproduced with permission from Springer Nature, CC BY 4.0.

hydrogen evolution with production rates of 118.9 mmol g^{-1} h^{-1} [21]. The carboxyl groups increased the intensity of the internal electric fields of PTA from the perylene center to the carboxyl border by 10.3 times to promote charge-carrier separation. The photogenerated electrons and holes directionally migrated to the edge and plane, respectively, to weaken charge-carrier recombination. Moreover, the PTA reduction potential increases from −0.47 to −1.13 V due to the decreased molecular conjugation and enhances the reduction ability. In addition, the carboxyl groups created hydrophilic sites. These works provide a strategy to engineer the molecular structures of future efficient photocatalysts.

It is worth mentioning PMI molecules, whose structure are mostly similar to PDI molecules. Unlike the PDI scaffold, PMI molecules bear only one part of the amide and side chain, but they can also form supramolecular self-assembly by π–π stacking behaviors, and π-electrons inside can migrate along the π–π stacking direction. The photocatalytic hydrogen production driven by hydrogel scaffold built with PMI self-assembled supramolecular was attempted by Stupp's group in 2014 [7]. They found that by screening different catalysts and electrolytes to prepare the PMI supramolecular nanoribbons for 3D architecture, the electronic coupling interaction in the PMI supramolecular containing Ni catalyst can be changed, thereby triggering different photocatalytic properties (Figure 9.6). In 2015, they separated the hydrophobic core and hydrophilic carboxylate headgroup with the linkage of different alkyl lengths, and upon the addition of salt, a series of PMI self-assembled nanostructures were constructed [22]. It was inferred that due to exciton splitting efficiencies originating from their higher orbital overlap, these hydrogel self-assemblies exhibited productive photocatalytic hydrogen amounts.

9.2.2 Porphyrin-based Supramolecular Photocatalysts for Hydrogen Production

Porphyrins, which are a group of heterocyclic organic compounds, are good absorbers of visible light. Porphyrin-based dyes have several intrinsic advantages, such as their rigid molecular structures with large absorption coefficients in the visible region and many reaction sites, i.e. four meso and eight β positions that are readily available for functionalization, which enables fine-tuning of the optical and physical properties of porphyrins. As a result of their excellent photophysical, photochemical, electrochemical, and structural properties, the distinguished π-conjugated porphyrins are of great interest [23]. The rich and extensive absorption features of porphyrins guarantee efficient use of the solar spectrum [24]. Additionally, their rigid and planar molecular skeleton and inherent aromatic electronic features facilitate their assembly into well-defined nanostructures with favorable optoelectronic properties [25]. Actually, the porphyrin self-assembled supramolecular is characterized by its synthetic versatility and morphology controllability. And since the self-assembly of porphyrins is mainly dependent on various intermolecular non-covalent interactions, porphyrin nanostructures with a certain size, shape, and function can be provided through careful molecular and supramolecular design. To date, porphyrin-based supramoleculars have been extensively applied

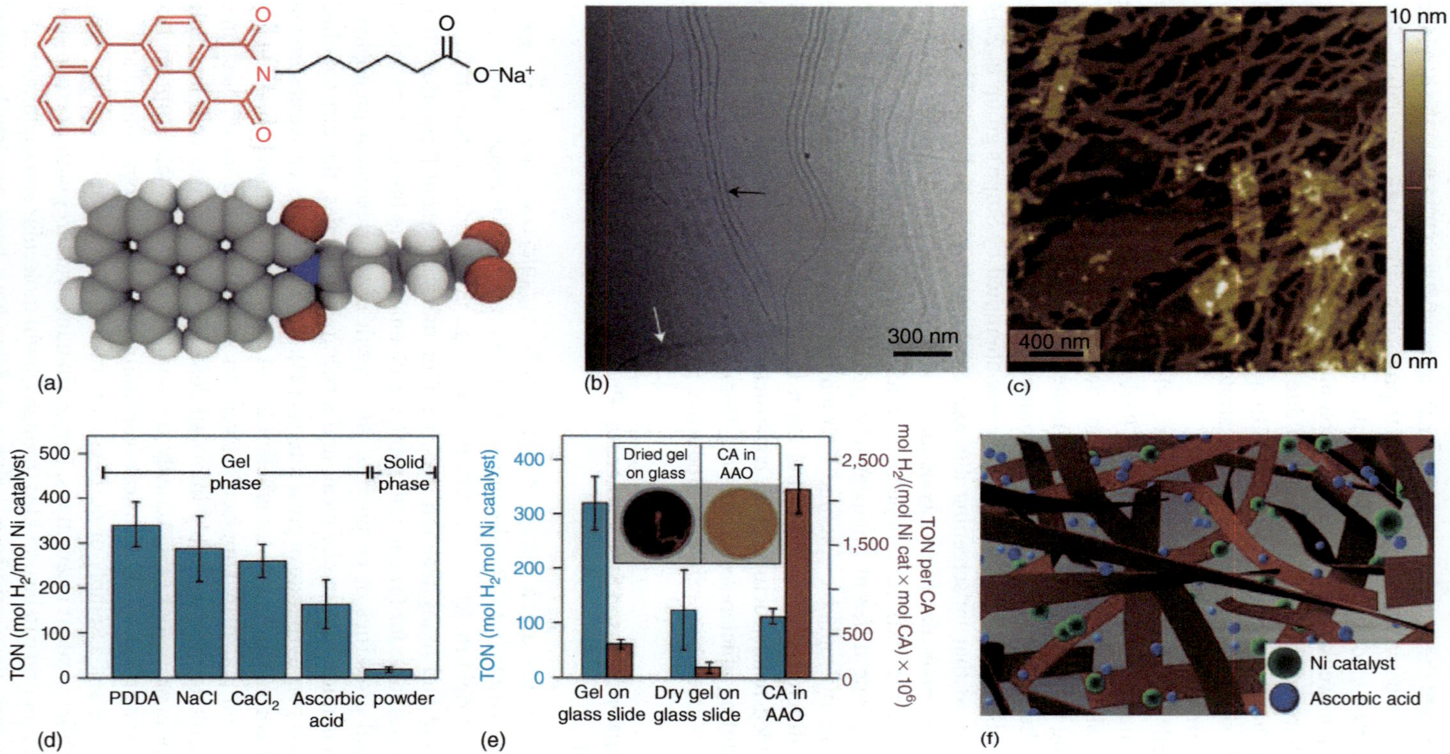

Figure 9.6 (a) Molecular structure PMI-based CA, (b) Cryo-TEM micrograph of CA solution, (c) AFM image of a dilute CA solution, (d) H_2 production histogram of CA gels prepared with NaCl, PDDA, $CaCl_2$, and ascorbic acid compared to insoluble protonated CA, (e) H_2 production from CA–PDDA gels, (f) Schematic of gel showing that CA nanoribbons trap solvent water molecules within a 3D architecture. Source: Weingarten et al. [7]/reproduced with permission from Springer Nature.

for visible-light photocatalysis [26]. However, the photocatalytic efficiency of the as-obtained porphyrin supramoleculars is still limited by the fast recombination of photoinduced electron-hole pairs. In this regard, we review here the recent advances in the design and fabrication of porphyrin-based supramoleculars and highlight their photocatalytic applications in hydrogen generation.

In 2019, our group constructed a full-spectrum (300–700 nm) responsive Tetra(4-carboxylphenyl)porphyrin (TCPP) supramolecular photocatalyst with a theoretical solar spectrum efficiency of 44.4% by the reprecipitation route (Figure 9.7) [27]. For the first time, hydrogen and oxygen evolution (40.8 and 36.1 $\mu mol\ g^{-1}\ h^{-1}$) is demonstrated by a porphyrin photocatalyst without the addition of any cocatalysts. The high photocatalytic reduction and oxidation activity arise from a strong built-in electric field due to molecular dipoles of electron-trapping groups and the nanocrystalline structure of the supramolecular photocatalyst. The appropriate band structure of the supramolecular photocatalyst adjusted via the highest occupied molecular orbital and lowest unoccupied molecular orbital energy levels of the porphyrin gives rise to the thermodynamic driving potential for H_2 and O_2 evolution under visible light irradiation. This work revealed that controlling the energy band structure of photocatalysts via the ordered assembly of structure-designed organic molecules could provide a novel approach to the design of organic photocatalysts in energy applications. However, further improvement of the photocatalytic activity is greatly limited by the high exciton binding energy and low charge carrier mobility of organic semiconductors, leading to inferior photogenerated charge separation efficiency.

The organic D–A structures have demonstrated wide applications in the field of organic electronics, such as transport, photoconductive, and ferroelectricity. Notably, electrons would prefer to transfer from D to A in the D–A structure, in favor of improving overall charge separation. Therefore, organic photocatalysts with a D–A structure have been developed based on covalent organic frameworks (COFs), conjugated polymers, and organic heterojunction. Subsequently, our group assembled a series of porphyrin-based supramolecular photocatalysts with a D–A structure, including tetraphenylporphyrin-C60, tetraphenylporphinesulfonate (TPPS)/fullerene, and TPPS/PDI [28]. As shown in Figure 9.8, the tetraphenylporphyrin-C60 is characterized by strong D–A interaction, inducing a significantly larger molecular dipole moment than the referential porphyrin molecule. As a consequence, a robust internal electric field (IEF) is built up in these D–A porphyrin-based supramolecular, contributing to ultrafast photogenerated charge separation and enhanced charge transport. Thus, they showed highly efficient photocatalytic hydrogen production performance with Pt nanoparticles as cocatalysts, and the maximum hydrogen evolution rate can reach 30.36 $mmol\ g^{-1}\ h^{-1}$.

Moreover, many metallated porphyrins have also been studied. These complex structures are ionically self-assembled by taking advantage of the electrostatic interactions between anionic and cationic porphyrins. The oppositely-charged pairs admit an enhanced photoinduced electron transfer across the porphyrin subunits, leading to higher photoactivity than individual porphyrins. Recently, our group also

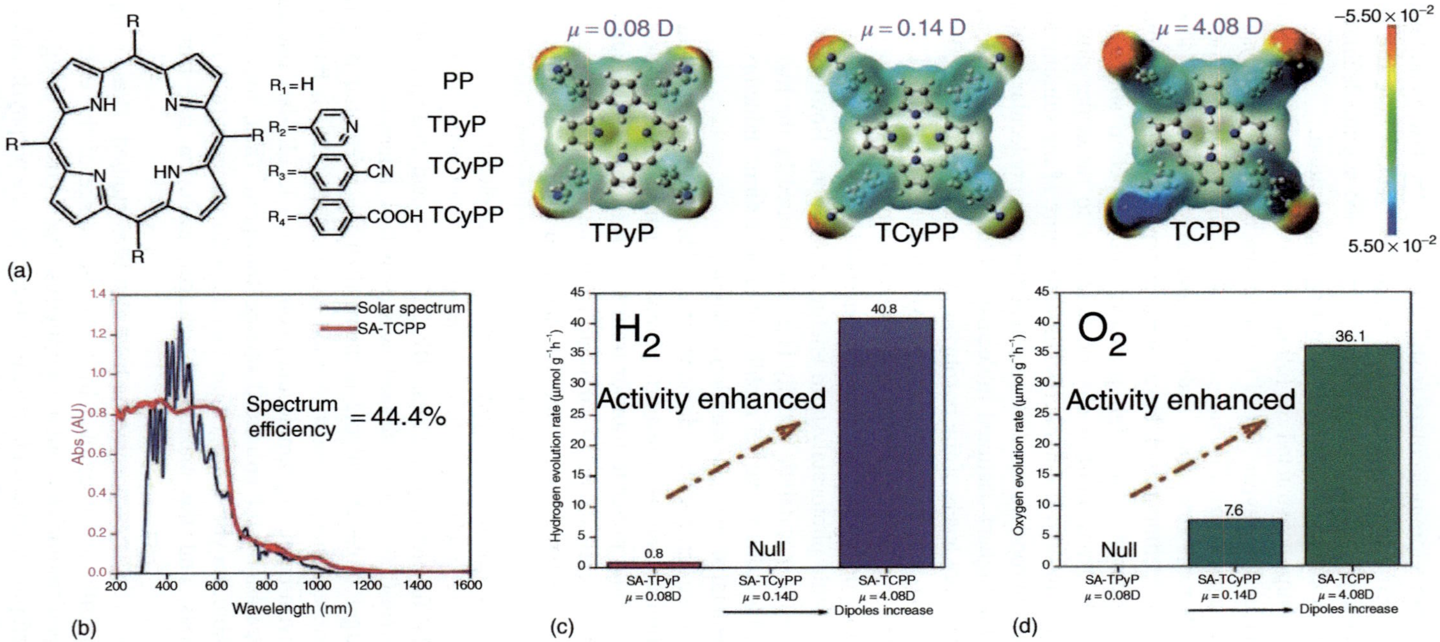

Figure 9.7 (a) Molecular structures, electrons distribution, and molecular dipoles of several porphyrin derivatives; (b) UV–vis diffuse reflection spectroscopy of SA-TCPP supramolecular photocatalyst, and solar spectrum observed by optical fiber spectrometer; (c) Photocatalytic hydrogen evolution of porphyrin supramoleculars without cocatalyst, (d) Photocatalytic oxygen evolution of porphyrin supramoleculars without cocatalyst. Source: Reproduced with permission from Zhang et al. [27]/John Wiley & Sons.

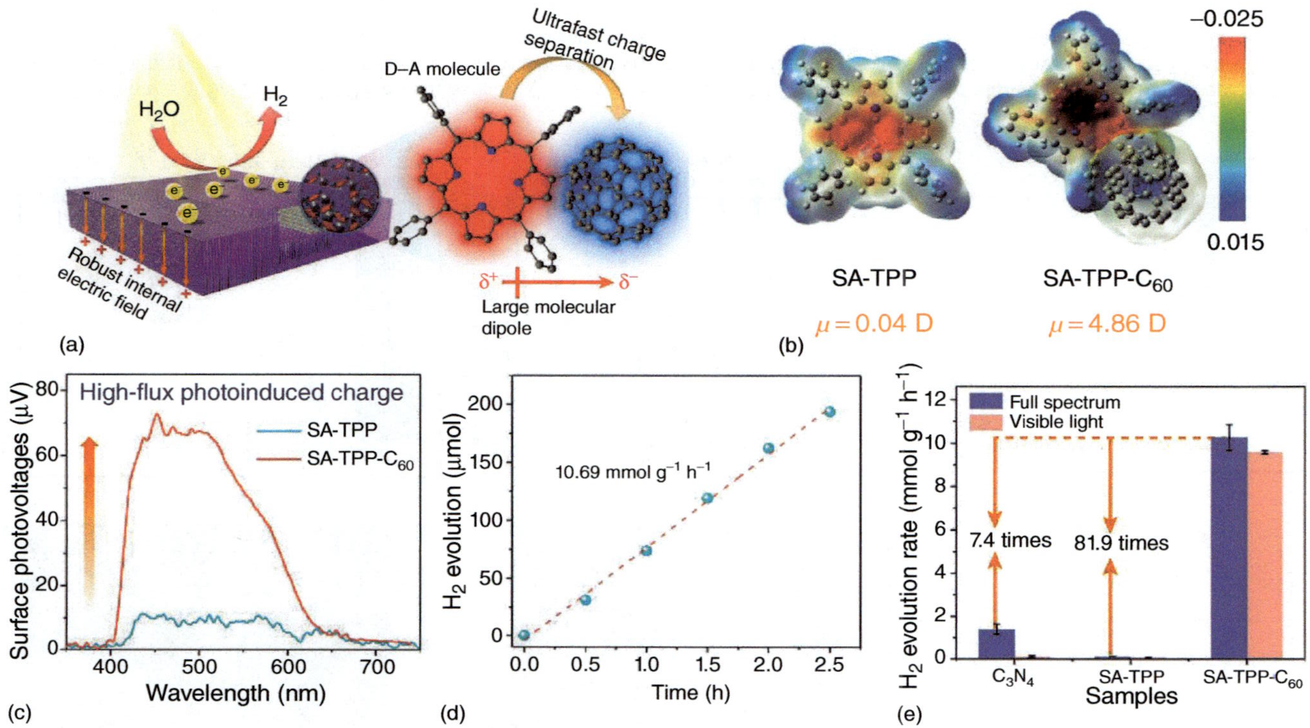

Figure 9.8 (a) Schematic mechanism of the photocatalytic reactions by SA-TPP-C60; (b) the calculated electrostatic potential maps, molecular dipoles; (c) the surface photovoltage of SA-TPP and SA-TPP-C_{60}; (d) the photocatalytic hydrogen evolution over SA-TPP-C60; (e) the comparison of hydrogen evolution rates. Source: Reproduced with permission from Liu et al. [28]/Elsevier.

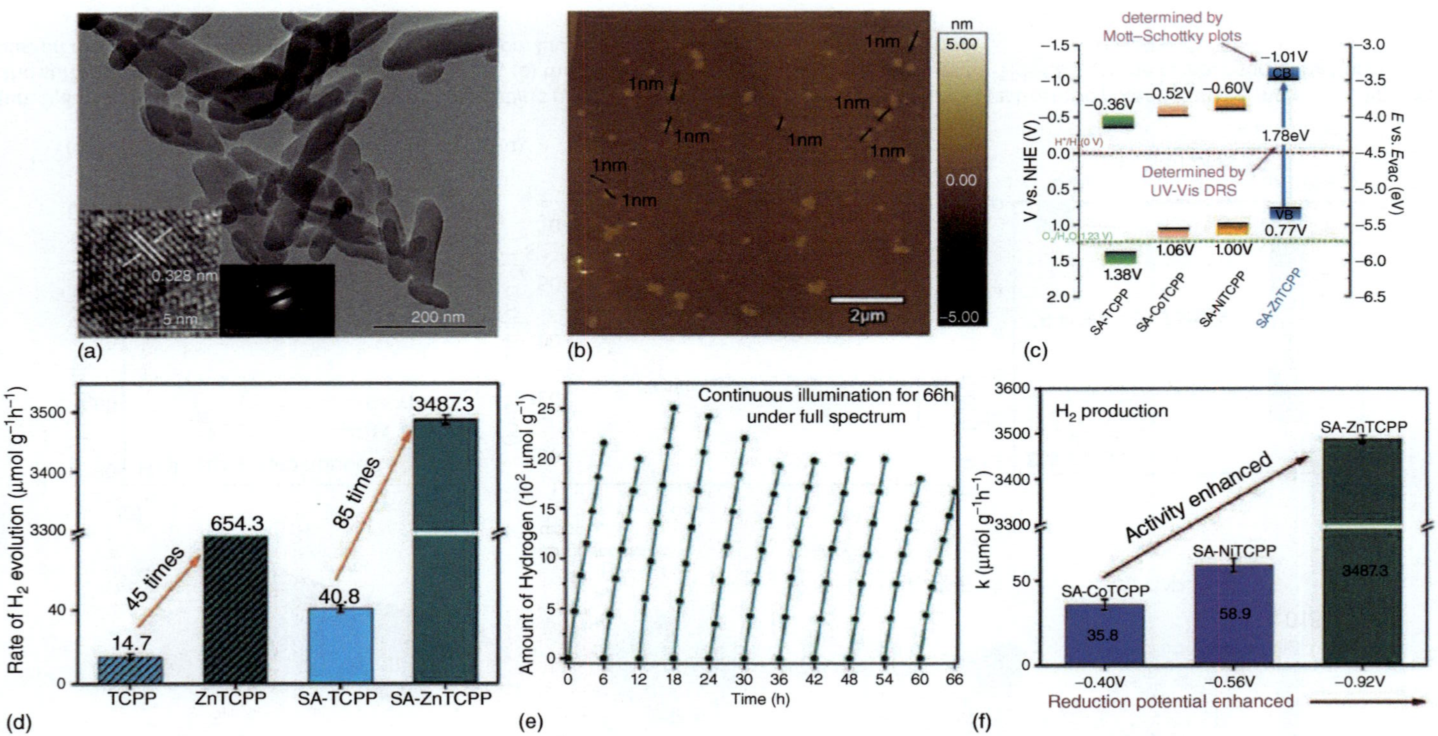

Figure 9.9 (a) TEM image of SA-ZnTCPP (HRTEM image inset); (b) AFM image of SA-ZnTCPP; (c) energy-band alignment diagram for supramolecular porphyrins; (d) photocatalytic H_2 evolution rate comparison of zinc porphyrin and metal-free porphyrin; (e) the repeated cycles of SA-ZnTCPP; (f) The increased photocatalytic activity of different supramolecular metalloporphyrins with the enhanced reduction potential. Source: Tian et al. [30]/reproduced with permission from Royal Society of Chemistry.

successfully fabricated ultrathin 2D self-assembled tetrakis (4-carboxyphenyl) zinc porphyrin (SA-ZnTCPP) supramolecular nanosheets for photocatalytic hydrogen production [29]. SA-ZnTCPP presents a deep CB (−1.01 V) contributed by the elevation of the orbital energy level of porphyrin after Zn^{2+} coordination, as shown in Figure 9.9. In addition, a strong built-in electric field in SA-ZnTCPP markedly boosts the rapid separation and mobility of photogenerated carriers. Consequently, the strong reduction driving force and faster electron transfer kinetics result in highly efficient photocatalytic hydrogen production of SA-ZnTCPP, and the H_2 evolution rate can reach 3487.3 $\mu mol\,g^{-1}\,h^{-1}$, which is 85 times higher than TCPP supramolecular. This self-metallization is also applicable to binary porphyrin structures for hydrogen evolution [30].

9.3 Conclusion and Perspectives

To summarize, this chapter has outlined the recent significant advances related to PDI-based and porphyrin-based supramoleculars for photocatalytic water splitting for hydrogen production under visible-light irradiation. Organizing PDI and porphyrin supramolecular nanostructures can benefit significantly from the improved photostability and light-harvesting properties based on aggregation. An increased charge transfer/transport process can be achieved by texture or crystal modification and interfacial heterostructure for final enhanced photocatalytic performance. The unique structure and tunable functionalization of PDI and porphyrins render their assembly or heterostructure easily tailored through various valent or non-valent interactions, to better activate the interface or surface for charge transfer. This is a clear advantage when compared to inorganic photocatalysts. On this basis, significant progress has been achieved, yet extended efforts are still required in various aspects to further advance the utilization of PDI- and porphyrin-based nanostructures for visible-light photocatalysis.

Among the issues requiring further effort, the first is how to bring together two different types of organic and inorganic semiconductors with a large and beneficial interface. Moreover, because electron transport through inorganic materials is very efficient, the overall electron flow in such junctions is mostly determined by the interface itself or by the lower mobility of the organic component. It thereby requires continued efforts to find better ways to improve the charge mobility of PDI and porphyrin supramoleculars. Further exploration is still needed into supramolecular/metal nanohybrids for plasmonic enhancement of photocatalytic efficiency. To gain a significant breakthrough in photocatalytic performance, an in-depth understanding of interface geometry, electronic properties, excitation dynamics, and charge transport properties by rational calculation and simulation would be useful for the design of efficient PDI and porphyrin supramolecular photocatalysts. Besides, it should be pointed out that, the IEF is a powerful driving force for the separation of photogenerated charge, which is closely related to photocatalytic performance. In our studies, the intrinsic IEF of materials is often improved by increasing the molecular dipole, thus improving the photocatalytic performance. However, improving the intrinsic IEF of a single component often involves very complex preparation processes and such an intrinsic IEF induced by a large molecular dipole is still limited in improving charge separation efficiency. As a consequence, exploring new strategies to obtain giant IEF is necessary to achieve a breakthrough in photocatalytic performance.

Alongside the heterogeneous photocatalysis, transferring or growing PDI-based and porphyrin-based supramoleculars onto an electrode by a proper deposition method is desired for PEC water splitting hydrogen evolution. Photoelectrodes have the advantage that an electric field can be created at the photoelectrode/electrolyte junction to manipulate the charge transfer reaction. Particularly in the case of flexible PEC devices, the high flexibility of the PDI and porphyrin structure and its compatibility with flexible and lightweight substrates may open broader applications for the next generation of optoelectronic devices. Future improvement of energy conversion efficiency calls for an effective alignment of these molecular assemblies at the macroscopic level aiming for better control of directional charge movement, as well as light absorption pathways in the nanostructure arrays.

References

1 Chen, X., Shen, S., Guo, L., and Mao, S.S. (2010). Semiconductor-based photocatalytic hydrogen generation. *Chem. Rev.* 110: 6503–6570.

2 Ma, G., Chen, S., Kuang, Y. et al. (2016). Visible light-driven Z-scheme water splitting using oxysulfide H_2 evolution photocatalysts. *J. Phys. Chem. Lett.* 7: 3892–3896.

3 Yanagida, S., Kabumoto, A., Mizumoto, K. et al. (1985). Poly (p-phenylene)-catalysed photoreduction of water to hydrogen. *Chem. Commun.* 8: 474–475.

4 Wang, X., Maeda, K., Thomas, A. et al. (2009). A metal-free polymeric photocatalyst for hydrogen production from water under visible light. *Nat. Mater.* 8: 76–80.

5 Zhang, J., Zhang, G., Chen, X. et al. (2012). Co-monomer control of carbon nitride semiconductors to optimize hydrogen evolution with visible light. *Angew. Chem. Int. Ed.* 51: 3183–3187.

6 Würthner, F., Saha-Möller, C.R., Fimmel, B. et al. (2016). Perylene bisimide dye assemblies as archetype functional supramolecular materials. *Chem. Rev.* 116: 962–1052.

7 Weingarten, A.S., Kazantsev, R.V., Palmer, L.C. et al. (2014). Self-assembling hydrogel scaffolds for photocatalytic hydrogen production. *Nat. Chem.* 6: 964–970.

8 Balakrishnan, K., Datar, A., Naddo, T. et al. (2006). Effect of side-chain substituents on self-assembly of perylene diimide molecules: morphology control. *J. Am. Chem. Soc.* 128: 7390–7398.

9 Bai, S., Debnath, S., Javid, N. et al. (2014). Differential self-assembly and tunable emission of aromatic peptide bola-amphiphiles containing perylene bisimide in polar solvents including water. *Langmuir* 30: 7576–7584.

10 Wei, W. and Zhu, Y. (2019). TiO_2@Perylene diimide full-spectrum photocatalysts via semi-core–shell structure. *Small* 15: 1903933.

11 Gong, Q., Xing, J., Huang, Y. et al. (2020). Perylene diimide oligomer nanoparticles with ultrahigh photothermal conversion efficiency for cancer theranostics. *ACS App. Bio Mater.* 3: 1607–1615.

12 Zang, L. (2015). Interfacial donor–acceptor engineering of nanofiber materials to achieve photoconductivity and applications. *Acc. Chem. Res.* 48: 2705–2714.

13 Wang, J., Shi, W., Liu, D. et al. (2017). Supramolecular organic nanofibers with highly efficient and stable visible light photooxidation performance. *Appl. Catal. B-Environ.* 202: 289–297.

14 Liu, D., Wang, J., Bai, X. et al. (2016). Self-assembled PDINH supramolecular system for photocatalysis under visible light. *Adv. Mater.* 28: 7284–7290.

15 Yang, Z., Fan, W., Zou, J. et al. (2019). Precision cancer theranostic platform by in situ polymerization in perylene diimide-hybridized hollow mesoporous organosilica nanoparticles. *J. Am. Chem. Soc.* 141: 14687–14698.

16 Zeng, L., Liu, T., He, C. et al. (2016). Organized aggregation makes insoluble perylene diimide efficient for the reduction of aryl halides via consecutive visible light-induced electron-transfer processes. *J. Am. Chem. Soc.* 138: 3958–3961.

17 Chen, S., Slattum, P., Wang, C., and Zang, L. (2015). Self-assembly of perylene imide molecules into 1D nanostructures: methods, morphologies, and applications. *Chem. Rev.* 115: 11967–11998.

18 Chen, Y., Li, A., Yue, X. et al. (2016). Facile fabrication of organic/inorganic nanotube heterojunction arrays for enhanced photoelectrochemical water splitting. *Nanoscale* 8: 13228–13235.

19 Chen, S., Wang, C., Bunes, B.R. et al. (2015). Enhancement of visible-light-driven photocatalytic H_2 evolution from water over g-C_3N_4 through combination with perylene diimide aggregates. *Appl. Catal. A* 498: 63–68.

20 Li, L. and Cai, Z. (2016). Structure control and photocatalytic performance of porous conjugated polymers based on perylene diimide. *Polym. Chem-UK* 7: 4937–4943.

21 Guo, Y., Zhou, Q., Nan, J. et al. (2022). Perylenetetracarboxylic acid nanosheets with internal electric fields and anisotropic charge migration for photocatalytic hydrogen evolution. *Nat. Commun.* 13: 1–10.

22 Weingarten, A.S., Kazantsev, R.V., Palmer, L.C. et al. (2015). Supramolecular packing controls H_2 photocatalysis in chromophore amphiphile hydrogels. *J. Am. Chem. Soc.* 137: 15241–15246.

23 Ng, K.K., Lovell, J.F., Vedadi, A. et al. (2013). Self-assembled porphyrin nanodiscs with structure-dependent activation for phototherapy and photodiagnostic applications. *ACS Nano* 7: 3484–3490.

24 Aratani, N. and Osuka, A. (2001). Monodisperse giant porphyrin arrays. *Macromol. Rapid Commun.* 22: 725–740.

25 Hasobe, T. (2012). Photo-and electro-functional self-assembled architectures of porphyrins. *Phys. Chem. Chem. Phys.* 14: 15975–15987.

26 Guo, P., Chen, P., and Liu, M. (2013). One-dimensional porphyrin nanoassemblies assisted via graphene oxide: sheetlike functional surfactant and enhanced photocatalytic behaviors. *ACS Appl. Mater. Interfaces* 5: 5336–5345.

27 Zhang, Z., Zhu, Y., Chen, X. et al. (2019). A full-spectrum metal-free porphyrin supramolecular photocatalyst for dual functions of highly efficient hydrogen and oxygen evolution. *Adv. Mater.* 31: 1806626.

28 Liu, L., Chen, X., Chai, Y. et al. (2022). Highly efficient photocatalytic hydrogen production via porphyrin-fullerene supramolecular photocatalyst with donor-acceptor structure. *Chem. Eng. J.* 444: 136621.

29 Jing, J., Yang, J., Zhang, Z., and Zhu, Y. (2021). Supramolecular zinc porphyrin photocatalyst with strong reduction ability and robust built-in electric field for highly efficient hydrogen production. *Adv. Energy Mater.* 11: 2101392.

30 Tian, Y., Martin, K.E., Shelnutt, J.Y.-T. et al. (2011). Morphological families of self-assembled porphyrin structures and their photosensitization of hydrogen generation. *Chem. Commun.* 47: 6069–6071.

10

Visible Light-responsive TiO_2 Thin-film Photocatalysts for the Separate Evolution of H_2 and O_2 from Water

Aswathy Rajan[1], Bernaurdshaw Neppolian[1], and Masakazu Anpo[2,3]

[1] *SRM Institute of Science and Technology, Department of Chemistry, Chennai, 603203, Tamil Nadu, India*
[2] *Fuzhou University, State Key Laboratory of Photocatalysis on Energy and Environment, Fuzhou, Fujian 350116, P.R. China*
[3] *Osaka Prefecture University (present, Osaka Metropolitan University), Department of Applied Chemistry, Sakai, Osaka 599-5831, Japan*

10.1 Introduction

The production of H_2 via solar water splitting by photocatalytic systems is an enticing approach to the global energy crisis and pollution issues. With this technology, freely accessible and limitless solar energy can be converted into chemical energy, resulting in the production of clean H_2 from water. H_2 releases only water through the combustion with O_2, therefore, H_2 is an ideal fuel [1]. For the commercialization of photocatalytic water splitting to produce H_2, the development of photocatalytic materials that effectively decompose water into H_2 and O_2 under sunlight irradiation is necessary. Titanium dioxide (TiO_2) has been a pioneering photocatalyst since Schrauzer and Guth reported the photolysis of H_2O and photoreduction of N_2 on TiO_2 particles in 1977 [2]. Prior to that, Honda and Fujishima reported a photosensitization effect of the TiO_2 electrode for the electrochemical photolysis of water at a TiO_2 electrode with a Pt counter electrode under UV light irradiation [3]. Following these reports, semiconducting powders or particles of metal oxides and sulfides have been widely utilized as photocatalysts. These powdered photocatalysts are easier to use, however, separation of the oxidation and reduction reactions is difficult, which results in efficient recombination of photogenerated H_2 and O_2 (so-called back reaction), leading to lower efficiency in the photocatalytic splitting of water. Furthermore, it is very difficult to recover these powder photocatalysts from the reaction systems. Also, in the solar spectrum, UV light accounts for only 3–5%, and therefore, designing solar light-responsive, so-called visible light-responsive photocatalysts is crucial.

To overcome these issues, visible light-responsive semiconducting thin-film photocatalysts have been desired since they are enabled to utilize sunlight efficiently

UV-Visible Photocatalysis for Clean Energy Production and Pollution Remediation: Materials, Reaction Mechanisms, and Applications, First Edition. Edited by Xinchen Wang, Masakazu Anpo, and Xianzhi Fu.
© 2023 WILEY-VCH GmbH. Published 2023 by WILEY-VCH GmbH.

and also separate the reduction and oxidation reactions [4, 5]. Various photocatalysts have been developed over the years, but their efficiencies are still low for practical application. Various strategic studies have also been widely carried out to improve the efficiency of H_2 production from water. With the addition of organic compounds such as alcohol or biomass into the reactant water, photocatalytic splitting of water has enhanced remarkably since these additives work as a sacrificial reagent, resulting in an enhancement of the H_2 evolution from water [6–8]. However, in those reactions, CO_2 has been produced as an oxidation product instead of O_2. Anpo et al. have proposed to use this produced CO_2 in an artificial LED light-type plant factory to enhance the photosynthesis of green vegetables leading to the rapid growth of vegetables. Such integration of photocatalytic splitting of water (so-called artificial photosynthesis) and natural photosynthesis of green plants is also introduced as promising, economical, sustainable, and environmentally friendly systems [9].

10.1.1 Fabrication of Visible Light-responsive TiO_2 Thin Films

Several deposition techniques such as spray pyrolysis, chemical vapor deposition, sol-gel, and sputtering have been applied for the fabrication of TiO_2 thin films [10, 11]. However, TiO_2 fabricated by these methods was found to function only under UV light irradiation (wavelength, $\lambda < 400$ nm) since it was a typical TiO_2 having a bandgap of 3.2 eV. Among them, it was accidentally found that the visible light-responsive TiO_2 thin-film photocatalysts with good stability and high surface areas have been fabricated using the radio frequency magnetic sputtering (RF–MS) deposition method at high substrate temperatures [11-13]. Figure 10.1A shows the advantages of a TiO_2 thin-film photocatalyst over a powdered photocatalyst. Figure 10.1B illustrates the schematic diagram of the RF–MS deposition method for the fabrication of a visible light-responsive TiO_2 thin-film photocatalyst.

For sputtering, a calcined TiO_2 plate was utilized as a target, and Ar was used as the sputtering gas. The distance between the target and substrate was fixed at 80 mm. The quartz glass plate or Ti-foil was placed on the substrate holder. The chamber was evacuated to a pressure less than 5.0×10^{-4} Pa before the introduction of Ar gas at 2.0 Pa. The sputtering parameters, such as the substrate temperature (T_S), the distance between target and substrate (D_{T-S}), and the pressure of Ar gas were found to greatly affect the characteristics of the fabricated TiO_2 thin films. Especially, the substrate temperature, which was changed between 473 and 873 K, greatly affected the optoelectronic property and photocatalytic activity of TiO_2 thin-film photocatalysts [12, 13].

10.1.2 Characteristics of the Visible Light-Responsive TiO_2 Thin Films Fabricated by RF–MS Deposition Method

Various spectroscopic techniques were employed to characterize and understand the optoelectronic and structural properties of the fabricated TiO_2 thin films.

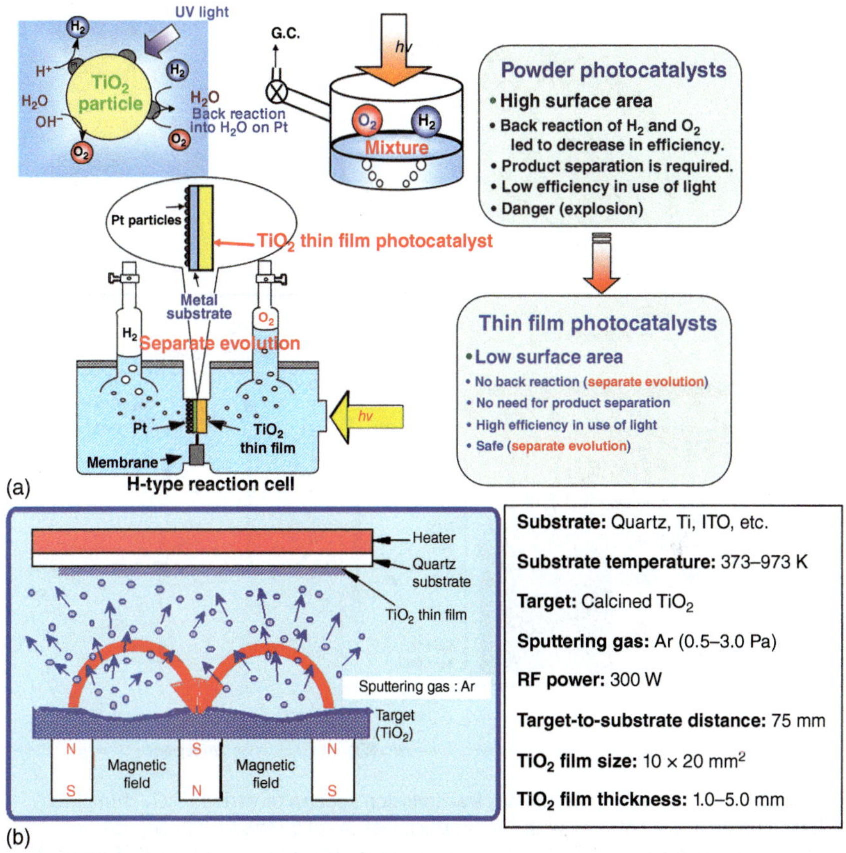

Figure 10.1 (A) Characteristic features of powder photocatalysts and thin-film photocatalysts. (B) Schematic illustration of the RF–MS deposition method and sputtering conditions to fabricate TiO_2 thin films by the RF–MS method. Source: Reproduced with permission from Matsuoka et al. [4]/Elsevier.

The different parameters that affect the photocatalytic activity of TiO_2 thin films are briefly explained below.

10.1.2.1 Effect of the Distance Between the Target and Substrate (D_{T-S}) and Substrate Temperature (T_S)

Figure 10.2A clearly exhibits that the UV–vis transmission spectrum of the TiO_2 thin film, which was fabricated on a quartz substrate, is remarkably affected by the substrate temperature. The TiO_2 thin film fabricated at 473 K substrate temperature (T_S) (UV-TiO_2) exhibits no absorption band in the wavelength regions longer than 380 nm. On the other hand, yellow-colored TiO_2 thin films fabricated at T_S higher than 673 K (Vis-TiO_2-673) exhibit significant absorption bands in the wavelength regions longer than 380 nm ($\lambda > 380$ nm), showing absorption of the visible light. Furthermore, a shift in the onset of the absorption band of the Vis-TiO_2 fabricated

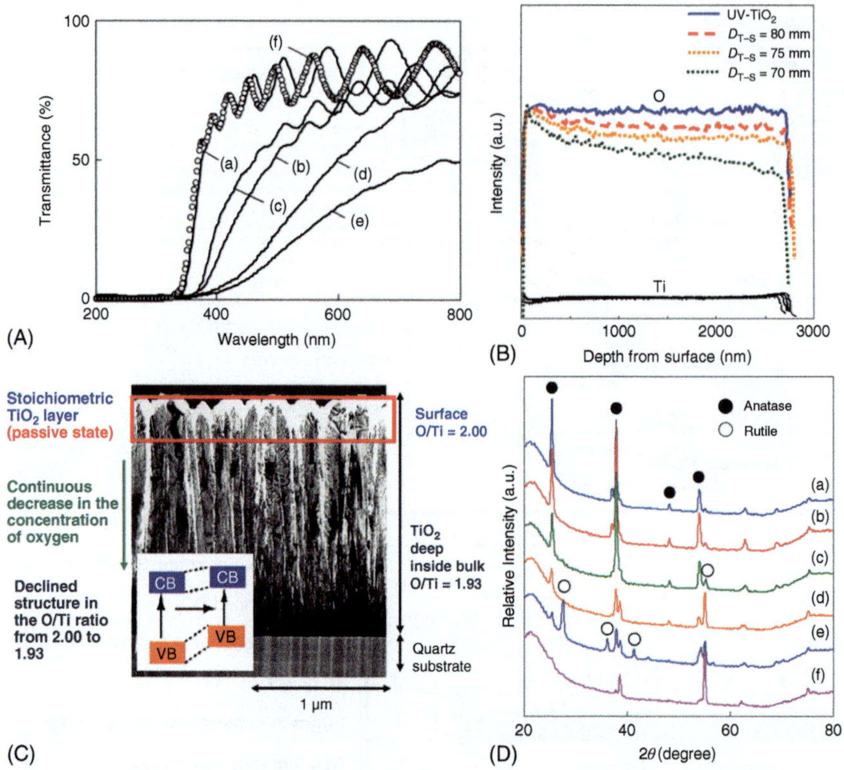

Figure 10.2 (A) Changes in the UV–vis transmission spectra of various TiO_2 thin films fabricated on quartz substrates with 1.2 μm thickness (a–e) by the RF-MS deposition method upon varying T_S and D_{T-S} in Ar atmosphere, T_S (K): 473 (a); 673 (b); 873 (c–f); D_{T-S} (mm): (c) 90; (a, b, d, and f) 80; (e) 70. Source: Reproduced with permission from Matsuoka et al. [4]/Elsevier. (B) SIMS depth profiles of ^{18}O and ^{48}Ti of UV-TiO_2 and Vis-TiO_2 by varying D_{T-S}. Source: Reproduced with permission from Ebrahimi et al. [13]/Springer Nature. (C) Cross-sectional TEM image of the Vis-TiO_2 thin film. Source: Matsuoka et al. [4]/ reproduced with permission from Elsevier. (D) XRD patterns of TiO_2 thin films fabricated at various substrate temperatures, T_s: 473 (a); 573(b); 673 (c), 773 (d), 873 (e), and 973 K (f). Source: Reproduced with permission from Takeuchi et al. [14]/Springer Nature.

at 873 K (Vis-TiO_2-873) is seen even at a wavelength longer than 600 nm, showing a big difference in the optical property in comparison to UV-TiO_2. Thus, it is clear that careful control of the substrate temperature is critically important for designing visible light-responsive TiO_2 thin-film photocatalysts. As shown in Figure 10.2A, with an increase in the substrate temperature, the efficiency of the absorption of visible light was also improved [12]. Moreover, the distance between a target and substrate (D_{T-S}) was also found to affect absorption properties. As D_{T-S} was reduced from 90 to 70 mm, the absorption spectra were found to shift toward longer wavelength regions (Figures 10.2A(c–e)). These phenomena are attributable to the fact that sputtering Ar atoms undergo continuous collisions on their way to reach the surface of the substrate, therefore, as D_{T-S} decreases, the kinetic energy of sputtering Ar atoms increases, leading to a difference in the optoelectronic property of TiO_2 thin films.

As shown in Figures 10.2B,C, the visible light-responsive TiO_2 thin films fabricated at different T_S and D_{T-S} exhibit characteristic features in their spectra of the secondary-ion mass spectrometry (SIMS) and transmission electron microscope (TEM) measurements. The intensity of SIMS due to the ^{18}O element of the TiO_2 thin film exhibits a steady decrease in the value from the top surface (O/Ti ratio = 2.00 ± 0.01) to the deep bulk inside of 1.93 for the Vis-TiO_2-873 thin film. On the other hand, for UV-TiO_2 thin films, noticeable variations in the O/Ti values were not observed. These findings clearly suggested that larger the kinetic energy of sputtering Ar atoms, lower the O/Ti ratios of the TiO_2 thin films. An anisotropic structure with a lower O/Ti ratio is explained by the substantial perturbation in its electronic properties and a significant shift in its absorption edge to the visible wavelength regions. As shown in Figure 10.2C, the cross-sectional TEM image indicates that top surface of the columnar TiO_2 crystal thin film is covered with the stoichiometric TiO_2 phase. This stable surface phase acts as a passive phase to protect the oxidation of the bulk inside the TiO_2 thin film, which maintains the stability of the visible light-responsive TiO_2 thin-film photocatalysts' efficient absorption and photocatalytic activity [12].

Figure 10.2D exhibits the X-ray diffraction (XRD) patterns of the TiO_2 thin films fabricated at temperatures below 673 K. A characteristic peak at 37.0° is attributed to the (004) phase of anatase. While the thin films fabricated at temperatures higher than 673 K exhibit well-defined peaks due to rutile also. The particle sizes of these TiO_2 thin films were calculated to be constantly around 20 nm by Scherrer's equation, irrespective of the fabricated temperatures [14]. The conversion of the crystal form of TiO_2 from anatase to rutile is known to occur by calcination at a higher temperature than 673 K, and simultaneously the particle size of TiO_2 increases while the surface area decreases. However, such a trend was not observed for TiO_2 thin films because the TiO_2 particles fabricated on Ti-foil receive interfacial stress from the substrates, and crystallization may be limited along their depth direction only.

10.1.2.2 Effect of the Pressure of Sputtering Ar Gas

As shown in Figure 10.3A,B, varying the pressure of sputtering Ar gas (PAr) from 0.5 to 5.0 Pa during the fabrication has a significant influence on their absorption spectra (Figure 10.3A) and XRD patterns (Figure 10.3B). With decreasing P_{Ar}, the absorption edge shifts to a longer wavelength region, and a broad absorption band is observed for the Vis-TiO_2 film ((c); P_{Ar} 1.5 Pa) in the 600–800 nm wavelength regions [13]. As seen in Figure 10.3B, varying the pressure of the sputtering Ar gas also affects significantly the intensity of the XRD peaks. As mentioned above, the sputtering Ar atoms have some kinetic energy depending on the pressure and temperature, which enables the promotion of the growth of crystals and the formation of the rutile phase. Furthermore, the collision of sputtering Ar atoms causes damage and degradation of

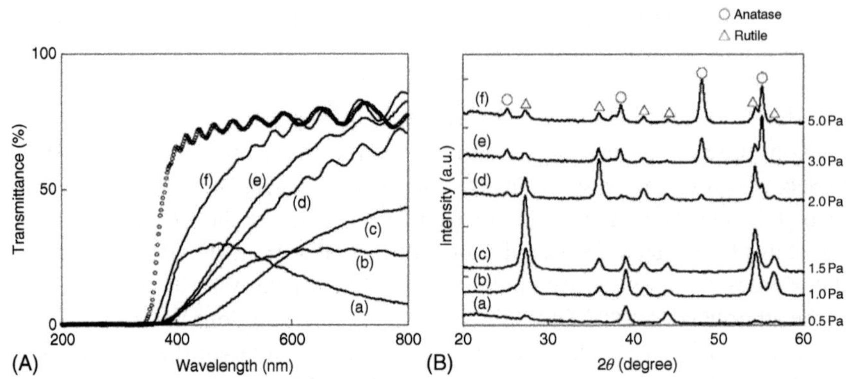

Figure 10.3 (A) UV–vis transmission spectra of TiO_2 thin films fabricated under various different pressure of Ar, P_{Ar} (solid line). Source: Reproduced with permission from Ebrahimi et al. [13]/Springer Nature. (B) XRD patterns of Vis-TiO_2-thin films fabricated under various different pressure of Ar, $P_{Ar.}$ In both figures of (A) and (B), P_{Ar} (Pa): (a) 0.5, (b) 1.0, (c) 1.5, (d) 2.0, (e) 3.0, and (f) 5.0. Source: Reproduced with permission from Ebrahimi et al. [13]/Springer Nature.

the crystallinity of the films. Thus, the kinetic energy of sputtering Ar atoms is one of the key factors in controlling the optical property of TiO_2 thin films fabricated by the RF-MS deposition method.

10.1.2.3 Effect of Surface Treatments on the TiO_2 Thin Films

As depicted in Figure 10.4A, the hydrothermal treatment of the Vis-TiO_2 with 10 M NaOH solution exhibits a decrease in the intensity of the XRD patterns assigned to the rutile phase, and simultaneously the peaks due to titanates are observed. Figure 10.4B exhibits SEM images of Vis-TiO_2 thin film treated with NaOH solution, and as the reaction progresses, the structure of nanowires becomes remarkable. The surface area of the nanowire TiO_2 was also found to remarkably increase with an increment in the time of hydrothermal reaction [15]. Such nanostructured TiO_2 films were found to exhibit effective charge separation of the photogenerated electrons and holes, which led to enhanced photocatalytic activity under visible light irradiation.

A chemical etching with HF was used to enlarge the surface area of the Vis-TiO_2 (HF-Vis-TiO_2). Figure 10.4C displays the XRD patterns of the Vis-TiO_2 and Vis-TiO_2 (HF-Vis-TiO_2). The intensity of the peak assigned to the rutile phase decreased after the HF treatment [16], while the content of the anatase phase increased from 10% to 25% after the HF treatment. Figure 10.4D shows the SEM image of the HF-Vis-TiO_2 and as seen, the surface roughness of the thin film increases, and the interspaces between columnar TiO_2 crystallites lengthen after the HF treatment. The results also indicate that the diffusion lengths of photogenerated electrons and holes to reach the liquid–solid interfaces of HF-Vis-TiO_2 become shorter. In fact, with an increase in the HF treatment time, the surface areas of the Vis-TiO_2/Ti thin film increase.

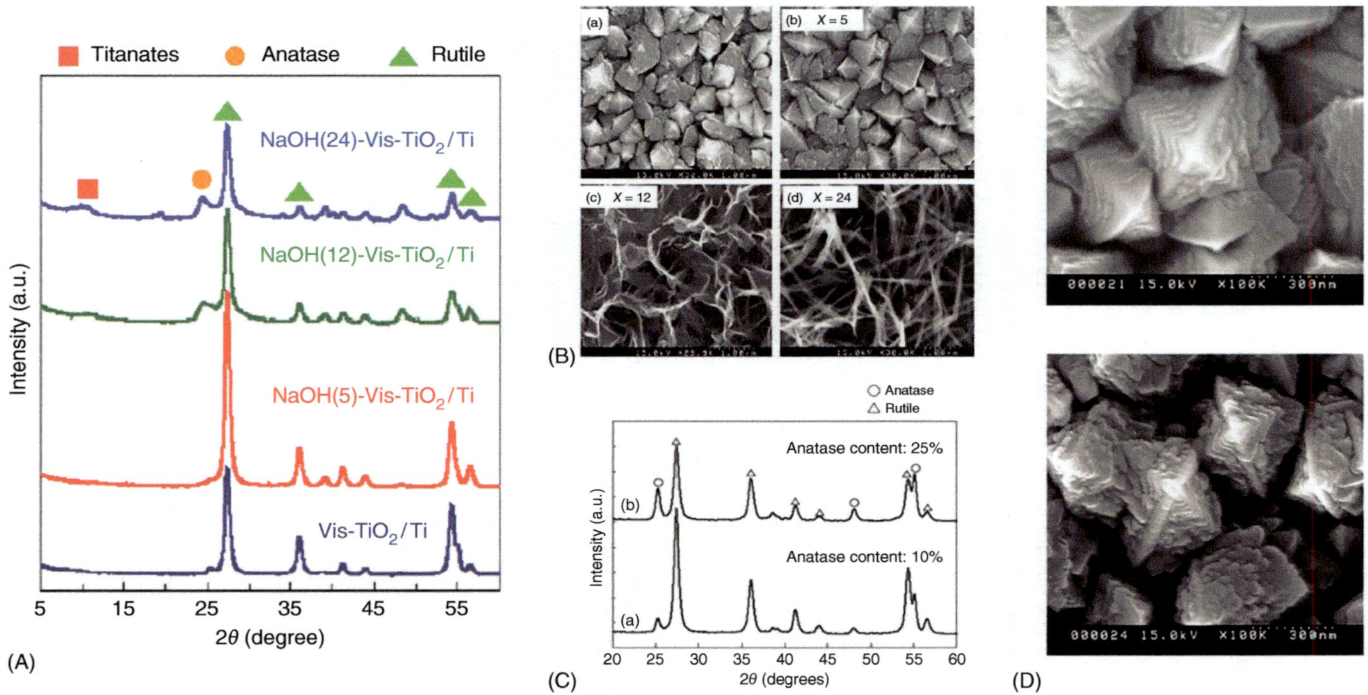

Figure 10.4 (A) XRD patterns of Vis-TiO_2/Ti and NaOH (X)-Vis-TiO_2/Ti, where X is the soaking time in NaOH. Source: Reproduced with permission from Matsuoka et al. [15]/Elsevier. (B) SEM images of Vis-TiO_2/Ti (a) and NaOH(X)-Vis-TiO_2/Ti (b–d). Source: Matsuoka et al. [15]/reproduced with permission from Elsevier. (C) XRD patterns of Vis-TiO_2/Ti (a) and HF-Vis-TiO_2/Ti (b). Source: Reproduced with permission from Kitano et al. [16]/Springer Nature. (D) SEM images of Vis-TiO_2/Ti (top) and HF (60)-Vis-TiO_2/Ti thin films (bottom). Source: Kitano et al. [16]/reproduced with permission from Springer Nature.

10.2 Photoelectrochemical Properties of TiO_2 Thin Films Fabricated by RF–MS Method

Photoelectrochemical measurements for the TiO_2 thin films were carried out on a standard three-electrode system with a 0.05 M aqueous NaOH solution. The photocurrent responses for the UV-TiO_2/Ti-foil and Vis-TiO_2/Ti-foil were recorded at the applied potential of +0.5 V vs. SCE (Figure 10.5A). A photoelectrochemical onset of about 520 nm was used to determine the bandgap to be 2.5 eV for the Vis-TiO_2 [17]. Furthermore, in order to evaluate the flat band potential (EFB) of the thin films, current-potential characteristics were recorded and a more negative zero current potential of −0.91 V vs. SCE was obtained for UV-TiO_2 compared to the −0.82 V for Vis-TiO_2 at pH 12.3. The obtained potential was found to be comparable to the EFB of polycrystalline semiconductors, and the conduction band edge (ECB) was also determined. Assuming the difference in ECB and EFB to be 0.2 eV for oxide semiconductors, the ECB of UV-TiO_2 and Vis-TiO_2 was calculated to be −1.11 V and −1.02 V vs. SCE at pH 12.3, respectively (Figure 10.5B) [1, 17]. The obtained valance and conduction band energy levels of the Vis-TiO_2 thin-film photocatalyst were found to be potential enough to induce the oxidation and reduction of water (splitting of water) to evolve H_2 and O_2. Thus, the possibility of photocatalytic splitting of water to produce H_2 and O_2 using Vis-TiO_2 thin-film photocatalysts fabricated by the RF–MS deposition method became clear, and its prospects to achieve the separate evolution of pure H_2 and O_2 from the water were further explored.

10.2.1 Setup the Reactor for Separate Evolution of H_2 and O_2 in the Photocatalytic Splitting of H_2O

There are several types of apparatus for the photocatalytic splitting of water [18, 19]. In the conventional setup of the single reactor, the efficiency of the splitting of water

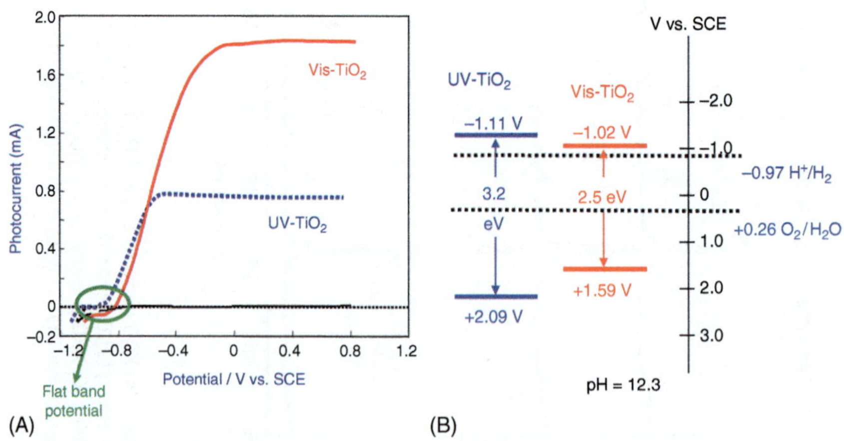

Figure 10.5 (A) Photocurrent curves of UV-TiO_2 and Vis-TiO_2 thin-film electrodes ($\lambda > 350$ nm). Ar-purged aqueous 0.05 M NaOH is used as the electrolyte. (B) Energy band levels of UV-TiO_2 and Vis-TiO_2 thin films against SCE. Source: Reproduced with permission from Kitano et al. [17]/Springer Nature.

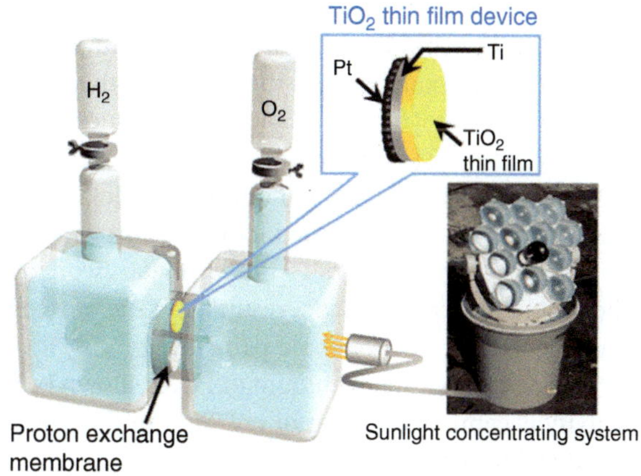

Figure 10.6 An H-type photocatalytic reactor for the separate evolution of pure H_2 and O_2 from water and a sunlight concentrating system utilized in the photocatalytic reaction under sunlight irradiation.

is greatly reduced by the back reaction or recombination of generated H_2 and O_2 to produce H_2O, and even worse, the cost of the separation of the evolved H_2 from a mixture with O_2 is quite high. To overcome these difficulties, the H-type reaction cell utilized in photoelectrochemical reactions was proposed for the separate evolution of H_2 and O_2 in the photocatalytic splitting of water. As shown in Figure 10.6, a visible light-responsive TiO_2 thin-film photocatalyst fabricated on Ti-foil has been utilized. Upon the absorption of UV-visible light, electrons and holes are generated in TiO_2 moieties, and then, electrons in the conduction band migrate to Pt particles through a Ti-foil. The electrons that reached Pt particles reduce the protons supplied from water to evolve H_2 molecules. The holes that remained in TiO_2 moieties oxidize OH^- supplied from water to evolve O_2 molecules from the TiO_2 thin film side. The efficiency of the separate evolution of H_2 and O_2 was expected to be strongly dependent on the work function of the metal substrate like a Ti-foil and also an external bias caused by the difference in the pH values in the solution of TiO_2 thin film side (O_2 evolution side) and Pt particle side (H_2 evolution side) as shown in Figure 10.6.

Anpo et al. designed an H-type quartz reactor for the photocatalytic splitting of water to achieve a separate evolution of pure H_2 and O_2 [12, 13]. As illustrated in Figure 10.6, a visible light-responsive TiO_2 thin-film photocatalyst and a proton exchange membrane were fixed in the middle of the H-type quartz cell. The H-type cell is separated through the membrane from aqueous solutions of different pH. The TiO_2 thin film was fabricated on one side of a Ti-foil with a 50 μm thickness, and Pt nanoparticles were deposited on the opposite side of the Ti-foil [17, 18].

10.3 Separate Evolution of Pure H_2 and O_2 Using a Visible Light-responsive TiO_2 Thin-film Photocatalyst Fabricated by RF–MS Deposition Method and the Factors Affecting the Efficiency

As mentioned above, the back reaction between the once evolved H_2 and O_2 makes photocatalytic splitting of water a particularly difficult reaction, and, therefore, avoiding the back reaction is a challenging issue to attain a high efficiency [1, 4, 5].

One approach is to utilize the so-called sacrificial reagents. The photocatalytic activity in the evolution of H_2 from water is improved by adding sacrificial reagents like alcohol or biomass into the water. In this case, the sacrificial reagent easily reacts with photogenerated holes and produces CO_2 instead of O_2, but enhances the evolution of H_2 from water. Such reactions are known as half-reactions of the splitting of water [7]. A small amount of metals, such as Pt, Au, Ag, and Ni have been loaded on the semiconductor photocatalyst since such metals enhance the photocatalytic evolution of H_2. These metals work as an efficient catalyst to reduce the proton to form hydrogen, resulting in the promotion of charge separation of the photogenerated electron–hole pairs. Thus, Vis-TiO_2 thin-film photocatalysts were also loaded with a small amount of Pt. As shown in Figure 10.7A, its photocatalytic activity is compared to that of UV-TiO_2 in the presence of methanol in water as a sacrificial reagent.

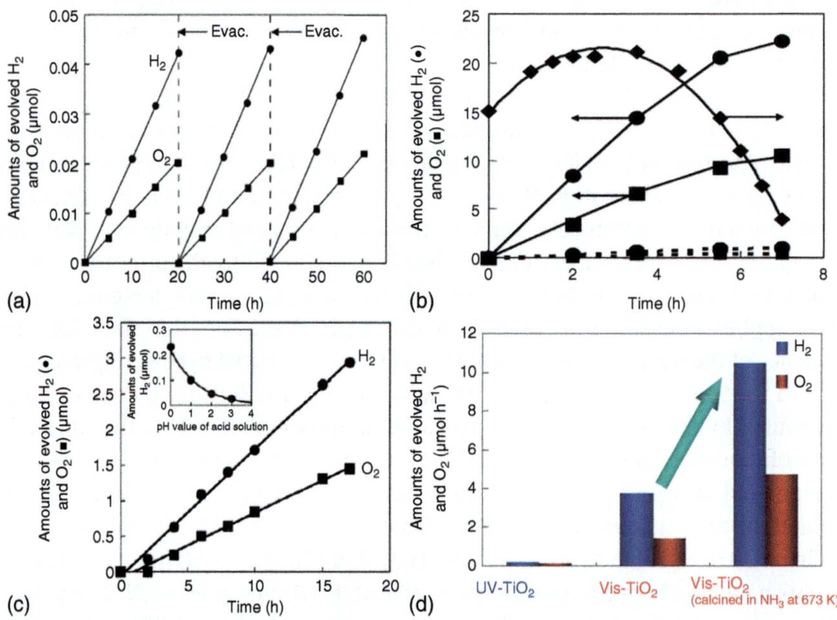

Figure 10.7 (A) Separate evolution of pure H_2 and O_2 from water using Pt-loaded Vis-TiO_2 thin-film photocatalyst. Source: Reproduced with permission from Kitano et al. [16]/ Springer Nature. (B) Separate evolution of pure H_2 and O_2 using a UV-TiO_2 thin-film photocatalyst (broken line) and a Vis-TiO_2-(873) thin-film photocatalyst (solid line). Source: Reproduced with permission from Kitano et al. [16]/Springer Nature. (C) Separate evolution of pure H_2 and O_2 from water using a Pt-loaded Vis-TiO_2 thin-film photocatalyst fabricated on Ti-foil (Vis-TiO_2/Ti/Pt) under visible light irradiation ($\lambda = 420$ nm). Inserted Figure shows the effect of the pH values of the aqueous solution on the Pt side. In these cases, the pH value of the aqueous solution on the TiO_2 thin film side.was always maintained to be 14.0. Source: Reproduced with permission from Kitano et al. [20]/Elsevier. (D) Effect of the calcination temperature of Vis-TiO_2 thin-film photocatalysts in NH_3 atmosphere on the yield of the H_2 evolution. Source: Reproduced with permission from Matsuoka et al. [21]/Springer Nature.

Table 10.1 Quantum yields for the separate evolution of H_2 and O_2 from water using various TiO_2 thin-film photocatalysts fabricated at various different T_S and D_{T-S} under UV light ($\lambda = 360$ nm) and visible light ($\lambda = 420$ nm) irradiation, respectively.

	Quantum yield (%)			
	$\lambda = 360$ nm		$\lambda = 420$ nm	
Photocatalyst	H_2 [a]	O_2 [b]	H_2 [a]	O_2 [b]
Pt/UV-TiO_2-(473,80)	26.2	12.6	0.00	0.00
Pt/vis-TiO_2-(673,80)	27.2	38.0	0.13	0.85
Pt/vis-TiO_2-(873,70)	5.2	8.2	0.05	0.18
Pt/vis-TiO_2-(873,80)	34.2	60.0	1.25	2.43
Pt vis-TiO_2-(873,90)	26.5	32.8	0.10	0.58

a) From 50 vol% aqueous methanol solutions.
b) From 0.05 M aqueous silver nitrate solution.
Source: Reproduced with permission from Matsuoka et al. [4]/Elsevier.

As explained in Section 10.2.2.1, the substrate temperature (T_S) and the distance between the target and substrate (D_{T-S}) significantly affect the absorption properties. It is clearly evident from Table 10.1 that the photocatalytic activity of the TiO_2 thin-film photocatalysts for the evolution of H_2 and O_2 changes significantly depending on the T_S and D_{T-S}. The evolution rates of H_2 and O_2 on Pt-Vis-TiO_2-(873) under visible light irradiation also remarkably increased with a decrease in D_{T-S} from 90 to 80 mm. Furthermore, the photocatalytic performance was found to further decrease with a further decrease in D_{T-S} to 70 mm. These results were attributed to the formation of the Ti^{3+} species, which work as the electron–hole recombination center, being created by the vigorous bombardment of sputtering atoms or Ar plasma with high-kinetic energy [12, 13]. Thus, it has been found that the regulation of the D_{T-S} is one of the key factors in the fabrication of highly active visible light-responsive thin-film photocatalysts.

The photocatalytic separate evolution of H_2 and O_2 was carried out in the H-shaped type cell with an aqueous NaOH (1 M) solution on the TiO_2 side and an aqueous H_2SO_4 (0.5 M) solution on the Pt side, respectively, to bring a chemical bias of 0.83 V to facilitate the electron transfer from TiO_2 moieties to Pt particles. As shown in Figure 10.6, upon the irradiation of the TiO_2 thin-film photocatalyst, the separate evolution of pure H_2 and O_2 was found on the Pt side and the TiO_2 side, respectively. As shown in Figure 10.7A, the Vis-TiO_2/Ti-foil/Pt thin-film photocatalyst leads to the separate evolution of H_2 and O_2 in a stoichiometric manner under visible light irradiation ($\lambda > 420$ nm). A chemical bias that was generated by two aqueous solutions with different pH values promotes the electron migration from the TiO_2 moieties to the Pt particle side. In fact, as shown in the inset figure in Figure 10.7C, it was found that the larger difference in the pH value between the two aqueous solutions on the TiO_2 side and the Pt side results in a higher rate of the evolution of H_2 and O_2 [12, 20].

The effect of calcination of Vis-TiO_2 thin films in air and NH_3 [(1.0×10^5 Pa) on the yields of the photocatalytic evolution of H_2 and O_2 from the water was studied at 673 K for 2 h. As shown in Figure 10.7D, it was observed that after calcination in NH_3, a remarkable increase in the yield of the evolution of H_2 and O_2 was observed. Further investigations were carried out to study the effect of the chemical etching of the TiO_2 thin films with HF solution on their photocatalytic activity for the decomposition of water. Figure 10.8A clearly depicts the photocatalytic activity of the Vis-TiO_2 thin-film photocatalyst and a remarkable increase in the yield by the HF treatment. Furthermore, Figure 10.8B revealed that the HF-Vis-TiO_2 has a greater photocurrent response than that of pristine Vis-TiO_2, whereas any response was hardly observed for UV-TiO_2. These results unequivocally demonstrated that the HF-Vis-TiO_2 thin film works as an effective photocatalyst for the independent production of H_2 and O_2 from water under visible light or sunlight irradiation. These enhancements in the photocatalytic activity were attributed to an increase in the surface area and surface roughness of the thin films due to the chemical

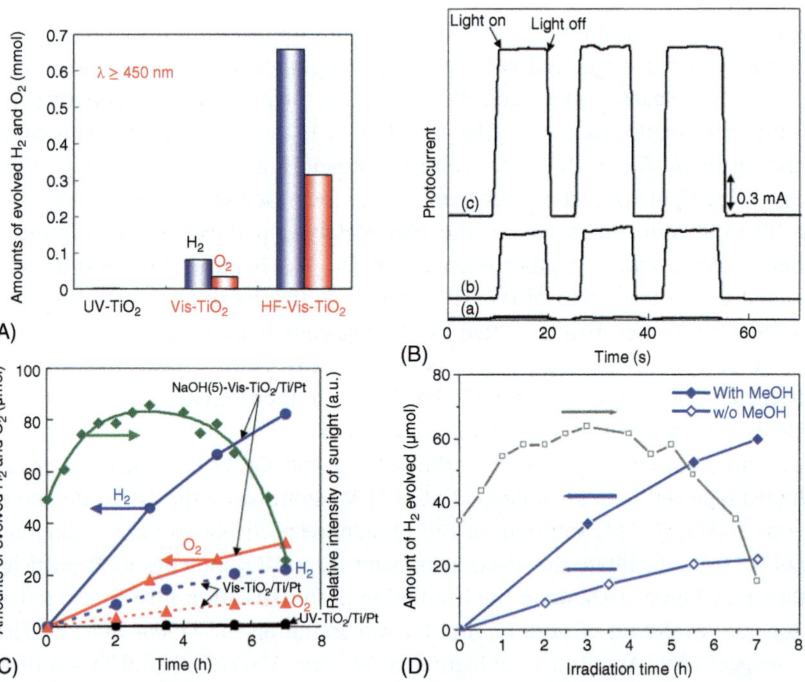

Figure 10.8 (A) The effect of HF etching of Vis-TiO_2 thin-film photocatalysts (HF-Vis-TiO_2) on the yields of photocatalytic separate evolution of H_2 and O_2. Source: Adapted from Kitano et al. [16]; © 2008 Springer. (B) Photocurrent responses of UV-TiO_2 (a), Vis-TiO_2 (b), and HF-Vis-TiO_2 (c). Source: Reproduced with permission from Kitano et al. [16]/Springer Nature. (C) The effect of NaOH treatment of TiO_2 thin films on the yield of the photocatalytic evolution of H_2 under visible light irradiation. Source: Reproduced with permission from Matsuoka et al. [15]/Elsevier. (D) Yields of photocatalytic evolution of H_2 from pure water (-o-) and water involving CH_3OH (-o-) using Vis-TiO_2. Source: Reproduced with permission from Horiuchi et al. [9]/Elsevier.

etching with the HF solution. In addition, the effect of NaOH treatment of the TiO_2 thin films (NaOH-Vis-TiO_2/Ti) on the yields of photocatalytic evolution of H_2 was also investigated [15]. In Figure 10.8C, the effect of the treatment of TiO_2 thin-film photocatalysts with an aqueous NaOH solution on the yields of the photocatalytic evolution of H_2 and O_2 under sunlight light irradiation through a sunlight-gathering apparatus (UV light is cut.) is visibly seen. The evolution rate of H_2 on the NaOH-Vis-TiO_2/Ti-foil/Pt photocatalyst was calculated to be about 15 mmol h^{-1} from the initial slope of the plots observed under sunlight irradiation [15].

The use of biomass (e.g. methanol) also significantly increased the rate of H_2 evolution. Thus, at a wavelength greater than 450 nm, the visible light catalytic performance of Vis-TiO_2 or UV-TiO_2 thin films was examined for the distinct evolution of H_2 in an aqueous solution of methanol. Figure 10.8 D shows a substantial increase in the H_2 evolution by adding 10 vol% methanol to 1.0 M NaOH aqueous solution on the TiO_2 side. The continuous charge separation caused by the effective movement of photo-formed holes and the current doubling effect by the ensuing organic radicals from organic hole scavengers can both be attributed to the acceleration of the reaction. Depending on the types of bio-related substances used, the degree of H_2 evolution changes. When several bio-related chemicals, in addition to methanol, were added to the electrolyte, elevated photocurrents were seen, demonstrating the beneficial role of these organic compounds as hole scavengers [9]. Thus, smaller organic compounds and polyols were identified to be preferred agents to speed up the process. Further photoelectrochemical studies also proved that CO_2 evolved as the byproducts in the photoreactor when biomass is used as a sacrificial agent.

10.4 Toward Greener Pathway: Integration of the Reaction System of the Photocatalytic Splitting of Water with an Artificial Plant Factory

The addition of organic compounds such as alcohol or biomass in water leads to a remarkable enhancement in the yield of photocatalytic evolution of H_2 from water with the production of CO_2 instead of O_2. Therefore, remedying this CO_2 problem is important to applying such a reaction system to produce H_2 from water using the photocatalyst. One possible approach has been reported: As shown in Figure 10.9, a visible light-responsive thin-film photocatalyst efficiently produces H_2 and CO_2 from water, involving biomass since biomass works as a sacrificial reagent. Thus, produced CO_2 is transferred into the plant factory in which green vegetables are cultivated under irradiation of artificial light of mainly red and blue LED lamps. The concentration of CO_2 in the plant factory is controlled to be about four times higher than that in the atmosphere, which enhances the photosynthesis of the vegetables, making vegetables grow rapidly and leading to a short cultivation period of about 1/3 of that in the outdoor farm. This proposal seems to be one of the most potential solutions to achieve photocatalytic H_2 production from water as a clean

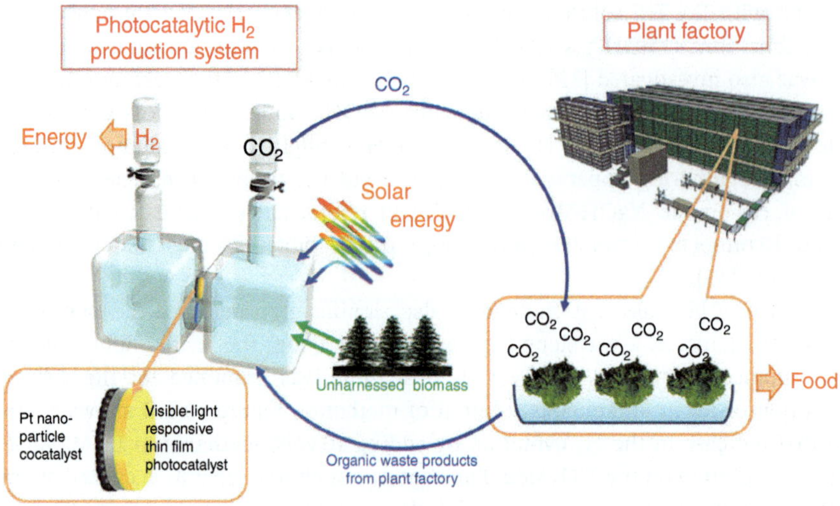

Figure 10.9 Conceptual illustrations for the integration of an efficient photocatalytic solar H_2 production system from water involving biomass by using a visible light-responsive thin-film photocatalyst and artificial light (LED)-type plant factory filled with a high concentration of CO_2 to enhance the growth of vegetables (enhance the photosynthesis) for sustainable and environmental-friendly greener technology. Source: Horiuchi et al. [9]/reproduced with permission Elsevier.

energy-supplying source for humankind since this integration system utilizes only solar energy, that is, the UV-visible light wavelength regions of sunlight are utilized for the photocatalytic evolution of H_2 and CO_2, and the long visible light wavelength and red light wavelength regions are utilized for solar cells to supply electric power to the LED lamps in the plant factory [9]. Thus, the hybridization or integration of the reaction system of a photocatalytic evolution of H_2 with an artificial LED lamp-type plant factory is considered a greener technology to supply clean H_2 energy and fresh vegetables.

10.5 Conclusion and Perspective

The present chapter summarizes the photocatalytic separate evolution of H_2 and O_2 from water using visible light-responsive TiO_2 semiconducting thin-film photocatalysts fabricated on Ti-foil by applying an RF–MS deposition method. The various factors that affect the photocatalytic activity of these thin-film photocatalysts were described in detail. Ways to improve the photocatalytic activity of these thin-film photocatalysts were also explained. The inevitable formation of H_2O from the already evolved H_2 and O_2 upon the surface of powdered catalysts has been consistently reported as a major hindrance to achieve good efficiency in the photocatalytic generation of H_2 and O_2. It was scientifically rationalized how the careful design and development of light-responsive TiO_2 thin films were considered better alternatives to these powered catalysts. The addition of alcohol or biomass into water led to an efficient photocatalytic evolution of H_2 from water but simultaneously produced CO_2 instead of O_2. In order to solve this dilemma, the integration of the efficient photocatalytic system that produces H_2 and CO_2 (artificial photosynthesis) separately with an artificial LED lamp-type plant factory (natural photosynthesis) was introduced as one of the future greener technologies.

References

1 Schneider, J., Matsuoka, M., Takeuchi, M. et al. (2014). Understanding TiO_2 photocatalysis: mechanisms and materials. *Chem. Rev.* 114: 9919–9986.

2 Schrauzer, G.N. and Guth, T.D. (1977). Photolysis of water and photoreduction of nitrogen on titanium dioxide. *J. Am. Chem. Soc.* 99: 7189–7193.

3 Fujishima, A. and Honda, K. (1972). Electrochemical photolysis of water at a semiconductor electrode. *Nature* 238: 37–38.

4 Matsuoka, M., Kitano, M., Takeuchi, M. et al. (2007). Photocatalysis for new energy production: recent advances in photocatalytic water splitting reactions for hydrogen production. *Catal. Today* 122: 51–61.

5 Matsuoka, M., Kitano, M., Takeuchi, M. et al. (2005). Photocatalytic water splitting on visible light-responsive TiO_2 thin films prepared by an RF magnetron sputtering deposition method. *Top. Catal.* 35: 305–310.

6 Shaheer, A.M., Vinesh, V., Lakhera, S.K., and Neppolian, B. (2021). Reduced graphene oxide as a solid-state mediator in $TiO_2/In_{0.5}WO_3$ S-scheme photocatalyst for hydrogen production. *Sol. Energy* 213: 260–270.

7 Lakhera, S.K., Rajan, A., Rugma, T.P., and Bernaurdshaw, N. (2021). A review on a particulate photocatalytic hydrogen production system: progress made in achieving high energy conversion efficiency and key challenges ahead. *Renewable Sustainable Energy Rev.* 152: 111694.

8 Hafeez, H.Y., Lakhera, S.K., Bellamkonda, S. et al. (2018). Construction of ternary hybrid layered reduced graphene oxide supported g-C_3N_4-TiO_2 nanocomposite and its photocatalytic hydrogen production activity. *Int. J. Hydrogen Energy* 43: 3892–3904.

9 Horiuchi, Y., Fukuda, H., Matsuoka, M., and Anpo, M. (2019). Integration of artificial photosynthesis (photocatalysis) and natural photosynthesis for the environmentally harmonious production of H_2 from H_2O involving biomass and vegetables. In: *Plant Factory Using Artificial Light* (ed. M. Anpo, H. Fukuda and T. Wada), 383–393. Elsevier.

10 Patil, M.K., Shaikh, S., and Ganesh, I. (2015). Recent advances on TiO_2 thin film-based photocatalytic applications (a review). *Curr. Nanosci.* 11: 271–285.

11 Eufinger, K., Poelman, D., Poelman, H. et al. (2008). TiO_2 thin films for photocatalytic applications. In: *Thin Solid Films: Process and Applications* (ed. S.C. Nam), 189–227. Transworld Research Network.

12 Kitano, M., Kikuchi, H., Hosoda, T. et al. (2006). The preparation of visible light-responsive TiO_2 thin films by applying an RF-magnetron sputtering

deposition method and their photocatalytic reactivity for the decomposition of water with a separate evolution of H_2 and O_2. *Key Eng. Mater.* 317: 823–826.

13 Ebrahimi, A., Kitano, M., Iyatani, K. et al. (2013). Effect of the sputtering parameters on the physical properties and photocatalytic reactivity of TiO_2 thin films prepared by an RF magnetron sputtering deposition method. *Res. Chem. Intermed.* 39: 1593–1602.

14 Takeuchi, M. and Anpo, M. (2010). Development of well-defined visible light-responsive TiO_2 thin film photocatalysts by applying an RF-magnetron sputtering deposition method. In: *Environmentally Benign Photocatalysts* (ed. M. Anpo and P.V. Kamat), 301–317. New York, NY: Springer.

15 Matsuoka, M., Kitano, M., Fukumoto, S. et al. (2008). The effect of the hydrothermal treatment with aqueous NaOH solution on the photocatalytic and photoelectrochemical properties of visible light-responsive TiO_2 thin films. *Catal. Today* 132: 159–164.

16 Kitano, M., Iyatani, K., Tsujimaru, K. et al. (2008). The effect of a chemical etching by HF solution on the photocatalytic activity of visible light-responsive TiO_2 thin films for solar water splitting. *Top. Catal.* 49: 24–31.

17 Kitano, M., Takeuchi, M., Matsuoka, M. et al. (2010). Applications of titanium oxide-based materials. In: *Environmentally Benign Photocatalysts* (ed. M. Anpo and P.V. Kamat), 545–560. Elsevier.

18 Selli, E., Chiarello, G.L., Quartarone, E. et al. (2007). A photocatalytic water splitting device for separate hydrogen and oxygen evolution. *Chem. Commun.* 47: 5022–5024.

19 Lo, C.C., Huang, C.W., Liao, C.H., and Wu, J.C. (2010). Novel twin reactor for separate evolution of hydrogen and oxygen in photocatalytic water splitting. *Int. J. Hydrogen Energy* 35: 1523–1529.

20 Kitano, M., Tsujimaru, K., and Anpo, M. (2006). Decomposition of water in the separate evolution of hydrogen and oxygen using visible light-responsive TiO_2 thin film photocatalysts: Effect of the work function of the substrates on the yield of the reaction. *Appl. Catal., A* 314: 179–183.

21 Matsuoka, M., Ebrahimi, A., Nakagawa, M. et al. (2009). Separate evolution of H_2 and O_2 from H_2O on visible light-responsive TiO_2 thin film photocatalysts prepared by an RF magnetron sputtering method. *Res. Chem. Intermed.* 35: 997–1004.

11

Development of Highly Efficient CdS-Based Photocatalysts for Hydrogen Production: Structural Modification, Durability, and Mechanism

Wei Li[1] *and Chuanyi Wang*[2]

[1] *Shaanxi University of Science & Technology, College of Chemistry and Chemical Engineering, Xi'an, Shaanxi 710021, China*
[2] *Shaanxi University of Science & Technology, School of Environmental Sciences and Engineering, Shaanxi Key Laboratory of Chemical Additives for Industry, Xi'an, Shaanxi 710021, China*

11.1 Introduction

Hydrogen energy is identified as an ideal alternative new energy in view of its advantages of low density, high calorific value, nontoxicity, green sustainability, easy storage, etc. More importantly, the combustion of hydrogen only produces a single product of H_2O, which effectively reduces greenhouse gas emissions. Thereby, the development of high-efficient hydrogen production strategies has attracted extensive attention in material chemistry and energy fields. However, conventional hydrogen production strategies generally have various limitations, such as high energy consumption, complicated operation, secondary pollution, or low efficiency. Photocatalytic water splitting to produce hydrogen is undoubtedly a promising candidate to convert solar energy into high-value chemical energy due to the advantages of simple operation, eco-friendliness, low cost, etc. [1].

During the past decades, a large number of semiconductor-based photocatalysts have been developed and applied in photocatalytic water splitting to produce hydrogen, such as TiO_2-based photocatalysts, bismuth compounds, copper compounds, transition metal sulfides (TMSs), silver compounds, and zinc compounds [2]. Furthermore, nonmetallic semiconductors (such as carbon nitride [C_3N_4], boron nitride [BN], and black phosphorus [BP]) have also been widely used to prepare high-efficient photocatalysts for hydrogen production. For instance, Shen and coworkers reported a B-doped N-deficient C_3N_4-based Z-scheme heterostructure for photocatalytic overall water splitting [3], and Wang and coworkers synthesized C-doped BN nanosheets for hydrogen or oxygen evolution under visible light illumination [4], and Tian et al. supported BP nanosheets on amorphous CoP for solar hydrogen production [5].

Beyond that, some novel non-semiconductor materials have also been extensively employed to construct high-efficient photocatalysts or improve the performance of traditional semiconductor photocatalysts, such as graphene, MXene, metal–organic

UV-Visible Photocatalysis for Clean Energy Production and Pollution Remediation: Materials, Reaction Mechanisms, and Applications, First Edition. Edited by Xinchen Wang, Masakazu Anpo, and Xianzhi Fu.
© 2023 WILEY-VCH GmbH. Published 2023 by WILEY-VCH GmbH.

frameworks (MOFs), and covalent organic frameworks (COFs). For example, in photocatalytic hydrogen production, Kudo and coworkers chose graphene oxide (GO) as a solid-state electron mediator to construct Z-schematic water splitting photocatalyst with metal sulfide [6], Zhao and coworkers anchored ultrathin $ZnIn_2S_4$ nanosheets on $Ti_3C_2T_X$ MXene [7], Xu and coworkers synthesized a MOFs-based catalysts [8], and Chen and coworkers activated the carbonyl oxygen sites in β-ketoenamine-linked COFs via cyano conjugation [9]. In this chapter, cadmium sulfide (CdS) is chosen as a model to discuss its merits/demerits, performance, and improvement in photocatalytic hydrogen production.

11.2 CdS-Based Photocatalysis

CdS, an important TMS with narrow a bandgap (~2.5 eV) and low work function, is widely used to prepare high-efficient photocatalysts or serves as a cocatalyst to improve the performance of other semiconductor-based photocatalysts due to its merits of excellent light-harvesting ($\lambda \geq$ ~580 nm), high-density photoinduced carriers generation and appropriate band structure for photocatalytic redox reactions (e.g. CO_2 reduction, nitrogen fixation, contaminant remediation, hydrogen/oxygen production, etc.) [10]. However, bare CdS semiconductor generally presents weak structural stability for easy photocorrosion under long-term illumination, and their photoinduced carriers are easily recombined, thus seriously restricting their photocatalytic performance. Meanwhile, nanomaterial toxicity due to heavy metal ions released in application also requires special attention. Therefore, extensive efforts have been devoted to enhancing the structural stability and boosting the photocatalytic performance of CdS semiconductor photocatalysts, such as heterojunction construction, plasma modification, structural regulation, and defect engineering, heteroatom doping. The heterointerface construction leads to the formation of a built-in electric field followed by improved optical responsivity, which will promote carriers' migration and restrain the recombination of photoinduced carriers. Thereby, the photocatalytic performance of semiconductor photocatalysts can be improved effectively.

11.2.1 Construction of p–n type Bi_xO_y/CdS Heterostructure

Bismuth oxide (Bi_xO_y), an important semiconductor with visible-light response, good structural stability, and nontoxicity, is identified as a potential candidate for photocatalysis. However, its poor broadband light-harvesting and easy electron–hole recombination generally led to unfavorable photoactivity. In order to overcome these drawbacks, Wang and coworkers chose CdS NPs as cocatalyst to introduce into the synthetic system of Bi_xO_y and construct the p–n type Bi_xO_y/CdS heterostructure via a solvothermal method [11]. The bare Bi_xO_y displays typical 2D nanosheet morphology with the size of several hundreds of nanometers (Figure 11.1a,b), but the Bi_xO_y/CdS composite displays a typical microsphere morphology assembled by abundant 2D nanosheets (Figure 11.1c,d). The HRTEM image in Figure 11.1e shows that plenty of nanoparticles are embedded into the 2D Bi_xO_y nanostructure,

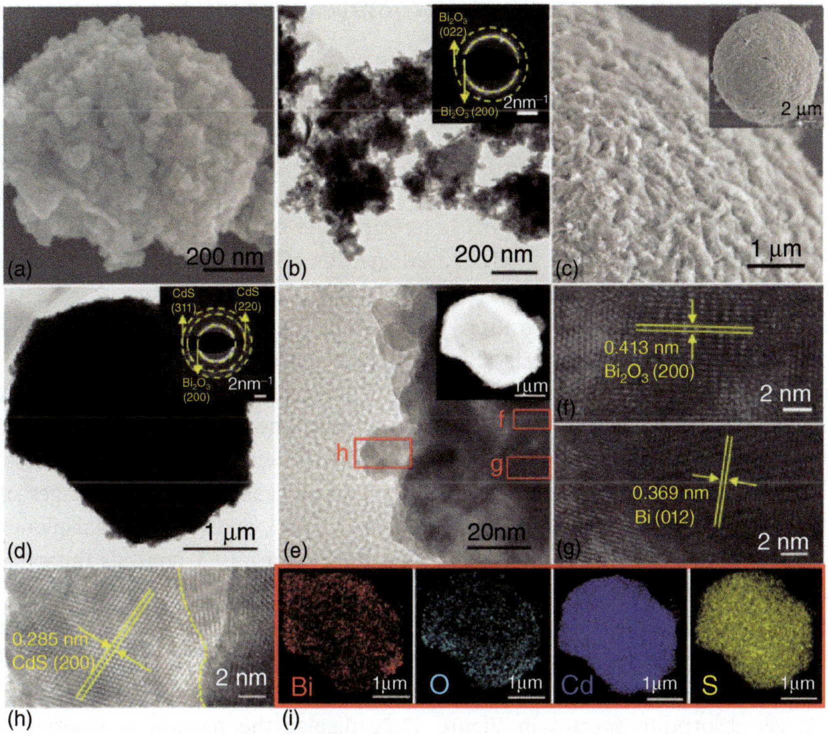

Figure 11.1 SEM and TEM images of Bi_xO_y (a, b) and optimized Bi_xO_y/CdS (c, d). (e–h) HRTEM images and (i) EDX mappings of Bi_xO_y/CdS. The insets of panel (b) and (d) are the corresponding SAED patterns, and the inset of panel (e) is the corresponding HAADF image. Source: Li et al. [11]/Reproduced with permission from Royal Society of Chemistry.

and a clear heterointerface can be observed. Meanwhile, the clear crystal fringes of $d = 0.413$, 0.369, and 0.285 nm can be observed (Figure 11.1f–h), which correspond to the (200) plane of Bi_2O_3, (012) plane of elemental Bi, and (200) plane of hexagonal CdS, respectively. Compared with the SAED pattern of Bi_xO_y in Figure 11.1b, the inset of Figure 11.1d shows the typical polycrystalline characteristic with high crystallinity, attributing to the introduction of CdS NPs to Bi_xO_y nanostructure. Also, the EDX mappings in Figure 11.1i reveal the uniform distribution of CdS NPs in the Bi_xO_y/CdS composite. Moreover, XPS peak shifts of both Bi 4f and Cd 3d chemical states suggest the strong interfacial interaction between CdS and Bi_xO_y, which attributes to the formation of the Bi_xO_y/CdS heterointerface.

Due to the typical II-type induced mechanism, the interfacial carriers' mobility and broadband light response were effectively improved, thereby the remarkably enhanced simulated sunlight (SSL)-induced photoactivity was achieved by the optimum Bi_xO_y/CdS composite in the absence of scavenger, which is ~34.2-fold greater than that of bare Bi_xO_y. Also, the promoted carrier transfer effectively suppresses the photocorrosion of CdS microstructure, resulting in desirable photostability under long-term photoirradiation.

11.2.2 Construction of CdS@h-BN Heterostructure on rGO Nanosheets

Hexagonal boron nitride (h-BN) with a strong covalent sp^2 bond is widely used in the electrocatalysis field due to its high mechanical strength, excellent medium resistance, thermal stability, etc. However, its wide bandgap feature (~4.5 eV) leads to weak conductivity and poor light-harvesting, seriously restricting its application in the photocatalysis field. To address this challenge, Wang et al. constructed a core–shell CdS@ h-BN heterostructure on reduced graphene oxide (rGO) nanosheets to prepare a ternary CdS@h-BN/rGO (CdS@BNG) composite photocatalyst via a structural reconstruction strategy [12].

The SEM and TEM characterizations demonstrate that the optimized CdS@BNG composite photocatalyst presents a typical porous structure assembled on rGO nanosheets, and the HRTEM image (Figure 11.2a) and EDX element mappings evince that the assembled porous structure is consisted by CdS NPs with thin h-BN coating, and the core–shell morphology is formed in the solvothermal process. The corresponding magnified HRTEM images display the clear lattice fringes of 0.206, 0.173, and 0.335 nm assigned to the (110), (201), and (002) crystal planes of hexagonal phase CdS (JCPDS 77-2306, Figure 11.2a,b). BET analysis indicates that the optimized CdS@BNG composite photocatalyst presents a typical porosity feature, and the type H1 hysteresis loop in the narrow relative pressure region (0.87–0.99) suggests the formation of uniform mesoporous structure originated from the structural reconstruction in solvothermal process.

UV–vis absorption spectra in Figure 11.2c display the redshifted absorption edge (607 nm) and improved light-harvesting property (near-infrared region)

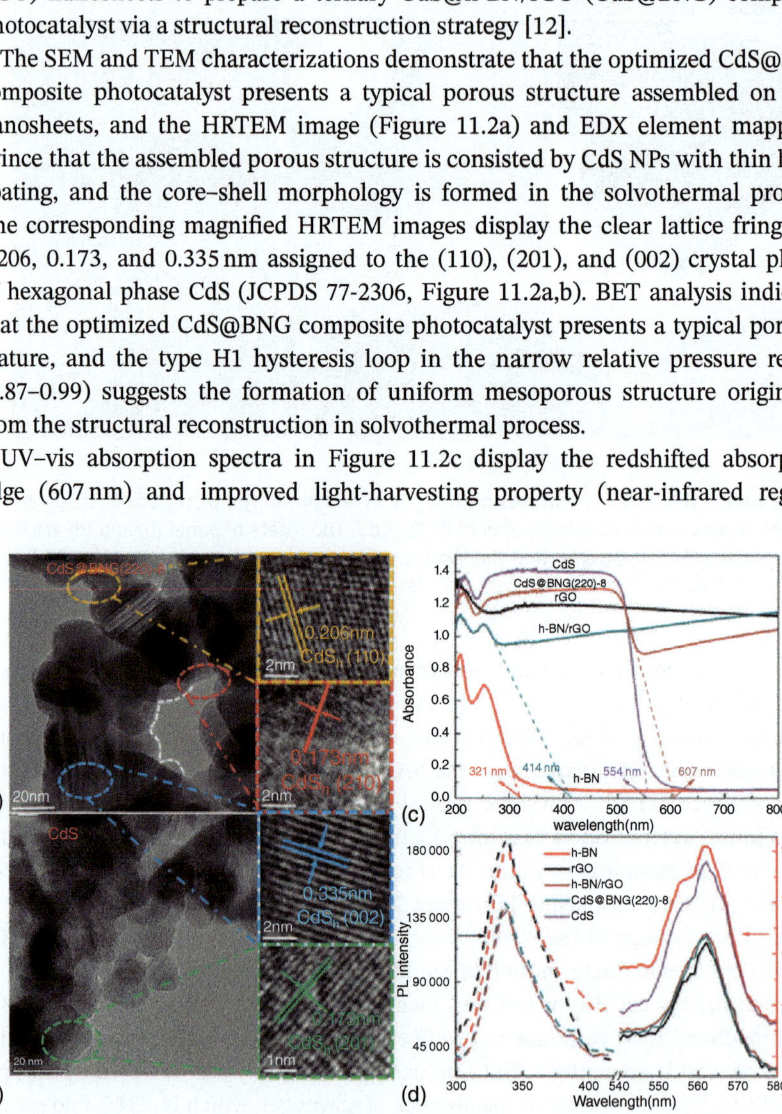

Figure 11.2 HRTEM images of optimized CdS@BNG (a) and bare CdS NPs (b). (c) UV–vis absorption spectra and (d) PL emission spectra of the samples. Source: Li et al. [12]/Reproduced with permission from Elsevier.

for the optimized CdS@BNG composite compared with that of bare CdS NPs (554 nm), which indicates that the formation of CdS@h-BN heterostructure on rGO nanosheets remarkably enhances the light-harvesting property and photoresponsivity. PL spectra in Figure 11.2d show that the optimized CdS@BNG composite photocatalyst presents lower PL intensity than that of bare h-BN and CdS NPs in the long-emission wavelength range, and the time-resolved PL decays show that the optimal CdS@BNG composite photocatalyst (1440 ps) presents slower PL decay than bare CdS NPs (1350 ps), indicating that the formation of h-BN/rGO and CdS@h-BN heterointerfaces remarkably promotes the separation of photoinduced carriers and effectively restricts their recombination.

The introduction of CdS cocatalyst to h-BN/rGO composite effectively enhances its photocatalytic HER performance. When the dosage of CdS cocatalyst is 8 mg, the optimized CdS@BNG composite achieves the highest HER activity, which is about 218-fold greater than h-BN/rGO hybrid and 11-fold greater than bare CdS NPs (Figure 11.3a). Furthermore, when the synthetic temperature and synthetic time are set as 220 °C and eight hours, the highest HER activity was achieved, and its apparent quantum yield (AQY) is about 19.8% under 420 nm of illumination. In detail, excess CdS cocatalyst will lead to the aggregation of CdS NPs, seriously breaking the CdS@h-BN heterointerface. Lower synthetic temperature and shorter synthetic time are insufficient to form an effective heterointerface, while higher synthetic temperature and longer synthetic time will also break the heterointerface, restraining the effective carriers' transfer. Therefore, decreased HER activities were observed. Although bare CdS possesses high HER activity under SSL, its photoactivity decreases remarkably after three hours of photoirradiation. However, the optimized CdS@BNG composite not only presents higher HER photoactivity than bare CdS but also presents linearly increased hydrogen production during 20 hours of continuous photoirradiation (Figure 11.3c). Excellent reproducibility was also

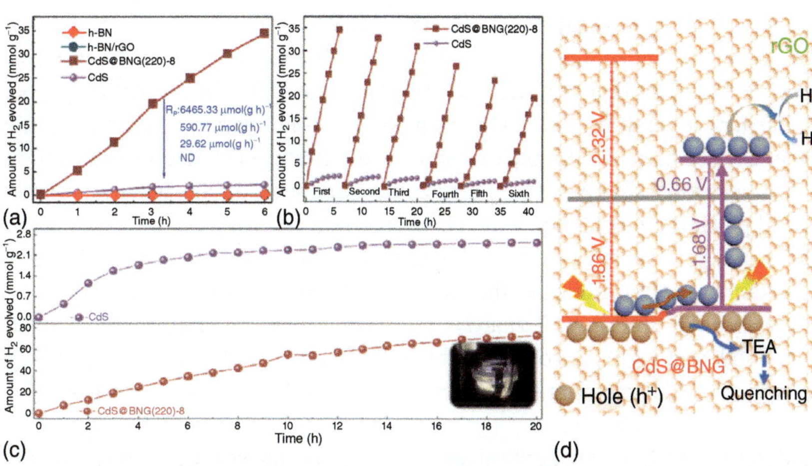

Figure 11.3 (a) HER kinetic curves of the samples. (b) Reproducibility and (c) durability of optimized CdS@BNG composite and bare CdS NPs. (d) Electron transfer mechanism of optimized composite. Source: Li et al. [12]/Reproduced with permission from Elsevier.

presented by this optimized composite (Figure 11.3b), indicating that the optimized CdS@BNG composite possesses excellent photostability. XPS, FTIR, Raman, and XRD analyses of the recovered sample further demonstrate the prominent structural stability of the optimized CdS@BNG composite.

Besides, the electrochemical measurements indicate that the optimized CdS@BNG composite holds excellent interfacial conductivity and remarkably improved photoexcited properties owing to the synergetic effect of h-BN/rGO and CdS@h-BN heterointerfaces. Based on the band structure and demonstrated electron transfer direction of the composite, the corresponding catalytic mechanism is proposed as displayed in Figure 11.3d. Due to the wide bandgap of h-BN (~4.18 eV), the electrons are hardly induced under SSL. However, for the formation of the CdS@h-BN heterointerface, the electrons in the valence band (VB) position of h-BN can be easily induced to the VB position of CdS NPs under photoirradiation. Then, the electrons can be induced to the conduction band (CB) position of CdS NPs from their VB position under photoirradiation. Because of the formation of the h-BN/rGO heterointerface, the photoinduced electrons can rapidly migrate to the catalyst surface, thereby the electron–hole recombination can be effectively restrained and hydrogen production can be efficiently promoted. In this process, the holes in the VB position of h-BN and CdS components are captured by scavenger (TEOA) to restrain the oxidation reaction.

11.2.3 N-doped CdS Nanocatalyst

In view of the practical application requirement, acquiring high-efficient photocatalytic performance by a simple and low-cost operation procedure is highly desired. Heteroatom doping is a simple and effective strategy to promote body carriers' migration and achieve high quantum yield due to the synergetic electronic effect of heteroatom sites in body structure. Compared with metal doping, nonmetallic atoms with high electronegativity can form strong coordination interaction with metal atoms in semiconductors, effectively modulating the band structure and improving the migration behavior of carriers, thereby it will dramatically boost body carriers' migration and hampering their recombination. Inspired by this, Wang et al. used urea as a doping precursor and introduced N heteroatoms into the lattice of hexagonal CdS NPs to synthesize the N-doped CdS nanocatalysts via wet chemical precipitation coupled with a hydrothermal process [13].

XRD analysis reveals that the N heteroatoms successfully enter the lattice of h-CdS. Raman analysis indicates the formation of coordination interaction between N atoms and Cd atoms. XPS analysis clearly presents the characteristics of the Cd—N bond at 404.4 eV. Also, the slight shifts of Cd 3d (0.1 eV) and S 2p (0.2 eV) characteristic peaks to higher binding energy further demonstrate the formation of Cd–N coordination interaction.

TEM image reveals that the optimized N–CdS nanocatalyst appears near-spherical micromorphology (Figure 11.4a), and its average particle size (~22.1 nm) is larger than that of bare CdS (~14.5 nm), which suggests that N doping benefits to the grain growth of CdS. The widened half-peak width of XRD patterns also well supports

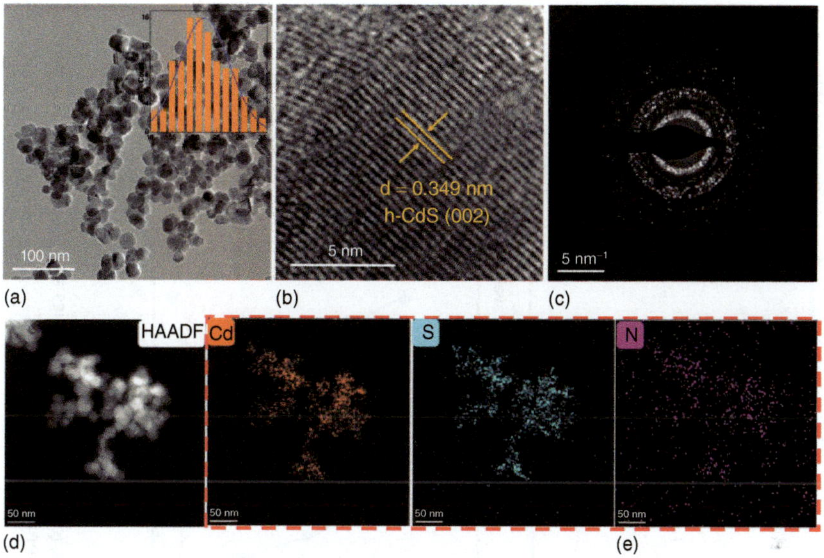

Figure 11.4 (a) TEM image, (b) HRTEM image, (c) SAED pattern, (d) HAADF image and (e) EDX mappings of optimized N–CdS. The insets of panel (a) is corresponding particle size distribution. Source: Li et al. [13]/Reproduced with permission from Elsevier.

this conclusion. Moreover, the HRTEM image in Figure 11.4b shows the widened lattice distance ($d = 0.349$ nm) compared with bare CdS ($d = 0.333$ nm), which provides powerful evidence for entering N heteroatoms to the lattice of CdS. The SAED pattern in Figure 11.4c reveals that N-doping increases the crystallinity of CdS, and it is consistent with the XRD results, which attributes to the stronger coordination ability of N heteroatoms. Additionally, the high-annular dark-field (HAADF) image in Figure 11.4d reveals the uniform phase state, and EDX mappings in Figure 11.4e display the uniform distribution of N elements in the body structure of CdS NPs.

In the presence of TEOA, the optimized N–CdS nanocatalyst presents a remarkably improved HER rate (3983.4 $\mu mol(h\,g)^{-1}$) during four hours of SSL irradiation (optimum pH = 14, Figure 11.5a), which raises to about ninefold greater than that of bare CdS NPs (442.6 $\mu mol(h\,g)^{-1}$, Figure 11.5b). Furthermore, Figure 11.5c displays the AQYs of the optimized N–CdS nanocatalyst for H_2 production under different fixed wavelengths of photoirradiation (500, 550, 600 nm), and about 32.4 % of AQY can be achieved at the wavelength of 500 nm, which is much higher than that of most reported CdS-based photocatalysts for H_2 production. The high AQY suggests the high-efficient utilization of photoexcited carriers, which is attributed to the promoted body carriers' migration due to the synergetic influence of heteroatom-semiconductor coordination (HSC) interaction. Moreover, the optimized N–CdS nanocatalyst holds a very stale and highly improved HER activity in a long irradiated time (15 hours, Figure 11.5d), and its HER activity presents a slower decline as cycle time increases (Figure 11.5e). XRD and XPS analyses confirm that the photocorrosion of bare CdS can be effectively restrained by N doping, thus the optimized N–CdS nanocatalyst not only possesses remarkably

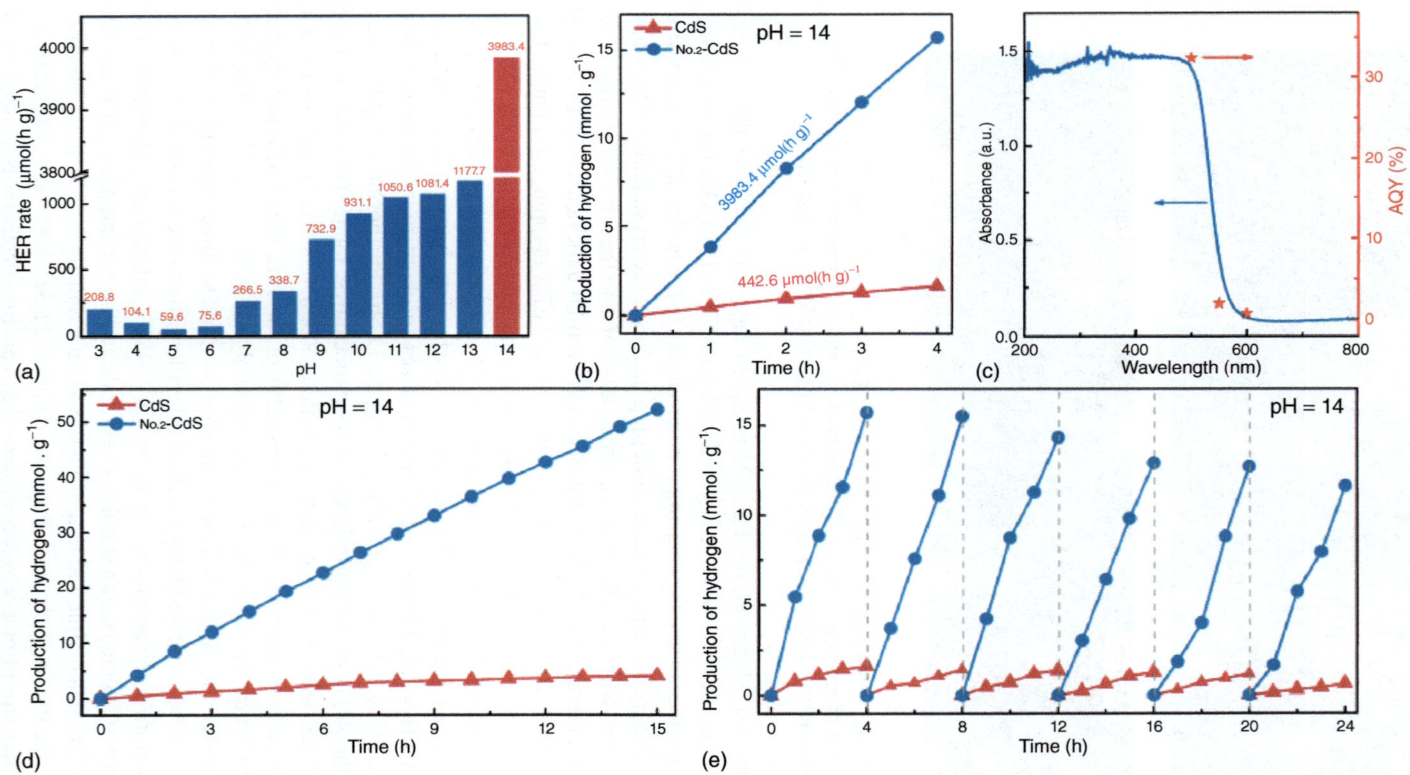

Figure 11.5 (a) pH value optimization of HER system (pH = 3–14), (b) SSL-induced HER kinetics, (d) HER durability kinetics and (e) HER recycling kinetics of CdS and N–CdS. (c) AQYs of N–CdS (500, 550, 600 nm). Source: Reproduced with permission from Li et al. [13]/Elsevier.

enhanced HER activity but also excellent photostability, even after experiencing long-term irradiation.

UV–vis absorption spectra suggest that the absorption edge of CdS appears to have a slight blue shift (578 nm → 568 nm) after N doping, and the corresponding bandgap broadens to 2.29 eV from the original 2.27 eV, which is very different from the commonly reported results. However, the doped sample still shows excellent light-harvesting and response characteristics in the visible-light range. Moreover, the photoinduced LSV curves in Figure 11.6b show that the optimized N–CdS nanocatalyst appears highly enhanced photocurrent signals than that of bare CdS in a wide range of initial potentials (−0.7 to −0.3 V), and the same result is obtained with transient photocurrent response curves recorded at the initial potential of −0.2 V (Figure 11.6c). Considering the broadened bandgap of N–CdS nanocatalyst, it is concluded that N doping effectively hinders the recombination behavior of photoexcited carriers in CdS body structure and results in remarkably increased photocurrent intensity (~2 times). PL spectra in Figure 11.6a show that the N–CdS nanocatalyst displays weakened PL intensity, and longer photoexcited carriers' retention time (2087 ps) than bare CdS (1903 ps), which provides direct evidence for restrained recombination behavior of photoexcited carriers, thereby more opportunities are ensured for them to participate in H_2 production. This attributes to the promoted carriers' mobility for the synergetic effect of N heteroatom sites in the body structure of CdS. Additionally, Tafel curves reveal that the N–CdS nanocatalyst appears to have a smaller polarized overpotential (999 mV) than that of bare CdS (1015 mV) at the same electrochemical test parameters, suggesting a higher reaction possibility for H_2 production. Besides, EIS data in Figure 11.6d demonstrates the remarkable promotion of body carriers' migration of CdS for N doping, which is the main reason for the restrained recombination behavior of photoexcited carriers. Therefore, the promoted separation and prolonged retention time of photoexcited carriers are the intrinsic driving forces for improving the HER activity of N–CdS nanocatalyst. Although the CB position of N–CdS is more positive than bared CdS, its Fermi level position (E_f) is more negative than bared CdS, which is the internal reason for its higher photoinduced reduction capacity.

Density functional theory (DFT) calculations provide theoretical evidence for promoted carriers' migration and formation of Cd–N–Cd coordination interaction. Furthermore, all the H_2O molecule adsorption (H_2O^*), transition state (TS), and H_2O molecule dissociation processes present remarkably reduced energy barriers in the N–CdS model compared with the CdS model, indicating that the formation of Cd–N–Cd coordination interaction thermodynamically enhances H_2O molecule's adsorption and accelerates its dissociation. Meanwhile, the N heteroatom site displays a much larger $|\Delta G_{H^*}|$ value (2.65 eV) than other sites, and the S site displays the smallest $|\Delta G_{H^*}|$ value (0.26 eV), which strongly demonstrates that S atoms are the main active sites for H_2 generation. Moreover, the $|\Delta G_{H^*}|$ value of S sites decreases obviously from the original 0.59 eV after N doping, suggesting remarkable promotion of H_2 generation on S sites owing to the synergetic HSC interaction.

Based on the results of DFT calculations, the HER mechanism on the N–CdS nanocatalyst can be illustrated in Figure 11.6e. Owing to the dramatically decreased

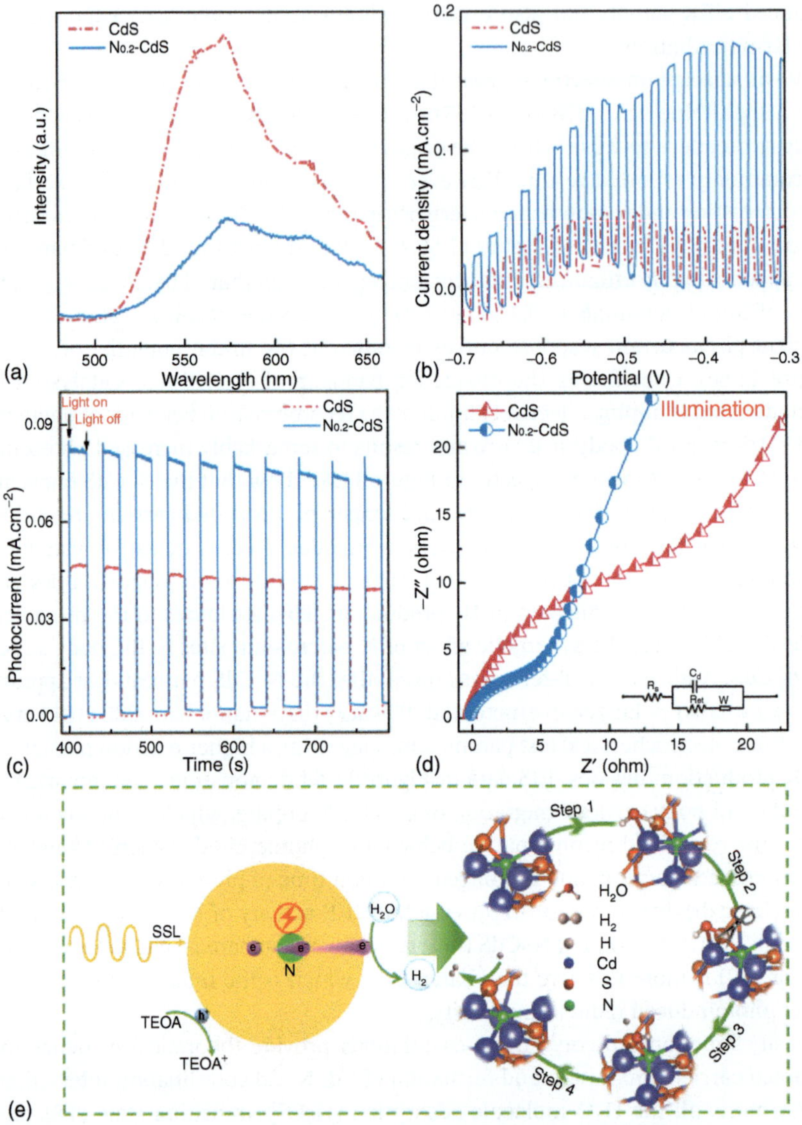

Figure 11.6 (a) PL spectra, (b) LSV curves, (c) transient photocurrent responses, and (d) EIS plots of CdS and N–CdS. (e) Synergetic promotion of body carriers migration and SSL-induced HER mechanism on N–CdS nanocatalyst. The LSV and EIS data were tested under photoirradiation. Source: Reproduced with permission from Li et al. [13]/Elsevier.

energy barriers of water cracking and H_2 generation on N–CdS nanocatalyst, highly improved HER activity can be achieved. Meanwhile, due to the synergetic promotion of body carriers' migration by HSC interaction, their recombination behavior can be effectively restrained, thus high quantum yield can be gained for high-efficient utilization of photoexcited carriers.

11.2.4 Pd Single-Atom Decorated CdS Nanocatalyst

In view of the unique electronic structure and unsaturated coordination environment of metallic single-atom sites (SASs), they are widely used to improve the catalytic performance of semiconductor catalysts for abundant active sites, exposing higher atomic availability. For instance, Tiwari et al. decorated Pd SASs on Co_2P NPs [14], and Wang et al. loaded Ir SASs on the NiO matrix [15]. Moreover, TiO_2, CdS, and C_3N_4-based catalysts were successively decorated by various metallic SASs. Additionally, Shi et al. decorated Co SASs on a porous carbon matrix [16], Zuo and Wang et al. loaded Pt and Cu SASs on MOFs [17, 18], and Huang and Sun et al. respectively anchored molybdenum and Pt SASs on N-doped graphene [19, 20]. Notably, SASs decoration can effectively improve the catalytic performance and enhance the structural stability of conventional catalysts, even realizing hydrogen production through overall water splitting under photoirradiation due to synergetic host–guest interaction. However, a great challenge remaining in metallic single-atom catalysts is stability, which is because single atom tends to aggregate due to their high surface energy. In view of this character, Wang et al. developed a strategy based on metal–semiconductor interaction to prepare stable single atomic Pd-decorated CdS (CdS–Pd) nanocatalyst via a surface photoinduced reduction strategy (Figure 11.8a) [21].

Aberration-corrected high-annular dark-field scanning TEM (AC HAADF-STEM) shows that Pd SASs uniformly distribute on the surface of CdS NPs (Figure 11.8b), which is further confirmed by corresponding EDX mappings. Moreover, the X-ray absorption near-edge structure (XANES) spectrum of the optimized CdS–Pd nanocatalyst is very different from that of Pd foil and PdO (Figure 11.7c), and its Pd K-edge locates at the right side of Pd foil (zero-valent state), indicating that Pd element exists in the positive-valent state. K-space spectra in Figure 11.7d show that the optimized CdS–Pd nanocatalyst does not appear obvious signals of metal–metal bonding at the high *k*-value region, suggesting the single-atom state of the Pd element in optimized CdS–Pd nanocatalyst. Also, compared with reference samples, its amplitude moves toward a higher *k* value, indicating the formation of Pd–S coordination interaction. Furthermore, the signals assigned to elemental Pd and Pd clusters are not presented in the extended X-ray absorption fine structure (EXAFS) spectrum (Figure 11.7e), and only the signal assigned to Pd–S coordination can be observed, which further demonstrates the formation of strong coordination interaction between CdS NPs and Pd single atoms. Its wavelet transform in Figure 11.7f also supports this conclusion. XPS analysis demonstrates the coexistence of Pd species on the surface of CdS NPs in two-valent states (Figure 11.7a). Raman spectra in Figure 11.7b show that the intensity ratio of the first-order longitudinal optical (1-LO) mode ($294\,cm^{-1}$) and second-order longitudinal optical (2-LO) mode ($593\,cm^{-1}$) of the optimized CdS–Pd nanocatalyst ($I_1/I_2 = 2.75$) is very different from that of bare CdS NPs ($I_1/I_2 = 2.80$), also implying the formation of strong CdS–Pd interaction, and this conclusion is consistent with XAFS analysis.

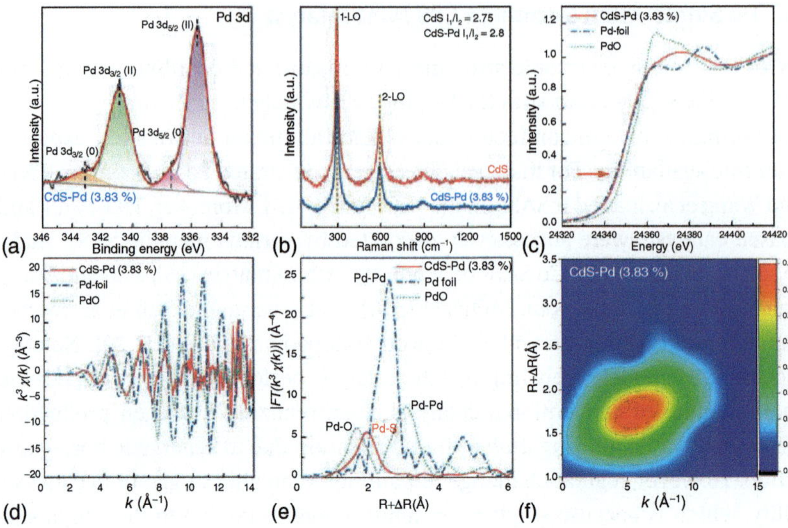

Figure 11.7 (a) Pd 3d XPS core-level spectrum, (b) Raman spectra of CdS and CdS–Pd, (c) XANES spectra, (d) K-space spectra, (e) EXAFS spectra, and (f) Wavelet transform of CdS–Pd. Source: Reproduced with permission from Li et al. [21]/Elsevier.

Due to the synergetic metal–semiconductor interaction, hydrogen production through the overall water splitting process can be realized under SSL irradiation, and about 56.88 μmol of hydrogen was evolved at 10 mg CdS–Pd nanocatalyst during six hours (Figure 11.8c), and its HER rate (~947.93 $\mu mol(g\,h)^{-1}$) is about 110-fold faster than that of bare CdS NPs (~8.64 $\mu mol(g\,h)^{-1}$). Its AQY is about 4.47% (420 nm) under 420 nm of photoirradiation (Figure 11.8e). If TEOA is added into the reaction system, about 440.15 μmol hydrogen was evolved at 10 mg CdS–Pd nanocatalyst during six hours (Figure 11.8d), and its HER rate (~8402.47 $\mu mol(g\,h)^{-1}$) is about 8.9-fold greater than overall water splitting performance and 25-fold greater than bare CdS NPs (~340.52 $\mu mol(g\,h)^{-1}$). Meanwhile, about 33.92% (420 nm) of AQY was achieved on this nanocatalyst (Figure 11.8e), which further confirms the high-efficient photoactivity of CdS–Pd nanocatalyst under broadband light illumination. Meanwhile, the sample bottle containing a very little amount of CdS–Pd nanocatalyst gives a large number of bubbles under outdoor sunlight irradiation, which provides visual evidence to demonstrate the high efficiency of CdS–Pd nanocatalyst for hydrogen production under broadband light irradiation. Moreover, the CdS–Pd nanocatalyst presented not only remarkably enhanced HER photoactivity but also excellent photostability in overall water splitting, which indicates that the photocorrosion of CdS NPs was highly restrained by Pd SASs decoration.

UV–vis absorption spectra show that the maximum absorption edge of CdS–Pd nanocatalyst redshifts to 630 nm from 587 nm, suggesting improved light-harvesting and optical-response properties. The PL spectra and the corresponding time-resolved transient decay PL spectrum show the remarkably

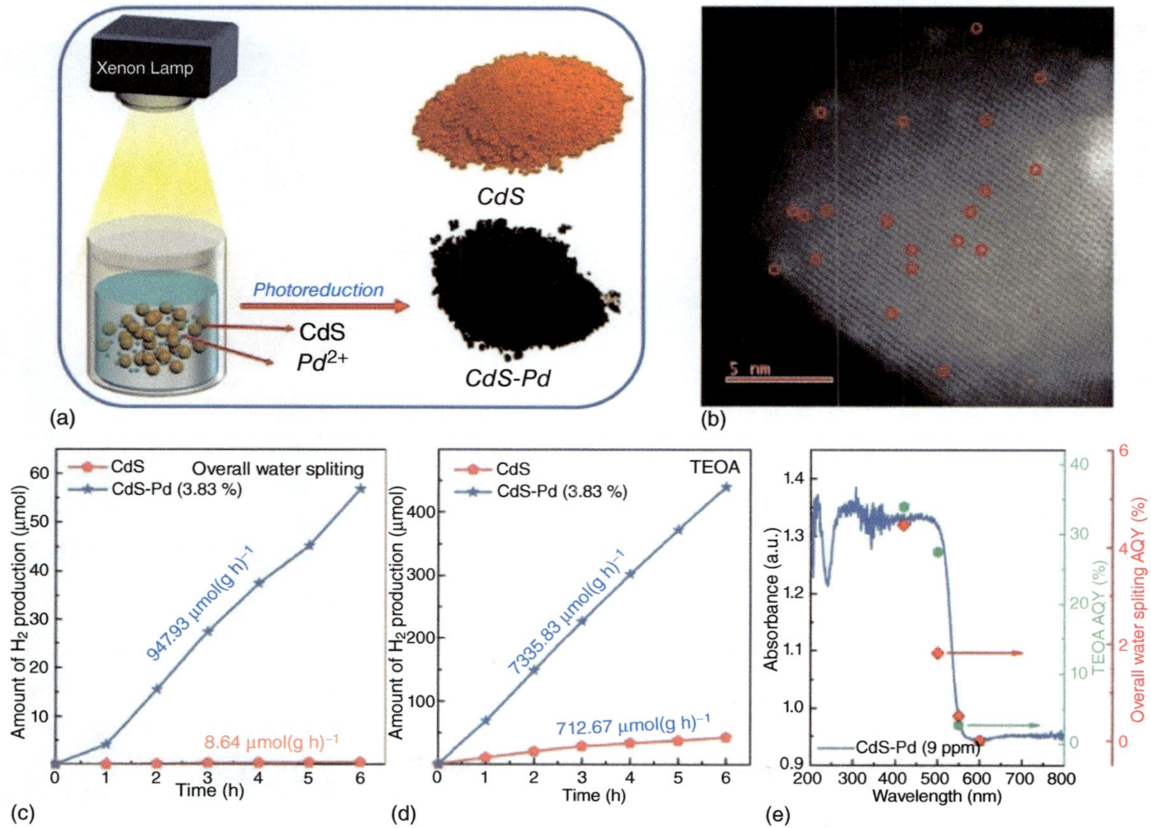

Figure 11.8 (a) Preparation schematic diagram and (b) AC HAADF-STEM image of CdS–Pd nanocatalyst. (c and d) HER kinetics of bare CdS NPs and CdS–Pd nanocatalyst without and with the addition of TEOA. (e) AQYs of CdS–Pd nanocatalyst at the presence and absence of TEOA. Source: Li et al. [21]/Reproduced with permission from Elsevier.

weakened emission peak intensity and conspicuously prolonged average lifetime compared with bare CdS NPs. It implies that electron–hole recombination was highly suppressed due to the synergetic CdS-to-Pd electron transfer, leading to more opportunities to induce water splitting for highly enhanced HER activity. The photoelectrochemical analysis suggests that the CdS–Pd nanocatalyst possesses reduced interfacial resistance compared with bare CdS NPs. When the ITO electrode was illuminated, its interfacial resistance was further reduced, indicating the promoted interfacial conductivity under photoirradiation. Furthermore, pure CdS NPs present a decreased photocurrent response under long-term photoirradiation, but CdS–Pd nanocatalyst keeps a very stable photocurrent response under long-term photoirradiation, attributing to the effective restraining effect to the photocorrosion of CdS NPs due to the promoted CdS-to-Pd electron transfer. Besides, Tafel curves show that the CdS–Pd nanocatalyst presents a larger polarized overpotential ($\eta = 1013$ mV) than bare CdS NPs ($\eta = 977$ V) in dark, but the opposite result ($\eta_{\text{CdS-Pd}}$ (992 mV) < η_{CdS} (1045 mV)) was obtained under photoirradiation, suggesting the enhanced photoelectric reduction ability of CdS–Pd nanocatalyst. Additionally, Mott–Schottky measurements demonstrate the formation of promoted CdS-to-Pd electron transfer.

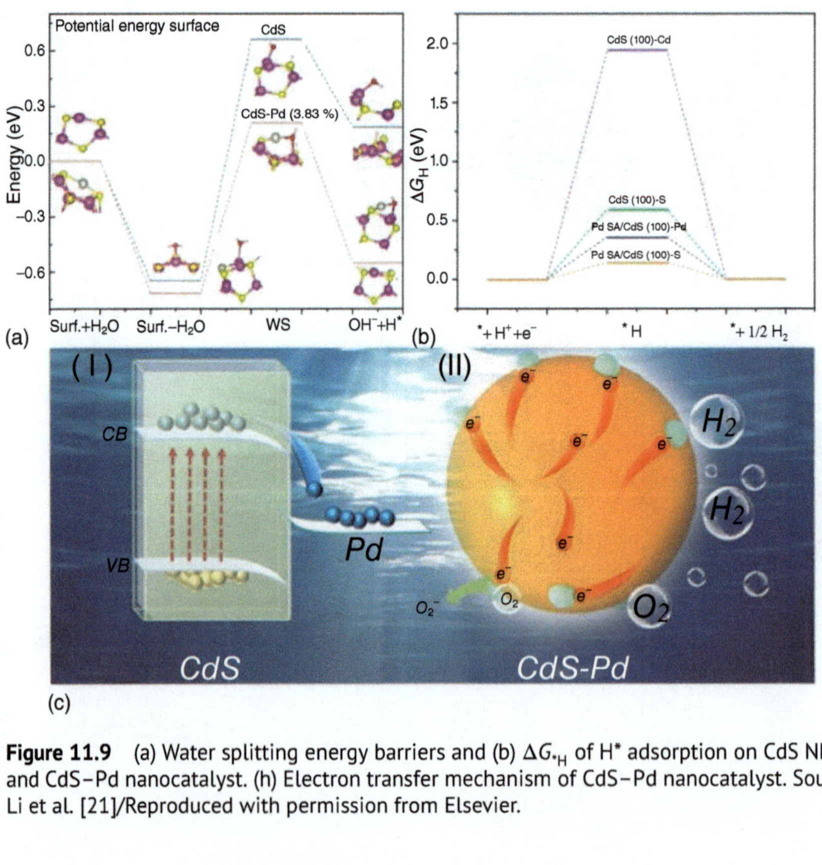

Figure 11.9 (a) Water splitting energy barriers and (b) ΔG_{*H} of H* adsorption on CdS NPs and CdS–Pd nanocatalyst. (h) Electron transfer mechanism of CdS–Pd nanocatalyst. Source: Li et al. [21]/Reproduced with permission from Elsevier.

DFT calculations further evince the formation of CdS-to-Pd electron transfer based on the S–Pd bonding interaction. Furthermore, the CdS–Pd model possesses a smaller activation barrier (0.93 eV) than the bare CdS model (1.31 eV) in the water splitting process (Figure 11.9a), suggesting easier water splitting on CdS–Pd nanocatalyst. Meanwhile, for the bare CdS, its Cd atom sites possess a larger ΔG_{*H} (1.94 eV) than its S atom sites (0.59 eV), revealing that its S atom sites are the main reactive sites for the HER process according to the Sabatier principle. After Pd SASs decoration, the ΔG_{*H} of S atom sites (0.14 eV) is further reduced (Figure 11.9b), and Pd atom sites also display much smaller ΔG_{*H} (0.36 eV) than S atom sites of bare CdS. Considering that the Pd atom only exists on the outer surface of CdS NPs, thereby S atom sites and Pd atom sites of CdS–Pd synergistically serve as the reactive sites for hydrogen production. Therefore, the overall reaction mechanism was proposed, as shown in Figure 11.9c, and the DFT result is consistent with what was concluded from experimental studies.

11.3 Summary and Prospect

Numerous studies have shown that the CdS semiconductor is indeed an outstanding material for synthesizing highly efficient hydrogen production photocatalysts. Apparently, improving the photocatalytic performance of CdS-based photocatalyst through a simple and green strategy has theoretical and practical significance. Through our work, we demonstrated that heteroatom doping and metallic surface decoration are identified as very effective strategies to promote the body carriers' migration and body-to-surface carriers' migration, which further improve the HER photoactivity of CdS-based photocatalyst. In future studies, the combined technology of heteroatom doping and metallic surface decoration or bimetallic decoration should be considered to synergistically promote the body-to-surface carriers' migration of CdS-based photocatalyst for higher efficient HER photoactivity under broadened light irradiation.

References

1 Richards, B.S., Hudry, D., Busko, D. et al. (2021). Photon upconversion for photovoltaics and photocatalysis: a critical review. *Chem. Rev.* 121: 9165–9195.

2 Xu, C., Anusuyadevi, P.R., Aymonier, C. et al. (2019). Nanostructured materials for photocatalysis. *Chem. Soc. Rev.* 48: 3868–3902.

3 Zhao, D., Wang, Y., Dong, C.-L. et al. (2021). Boron-doped nitrogen-deficient carbon nitride-based Z-scheme heterostructures for photocatalytic overall water splitting. *Nat. Energy* 6: 388–397.

4 Huang, C., Chen, C., Zhang, M. et al. (2015). Carbon-doped BN nanosheets for metal-free photoredox catalysis. *Nat. Commun.* 6: 7698.

5 Tian, B., Tian, B., Smith, B. et al. (2018). Supported black phosphorus nanosheets as hydrogen-evolving photocatalyst achieving 5.4% energy conversion efficiency at 353 K. *Nat. Commun.* 9: 1397.

6 Iwashina, K., Iwase, A., Ng, Y.H. et al. (2015). Z-schematic water splitting into H_2 and O_2 using metal sulfide as a hydrogen-evolving photocatalyst and reduced graphene oxide as a solid-state electron mediator. *J. Am. Chem. Soc.* 137: 604–607.

7 Zuo, G., Wang, Y., Teo, W.L. et al. (2020). Ultrathin $ZnIn_2S_4$ nanosheets anchored on $Ti_3C_2T_X$ MXene for photocatalytic H_2 evolution. *Angew. Chem. Int. Ed.* 59: 11287–11292.

8 Zhu, B., Zou, R., and Xu, Q. (2018). Metal-organic framework based catalysts for hydrogen evolution. *Adv. Energy Mater.* 8: 1801193.

9 Wang, L., Zhang, L., Lin, B. et al. (2021). Activation of carbonyl oxygen sites in β-ketoenamine-linked covalent organic frameworks via cyano conjugation for efficient photocatalytic hydrogen evolution. *Small* 17: 2101017.

10 Cheng, L., Xiang, Q., Liao, Y., and Zhang, H. (2018). CdS-based photocatalysts. *Energy Environ. Sci.* 11: 1362–1391.

11 Li, W., Liu, X., Chu, X. et al. (2021). Fast Cr(vi) wastewater remediation on a Bi_xO_y/CdS heterostructure under simulated solar light induction. *Environ. Sci.: Nano* 8: 3655–3664.

12 Li, W., Wang, X., Ma, Q. et al. (2021). CdS@h-BN heterointerface construction on reduced graphene oxide nanosheets for hydrogen production. *Appl. Catal. B-Environ.* 284: 119688.

13 Li, W., Wang, F., Liu, X. et al. (2022). Promoting body carriers migration of CdS nanocatalyst by N-doping for improved hydrogen production under simulated sunlight irradiation. *Appl. Catal. B-Environ.* 313: 121470.

14 Tiwari, J.N., Dang, N.K., Park, H.J. et al. (2020). Remarkably enhanced catalytic activity by the synergistic effect of palladium single atoms and palladium-cobalt phosphide nanoparticles. *Nano Energy* 78: 105166.

15 Wang, Q., Huang, X., Zhao, Z.L. et al. (2020). Ultrahigh-loading of Ir single atoms on NiO matrix to dramatically enhance oxygen evolution reaction. *J. Am. Chem. Soc.* 142: 7425–7433.

16 Shi, R., Tian, C., Zhu, X. et al. (2019). Achieving an exceptionally high loading of isolated cobalt single atoms on a porous carbon matrix for efficient visible-light-driven photocatalytic hydrogen production. *Chem. Sci.* 10: 2585–2591.

17 Zuo, Q., Liu, T., Chen, C. et al. (2019). Ultrathin metal-organic framework nanosheets with ultrahigh loading of single Pt atoms for efficient visible-light-driven photocatalytic H_2 evolution. *Angew. Chem. Int. Ed.* 58: 10198–10203.

18 Wang, G., He, C.-T., Huang, R. et al. (2020). Photoinduction of Cu single atoms decorated on UiO-66-NH_2 for enhanced photocatalytic reduction of CO_2 to liquid fuels. *J. Am. Chem. Soc.* 142: 19339–19345.

19 Huang, P., Cheng, M., Zhang, H. et al. (2019). Single Mo atom realized enhanced CO_2 electro-reduction into formate on N-doped graphene. *Nano Energy* 61: 428–434.

20 Sun, M., Ji, J., Hu, M. et al. (2019). Overwhelming the performance of single atoms with atomic clusters for platinum-catalyzed hydrogen evolution. *ACS Catal.* 9: 8213–8223.

21 Li, W., Chu, X., Wang, F. et al. (2022). Pd single-atom decorated CdS nanocatalyst for highly efficient overall water splitting under simulated solar light. *Appl. Catal. B-Environ.* 304: 121000.

12

Theoretical Studies on Photocatalytic H_2 Production from H_2O

Kangkang Lian, Zhonghui Wang, and Sen Lin

Fuzhou University, State Key Laboratory of Photocatalysis on Energy and Environment, Department of Chemistry, Wulong Jiangbei Avenue, Fuzhou, Fujian, P.R. China

12.1 Introduction

Due to the shortage of fossil fuels and the environmental pollution caused by their combustion, the search for a clean and renewable alternative energy source is urgent. Hydrogen is an excellent candidate, producing only water when burned. Over the last few decades, attempts have been made toward renewable hydrogen [1]. As a result, sunlight-driven water decomposition has emerged as a promising method for hydrogen production.

Since the discovery of the photosensitization effect of the TiO_2 electrode for the electrochemical photolysis of water at a TiO_2 electrode with a Pt counter electrode under UV light irradiation by Fujishima and Honda in 1972 [2], increasing attention has been paid to various semiconducting materials. The last few decades have seen an explosion of three-dimensional (3D) materials for visible light photocatalytic water splitting including oxides, sulfides, nitrides, and other materials [1], which are promising photocatalysts thanks to their high efficiency, photochemical stability, and natural content [1].

Novoselov and coworkers prepared two-dimensional (2D) graphene materials in 2004 [3]. Since then, 2D materials with excellent electrical, thermal, and mechanical properties have attracted widespread attention and provided a new direction for the application of 2D materials for photocatalytic water splitting. The advantage of 2D materials is that they have a high specific surface area for photocatalytic reactions. In addition, 2D materials shorten the distance of photogenerated carriers (electrons and holes) to the catalyst-water interface, thus reducing the recombination of photogenerated carriers and increasing the carrier lifetime [4]. Despite the fascinating properties exhibited by 3D and 2D materials for photocatalytic water splitting, they still do not meet the needs of industrial hydrogen production. Thus, efforts have been made to improve the overall photocatalytic water splitting activity of 3D and 2D photocatalysts through a variety of different physicochemical means.

Photocatalysts suitable for water splitting to produce H_2, whether 3D or 2D materials, need to fulfill three main characteristics [4]: (i) a sufficiently wide

UV-Visible Photocatalysis for Clean Energy Production and Pollution Remediation: Materials, Reaction Mechanisms, and Applications, First Edition. Edited by Xinchen Wang, Masakazu Anpo, and Xianzhi Fu.
© 2023 WILEY-VCH GmbH. Published 2023 by WILEY-VCH GmbH.

bandgap ($E_g > 1.23$ eV) but with visible light absorption is needed; (ii) suitable band edge positions are needed. The conduction band minimum (CBM) should be higher than the reduction potential of hydrogen for hydrogen production. If overall water splitting is to be achieved, then the valence band maximum (VBM) should be simultaneously below the oxidation potential of water for oxygen production. (iii) They need low recombination rates of photogenerated electrons and holes, namely high carrier lifetimes. Scientists have therefore carried out experimental work based on the existing photocatalysts, to address these issues, but most require a trial-and-error approach that is both time-consuming and labor-intensive.

Theoretical calculations can predict and model the energy band structure, carrier separation, and other important properties of photocatalysts. They are indispensable in the design and development of photocatalysts. Arguably, theoretical calculations are more economical, efficient, and accurate than experimental trial-and-error methods. Predicting efficient photocatalysts from theoretical calculations could help experimental scientists make breakthroughs in overall photocatalytic water splitting for hydrogen production.

In this chapter, we discuss recent advances in photocatalytic water splitting on 3D and 2D materials from a computational chemistry perspective. We discuss several examples, focusing on band structure engineering and carrier separation. Conclusions and a summary are given at the end.

12.2 3D Photocatalysts

In this section, we focus on theoretical and computational advances in the field of 3D materials for photocatalytic water splitting based on the modulation of energy band structures and carrier separation.

12.2.1 Band Structure Engineering

Several examples of the modulation of the energy band structure in terms of crystal planes, chemical doping, and solid solution formation are discussed here.

TiO_2 is one of the most studied catalysts for photocatalytic water splitting. However, all three of its crystalline forms have relatively wide bandgaps (3.0 eV for rutile, 3.2 eV for anatase, and 3.4 eV for brookite) [5, 6] and thus are not suitable for direct use in the photodissociation of water. As a result, substantial theoretical work has been devoted to the study of tuning the energy band structure of TiO_2. Based on density functional theory (DFT) calculations, Harb et al. [7] revealed the facet effect of TiO_2 on the photocatalytic water splitting activity. Figure 12.1 shows the calculated electronic structures (band edge positions) on different facets (110, 101, 001, and 100) of anatase and rutile by Heyd–Scuseria–Ernzerhof (HSE06) [8]. The results show that the 001 surfaces of anatase and rutile have a higher level of CBM, in that the 001 surfaces of anatase and rutile are more suitable for photocatalytic water splitting to produce hydrogen than others. The theoretical calculations presented above help in guiding the experiments on how to improve the activity of photocatalytic water splitting by exposing specific crystalline facets of catalysts.

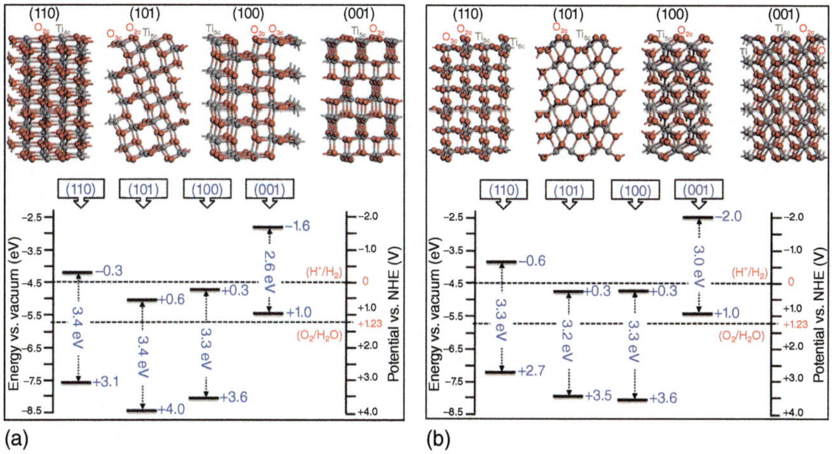

Figure 12.1 (a) Four possible low-index (110), (101), (100), and (001) anatase TiO_2 surfaces, and the corresponding electronic structures. (b) Four possible low-index (110), (101), (100), and (001) rutile TiO_2 surfaces and the corresponding electronic structures. VBM/CBM positions relative to vacuum and the normal hydrogen electrode (NHE) scales are also provided. Color legend: Ti is white, O is red. Source: Reproduced with permission from Harb et al. [7]/American Chemical Society.

Chemical doping can also modify the electronic structure of semiconductor materials [9]. In 2011, Wei and coworkers [10] used first-principle calculations to prove that the phosphorus (P)-doped anatase phase of TiO_2 can narrow the bandgap and improve its visible light absorption capacity without changing the CBM. As shown in Figure 12.2a, P impurities tend to replace Ti atoms after doping into TiO_2. The density of states (DOSs) and band edge positions of TiO_2 under different doping conditions were then calculated (Figure 12.2b,c). As the concentration of P doping increased, the valence band increased significantly; meanwhile, the conduction band remained nearly unchanged. Therefore, P doping can change the bandgap size of TiO_2. These results indicate that the spectral response range of the material has been expanded and increased the energy conversion efficiency, but the CBM position was not changed. Thus, this method can promote the efficiency of HER.

Ta_3N_5 and $Ti_3O_3N_2$ (Figure 12.3a,b) have sufficiently large bandgaps as reported by previous studies and high-throughput predictions, but they are difficult to commercialize due to their low-energy conversion efficiency [11]. Due to the band-bow effect, the solid solution can have a lower bandgap than the original two materials. $Ti_3O_3N_2$ and Ta_3N_5 have the same orthorhombic crystal structure (Figure 12.3) and similar volumes and the solid solution composed of xTa_3N_5 and $(1-x)Ti_3O_3N_2$ is denoted $Ta_{3x}Ti_{3-3x}O_{3-3x}N_{2+3x}$. Based on the above scenario, Wu and Ceder [11] proposed a new and promising Ta_3N_5:$Ti_3O_3N_2$ (Figure 12.3c) solid solution photocatalyst for water splitting. To determine the bandgap of a solid solution, band structures were calculated by the Δ-sol method, which is an efficient method for predicting bandgaps for solids using computationally efficient local and semilocal functionals [12]. Meanwhile, the HSE06 method was used to verify the band structure

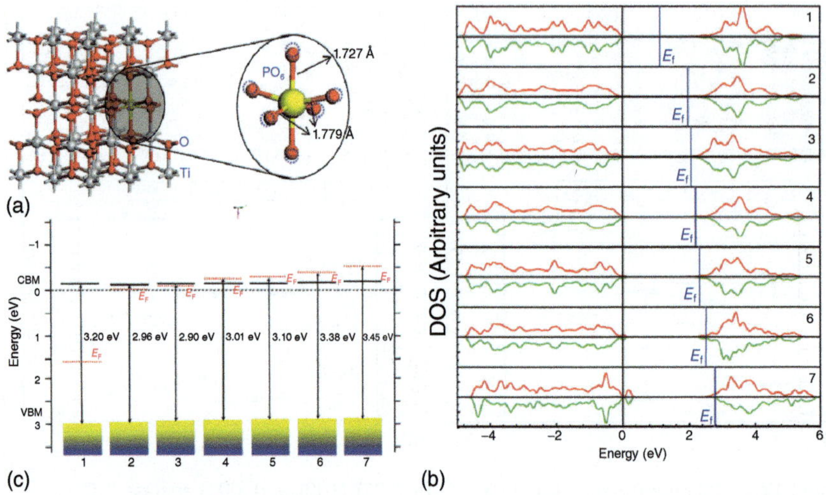

Figure 12.2 (a) Schematic of the phosphorus cation-doped anatase supercell. The red spheres represent O atoms, the silver spheres represent Ti atoms, and the yellow sphere represents the P atom. The dotted circles denote, schematically, the positions of oxygen atoms before relaxation. (b) Density of states (DOSs) for (1) pure anatase and P-doped configurations with various doping levels at (2) 0.3, (3) 0.7, (4) 1.0, (5) 1.4, (6) 2.1, and (7) 3.1 atm%. The dotted line represents the Fermi level, whereas the solid line represents the valence band maximum of undoped anatase TiO_2. (c) Electron transition energy for the configurations of (1) pure anatase and various P doping levels at (2) 0.3, (3) 0.7, (4) 1.0, (5) 1.4, (6) 2.1, and (7) 3.1 atm%. Source: Reproduced with permission from Peng et al. [10]/American Chemical Society.

calculated by the Δ-sol method. Figure 12.3d shows that the bandgap of the solid solution has a convex shape, that is, the bandgap is the smallest when X is 0.5 and becomes larger as X increases or decreases. The band edge position (Figure 12.3e) shows that the band edge position of the solid solution can satisfy the redox potential for photocatalytic water splitting when X is 0.5 and no bias is required. The reason for the reduction in the bandgap is that the energy of the 3d orbital of Ti is lower than that of the 5d orbital of Ta and the energy of the 2p orbital of N is higher than that of the 2p orbital of O; the CBM and VBM positions are due to the 3d orbital of Ti and the 2p orbitals of N. In summary, the $Ta_{3x}Ti_{3-3x}O_{3-3x}N_{2+3x}$ solid solution predicted by the theoretical calculations has good potential for efficient photocatalytic water splitting under solar illumination, which has yet to be verified experimentally in the future.

12.2.2 Carrier Separation

Carrier separation can also enhance the photocatalytic activity of photocatalysts. Liang and coworkers [13] presented a protocol based on a first-principles approach to investigate the photocatalytic activity of mixed-phase TiO_2 composed of anatase and rutile. The most stable mixed-phase TiO_2 was considered a heterojunction. The calculation results show that such heterojunction has a bandgap of 3.0 eV, and the VBM and CBM of rutile are 0.52 and 0.22 eV higher than those of anatase, respectively

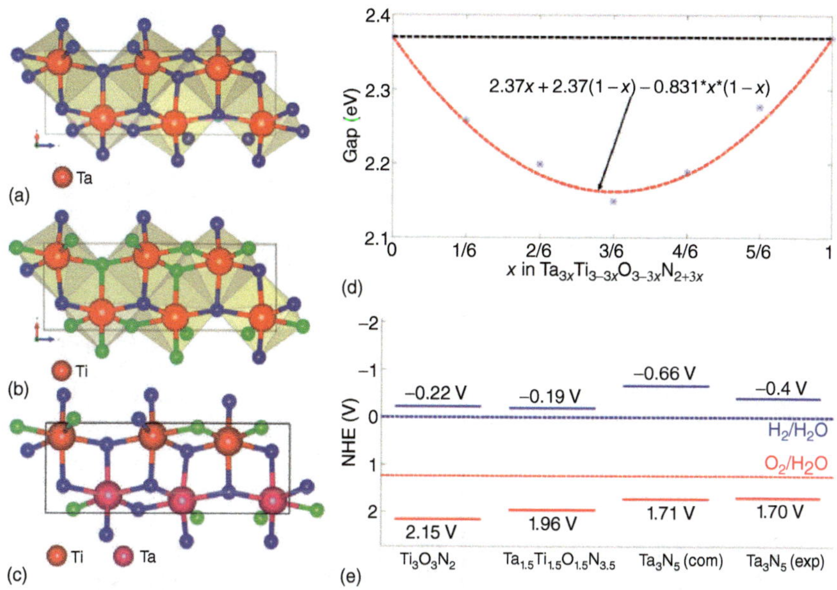

Figure 12.3 (a) Crystal structures of Ta_3N_5 and (b) $Ti_3O_3N_2$. (c) Crystal structures of Ta_3N_5:$Ti_3O_3N_2$. (d) Bandgap of the solid solution $Ta_{3x}Ti_{3-3x}O_{3-3x}N_{2+3x}$. Blue dots represent the bandgaps determined by the Δ-sol method. (e) Band edge positions of $Ti_3O_3N_2$, Ta_3N_5, and their solid solution $Ta_{1.5}Ti_{1.5}O_{1.5}N_{3.5}$ vs. NHE reference. The green and blue spheres represent O and N atoms, respectively. Source: Reproduced with permission from Wu and Ceder [11]/American Chemical Society.

(Figure 12.4a). The formation of heterojunctions leads to charge rearrangement and creates an electric field at the heterojunction interface. These induced electric fields thus facilitate the separation of photogenerated electrons and holes (Figure 12.4b), which promotes the photocatalytic HER activity of the TiO_2 heterojunction.

Pacchioni and coworkers [14] used DFT to study heterojunctions with co-exposed anatase (001) and (101) planes (Figure 12.4c). The energy band matching results of the formed heterojunction show that the difference between CBM and VBM is 0.54 and 1.02 eV at the crystal interface (Figure 12.4d). The heterojunction promotes carrier separation and behaves as the holes positioned on the (001) side and electrons positioned on the (101) side (Figure 12.4e), thus preventing carrier recombination and enhancing photocatalytic activity. This also leads to promoted photocatalytic HER.

Carrier separation can also be effectively promoted via surface distortion. A previous experiment by Chen et al. [15] found that by introducing structural distortion in the surface of TiO_2 nanocrystals, the photocatalytic activity of TiO_2 was significantly enhanced and the H_2 yield was two orders of magnitude higher than normal TiO_2. Subsequently, Huang and coworkers [16] used TiO_2 as a model and investigated the effect of surface distortion on carrier separation using DFT+U. Figure 12.5 shows the distribution of photogenerated electrons for TiO_2(101) and TiO_2(001) with surface distortion. Photogenerated electrons are transferred to

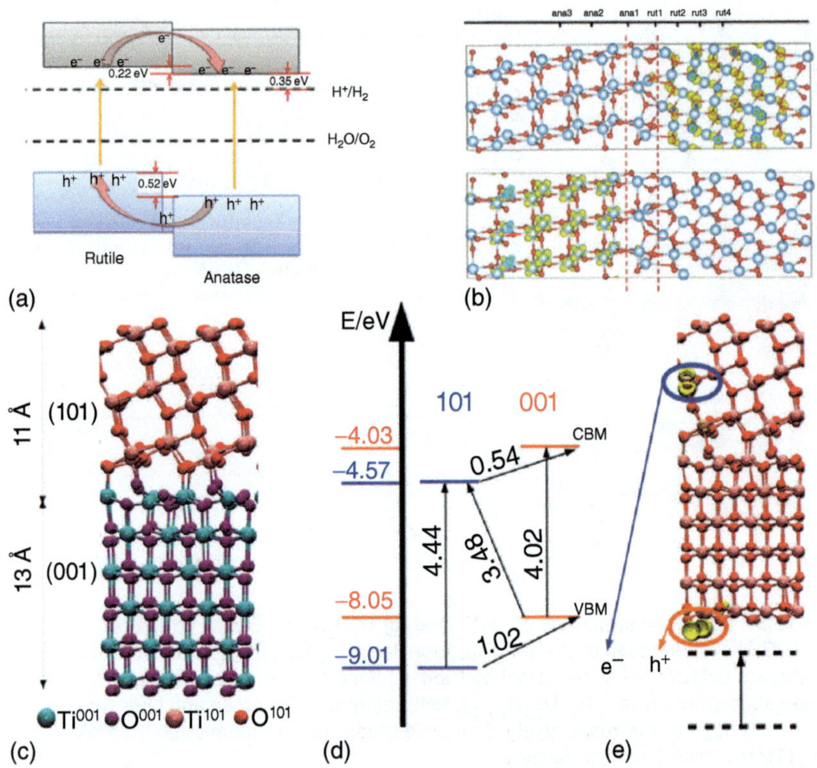

Figure 12.4 (a) The band offsets of heterojunction. (b) States of the valence band top and the conduction band bottom; red and blue spheres represent O and Ti atoms, respectively. Source: Reproduced with permission from Ju et al. [13]/American Chemical Society. (c) Side view of the optimized (001)-(101) anatase heterojunction model. (d) Band alignment of coexposed (001)-(101) anatase heterojunction. (e) Distribution of photogenerated electrons and holes. Source: Reproduced with permission from Di Liberto et al. [14]/American Chemical Society.

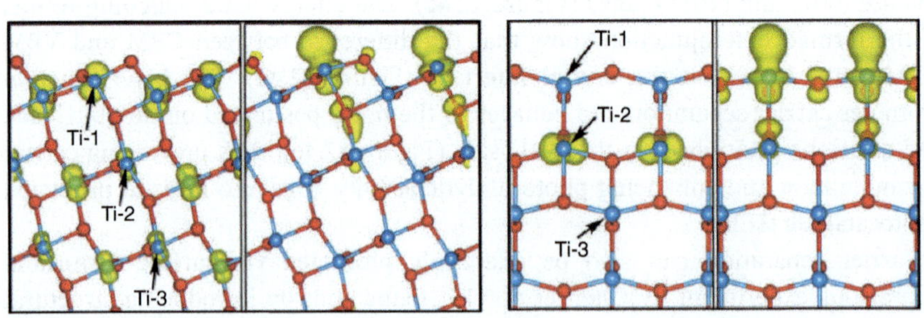

Figure 12.5 Distribution of the wave functions of photogenerated electrons in the (101) and (001) surface. (a) Electrons in the original (101) surface. (b) Electrons in the distorted (101) surface. (c) Electrons in the original (001) surface. (d) Electrons in the distorted (001) surface. The red and blue spheres represent O and Ti atoms, respectively. Source: Reproduced with permission from Ma et al. [16]/American Chemical Society.

the surface when the $TiO_2(101)$ surface is distorted (Figure 12.5a,b), while holes are prevented from transferring to the surface, which greatly reduces the carrier recombination rate. $TiO_2(001)$ behaves similarly (Figure 12.5c,d). The main reason for this is that the new metal–oxygen bonds within the distorted crystal plane form new antibonding orbitals, and molecular orbital theory suggests that the low-energy antibonding orbitals provide an energy gradient for electron transfer to the surface. Carriers are clearly separated effectively in the presence of surface distortion, thus facilitating hydrogen production from photocatalytic water splitting and confirming the experimental phenomenon [15].

The water splitting activity of the photocatalysts can also be improved via loading with metal and creating surface defects [17]. Due to the limited space of the chapter, doping/defects are not discussed further here.

12.3 2D Photocatalysts

In 2009, Wang et al. [18] discovered 2D g-C_3N_4 for photocatalytic water splitting to produce hydrogen and oxygen, which opened new directions for the application of 2D materials in photocatalytic water splitting. Furthermore, a growing number of experiments have shown that 2D materials can be stripped from bulk crystals or synthesized by chemical vapor deposition or other thin film growth techniques [4], thus making them candidates for photocatalytic water splitting.

12.3.1 Band Structure Engineering

Based on the previous DFT studies of hexagonal boron nitride (h-BN) doped with carbon atoms [19, 20] and graphdiyne (GDY) doped with B, N pairs [21], Lin's group investigated the energy band structure of halogen (F, Cl, and Br)-functionalized GDY via hybrid DFT [22]. The structure of halogen-functionalized GDY is shown in Figure 12.6a. Figure 12.6b shows that the bandgap energy of the modified GDY increases with increasing halogen atom pairs; meanwhile, the position of the VBM is influenced by the electronegativity of the halogen atoms. Greater electronegativity of the halogen atoms leads to a deeper VBM of the modified GDY (when the number of halogen atoms on the surface of GDY is the same). The calculations also show that the bandgap of GDY can be effectively adjusted via a suitable mixture of different halogen atoms for modification, and the valence band top and conduction band bottom of GDY, which were modified by a mixture of different halogen atoms can well match the redox potential of water. Thermodynamic analysis (Figure 12.6c–e) showed that the GDY modified with a mixture of different halogen atoms exhibited better photocatalytic water splitting activity than that of GDY modified with a single halogen.

2D phosphorus with remarkable photovoltaic properties has received much attention in recent years and is a very promising candidate for metal-free photocatalysts. Yang and coworkers [23] reported isomers of 2D phosphorus with porous structures (Figure 12.7a–e) using topological modeling methods and first

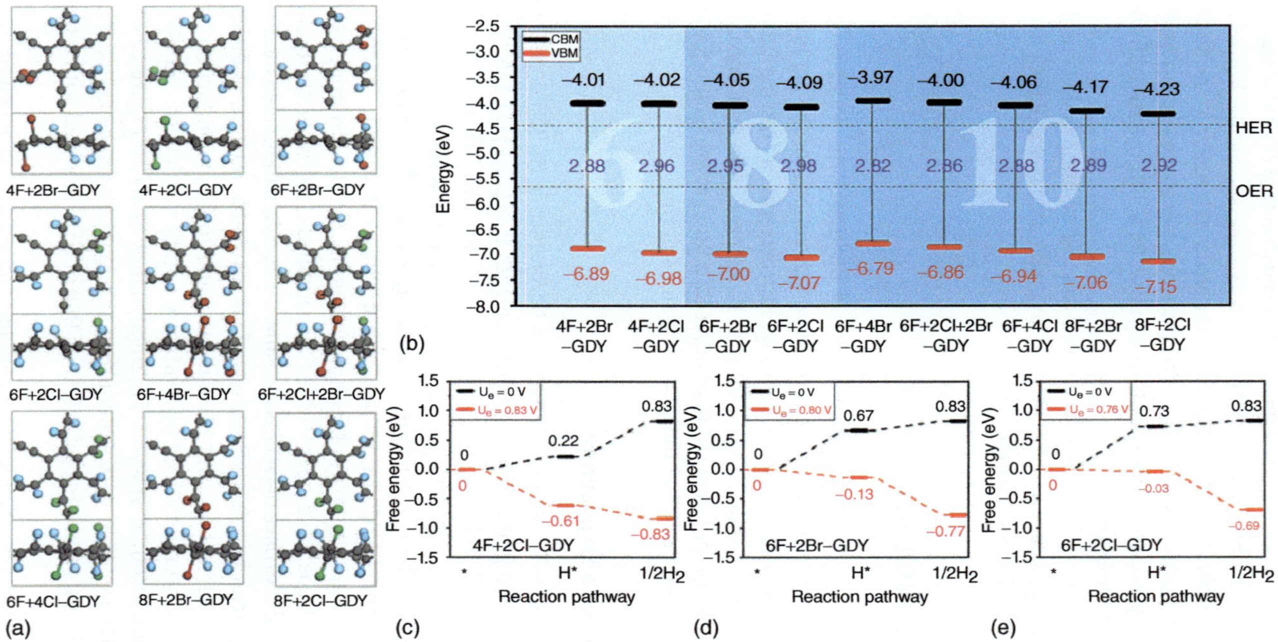

Figure 12.6 (a) Optimized structures of GDYs with mixed types of halogen atoms. The total number of halogen atoms is 6 (4F+2Br-GDY and 4F+2Cl-GDY), 8 (6F+2Br-GDY and 6F+2Cl-GDY), and 10 (6F+4Br-GDY, 6F+2Br+2Cl-GDY, 6F+4Cl-GDY, 8F+2Br-GDY, and 8F+2Cl-GDY). The gray, blue, green, and red spheres represent the C, F, Cl, and Br atoms, respectively. (b) Band structure for n1X_1+n2X_2+n3X_3-GDY. Horizontal dashed lines represent the redox potentials of water at pH = 0. (c)–(e) Calculated Gibbs free energy changes for the intermediate states involved in the HER processes on 4F+2Cl-GDY, 6F+2Br-GDY, and 6F+2Cl-GDY at pH = 7. OER, oxygen evolution reaction. Source: Reproduced with permission from Wang et al. [22]/AIP Publishing.

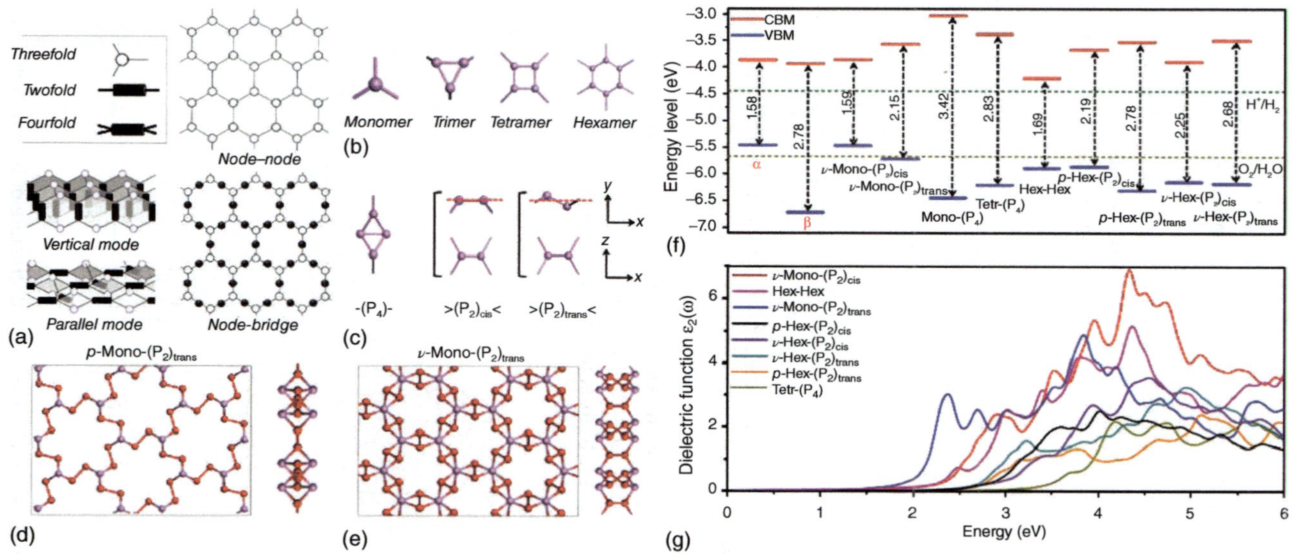

Figure 12.7 (a) Strategy to assemble "nodes" with threefold coordination and "bridges" with either twofold or fourfold coordination into 2D topological networks. (b) Four kinds of "nodes": monomer, trimer, tetramer, and hexamer. (c) Tetramers and dimers were chosen as twofold- and fourfold-coordinated "bridges." The "cis" and "trans" indicate two isomerized structures of dimer. (d, e) Two 2D phosphorus networks were obtained by assembling with monomer "node" and $>(P_2)_{trans}<$ "chain" in parallel or vertical methods. Both purple and red balls here denote the phosphorus atoms. The red balls illustrate the phosphorus atoms contributed by the "chain." (f) The band alignment of 2D phosphorus allotropes. Vacuum level is set as 0 eV and the chemical reaction potentials for H^+/H_2 and O_2/H_2O are plotted with dashed lines. (g) The imaginary part of dielectric function of 2D phosphorus allotropes. Source: Reproduced with permission from Zhuo et al. [23]/American Chemical Society.

principles. The electronic band structures calculated using the HSE06 indicate that the new structures are semiconductors with wide bandgaps ranging from 1.58 to 3.42 eV (Figure 12.7f); they could be tuned via a combination of elementary units in the network. Of these, eight isomers of 2D phosphorus possess suitable band gaps. Importantly, the positions of the conduction and valence band edges of certain isomers are well-matched to the chemical reaction potentials of H_2/H^+ and O_2/H_2O and can be used as photocatalysts for visible light-driven water splitting. Figure 12.7g shows that optical absorption spectroscopy predicts that porous 2D phosphorus materials can exhibit strong absorption in the visible light region. This theoretical study provides design principles for single-element 2D materials for photocatalytic water splitting.

Bulk GeTe is a semiconductor material with an indirect bandgap of 0.68 eV (Figure 12.8a–c), which is too narrow for photocatalysis. Li and coworkers [24] predicted the photocatalytic activity of 2D monolayer GeTe by DFT. The HSE06 calculations showed that monolayer GeTe has an indirect bandgap of 2.35 eV (Figure 12.8d), which is much larger than that of bulk GeTe, and the position of the band edges well matched the redox potential of water. Interestingly, when a 1% tensile strain is applied, the position of the CBM barely changes, while the position of the VBM decreases, thus resulting in a widening of the bandgap to 2.45 eV (Figure 12.8d). In contrast, when the strain is greater than 1%, the position of the CBM gradually shifts downwards with increasing strain, while the position of the VBM is less responsive to strain, thus resulting in a decrease in the bandgap with increasing strain. Figure 12.8e shows that a monolayer of GeTe can absorb visible and UV light. The results indicate that GeTe monolayers can be used for photocatalytic water splitting with high activity toward HER. This work provides

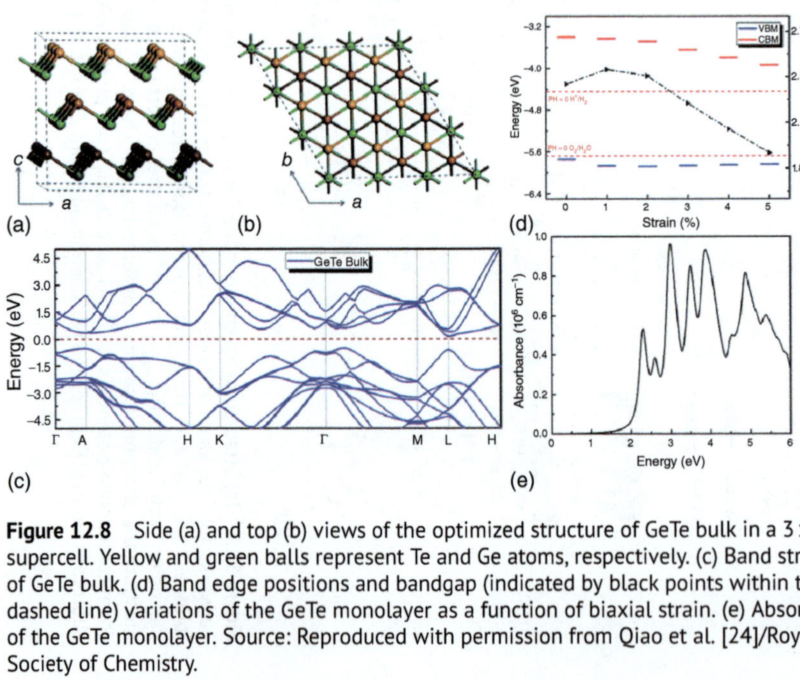

Figure 12.8 Side (a) and top (b) views of the optimized structure of GeTe bulk in a 3 × 3 × 1 supercell. Yellow and green balls represent Te and Ge atoms, respectively. (c) Band structure of GeTe bulk. (d) Band edge positions and bandgap (indicated by black points within the dashed line) variations of the GeTe monolayer as a function of biaxial strain. (e) Absorbance of the GeTe monolayer. Source: Reproduced with permission from Qiao et al. [24]/Royal Society of Chemistry.

important theoretical insight into the use of stress to modulate the energy band structure of 2D materials.

The energy band structure of 2D materials can also be modified using single-atom loading, such as g-C_3N_4 loaded with metal single atoms [25]. C doping also shows potential for the modulation of the energy band structure of boron nitride [26, 27].

12.3.2 Carrier Separation

As we mentioned earlier, g-C_3N_4 is a promising catalyst for photocatalytic water splitting to produce hydrogen, but usually suffers from easy recombination of photogenerated electrons and holes. Yang and coworkers [28] found that photogenerated electron and hole separation was significantly promoted by bending the 2D g-C_3N_4 to form a corrugated monolayer structure. Notably, this strategy does not employ any chemical modification. The bandgap of the rippled g-C_3N_4 in 2D did not change much vs. the previous sample via first-principle calculations. Nevertheless, it is interesting to note that the highest occupied molecular orbital (HOMO) and lowest unoccupied molecular orbital (LUMO) in rippled g-C_3N_4 are spatially separated from previous samples (Figure 12.9a–c). Furthermore, the separation of HOMO and LUMO and partial DOS analysis (Figure 12.9d) indicate that a type II heterojunction was formed inside the corrugated g-C_3N_4 and separated the charge carriers under the electric field so that photogenerated electrons in the conduction band could move toward the trough, and holes in the valence band could move toward the peak. As a result, the separation of photogenerated electrons and holes means that the probability of their recombination becomes smaller, namely increasing the lifetime of the carrier. This work provides a novel strategy for achieving carrier separation in 2D catalyst materials for photocatalytic water splitting.

In recent years, van der Waals (vdW) heterostructures have been proposed to alter the properties of 2D materials [29]. Lu and coworkers [30] investigated the ZrS_2-based vdW heterojunctions. Both h-BN/ZrS_2 and g-C_3N_4/ZrS_2 vdW heterojunctions were found to satisfy the potential for photocatalytic water splitting (Figure 12.10a,b). Their electronic properties were calculated to reveal whether bilayer vdW heterostructures based on ZrS_2 monolayers could enhance the visible light response of these 2D materials. The electronic structure of vdW heterojunctions was calculated by HSE06. The 2D monolayer material opens the bandgap under the action of the vdW interaction with the monolayer ZrS_2 and the applied stress. Figure 12.10c,d then show that the CBs are contributed by the ZrS_2 layer while the VBs are contributed by the other layer in all four heterostructures. This is due to the formation of the type II heterojunction. A Type II heterojunction is formed at the interface, and the carriers are separated by the action of the electric field. In this way, lower bandgaps and good carrier separation are achieved. In addition, the diffusion distance of charges from the bulk to the surface is reduced, thus reducing the possibility of recombination during migration. For photocatalytic water splitting, h-BN/ZrS_2 and g-C_3N_4/ZrS_2 heterostructures can satisfy basic requirements: suitable CBM and VBM sites, the sufficient driving force for electron, and hole transfer, as well as efficient charge separation, and good light absorption

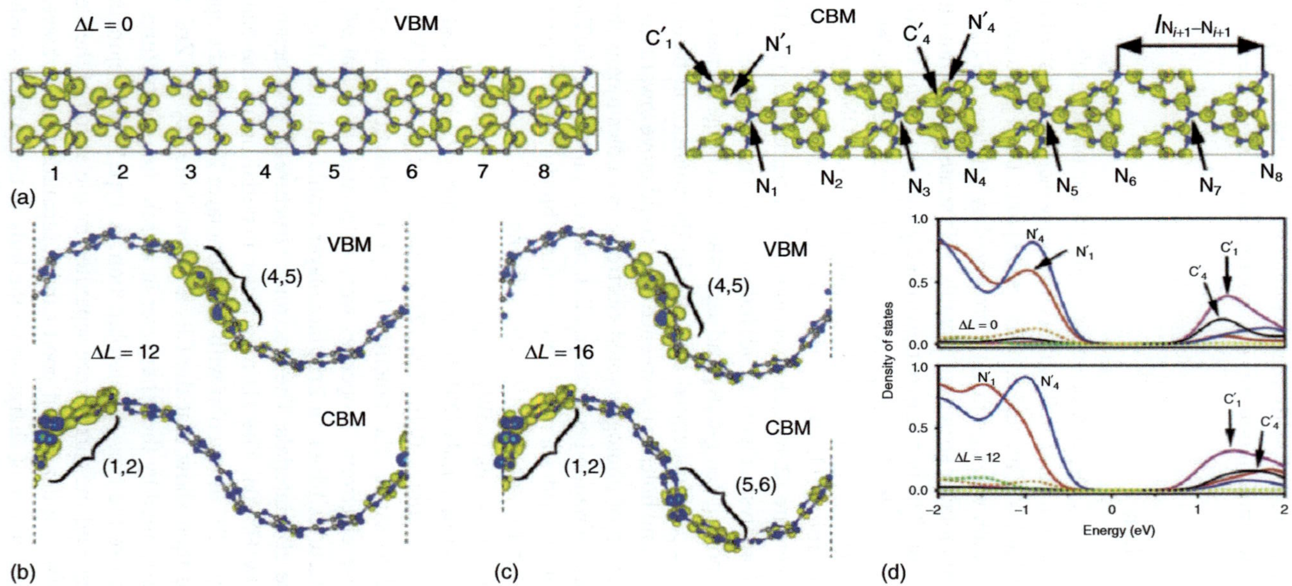

Figure 12.9 The electronic structure of g-C_3N_4-ΔL. Here, g-C_3N_4-ΔL means that the length of the supercell g-C_3N_4 shrunk ΔL angstrom. (a) Partial charge distributions (PCDs) for VBM and CBM in g-C_3N_4-0. The numbers represent the number of primitive cells in the supercell in the left part. The N_i ($i = 1-8$) represents the nodal N atom in the supercell in the right part. (b, c) The PCDs for VBM and CBM in g-C_3N_4-12 and g-C_3N_4-16, respectively. (d) The DOS of selected C and N atoms in g-C_3N_4-0 and g-C_3N_4-12, respectively. The gray and blue spheres represent C and N atoms, respectively. Source: Reproduced with permission from Sun et al. [28]/John Wiley and Sons.

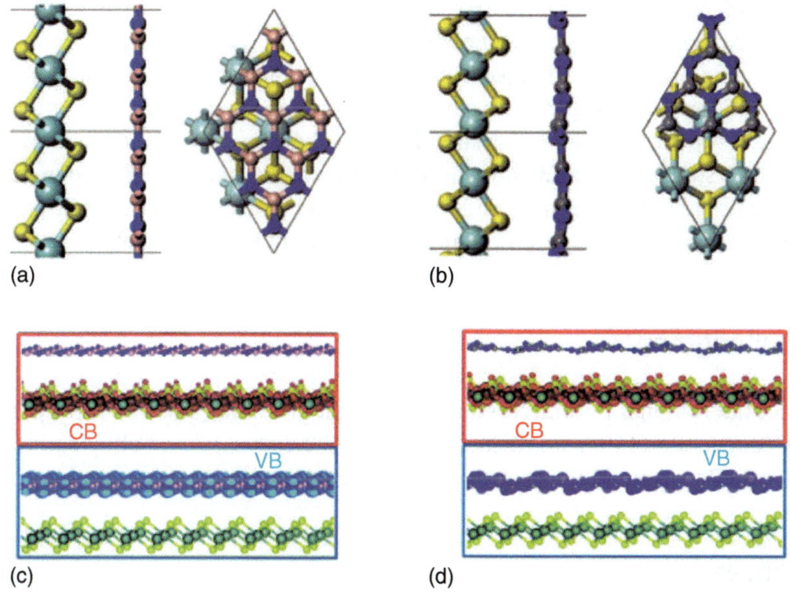

Figure 12.10 (a) h-BN/ZrS_2 vdW heterostructures. (b) *g*-C_3N_4/ZrS_2 vdW heterostructures. (c) The charge densities of the VB (blue) and CB (red) for h-BN/ZrS_2 with an isovalue of 0.02 e Å^{-3}. (d) The charge densities of the VB (blue) and CB (red) for *g*-C_3N_4/ZrS_2 with an isovalue of 0.02 e Å^{-3}. The yellow, green, pink, gray, and blue spheres represent S, Zr, B, C, and N atoms, respectively. Source: Reproduced with permission from Zhang et al. [30]/Royal Society of Chemistry.

under visible light. This research on the photocatalytic applications of layered vdW heterojunctions provides relevant theoretical and experimental efforts to find more efficient photocatalytic materials.

12.4 Summary and Perspectives

In this chapter, we describe several theoretical studies on photocatalytic water splitting to produce H_2 with 3D and 2D photocatalytic materials. Both the energy band structure and carrier separation of 3D and 2D semiconductor materials are important means of regulating the enhancement of photocatalytic water splitting. Theoretical calculations show that by exposing specific crystallographic surfaces, chemical doping and solid solution construction can modulate the bandgap of materials and adjust the position of the band edges to provide sufficient driving force for hydrogen production. Furthermore, surface distortion and heterojunction construction can also effectively improve the separation of photogenerated carriers, thus increasing the photocatalytic water splitting activity of the photocatalysts.

DFT is a very powerful tool for predicting the water splitting performance of photocatalysts. In contrast to experimental trial and error, the DFT can successfully simulate the energy band structure, carrier separation, and other important properties of photocatalysts with less time and effort, which in turn offers a theoretical basis for the development of photocatalysts for water splitting.

Despite rapid progress in the theoretical study of photocatalytic water splitting, there are still many challenges; for example, the effect of aqueous solvation on the energy band structure of photocatalysts and the catalytic process of water splitting; Previous DFT studies have paid only a little attention to the dynamic effects of reactant molecules and intermediates on the energy band structure of photocatalysts. New theories can lead to better insight into the important issues listed above.

Acknowledgments

We thank the National Natural Science Foundation of China (21973013), Natural Science Foundation of Fujian Province of China (2020J02025), and "Chuying Program" for the Top Young Talents of Fujian Province.

References

1 Wang, Q. and Domen, K. (2020). Particulate photocatalysts for light-driven water splitting: mechanisms, challenges, and design strategies. *Chem. Rev.* 120: 919–985.

2 Fujishima, A. and Honda, K. (1972). Electrochemical photolysis of water at a semiconductor electrode. *Nature* 238: 37.

3 Geim, A.K., Novoselov, K.S., Morozov, S.V. et al. (2004). Electric field effect in atomically thin carbon films. *Science* 306: 666.

4 Singh, A.K., Mathew, K., Zhuang, H.L., and Hennig, R.G. (2015). Computational screening of 2D materials for photocatalysis. *J. Phys. Chem. Lett.* 6: 1087–1098.

5 Ge, M., Cai, J., Iocozzia, J. et al. (2017). A review of TiO_2 nanostructured catalysts for sustainable H_2 generation. *Int. J. Hydrogen Energy* 42: 8418–8449.

6 Zhang, X., Zhou, H., Hao, Y.N. et al. (2020). Polaron-induced deep defect levels in brookite TiO_2: a many-body green's function theory study. *J. Phys. Chem. C* 124: 19024–19032.

7 Harb, M., Jeantelot, G., and Basset, J.M. (2019). Insights into the most suitable TiO_2 surfaces for photocatalytic O_2 and H_2 evolution reactions from DFT calculations. *J. Phys. Chem. C* 123: 28210–28218.

8 Jain, M., Chelikowsky, J.R., and Louie, S.G. (2011). Reliability of hybrid functionals in predicting band gaps. *Phys. Rev. Lett.* 107: 216806.

9 Schneider, J., Matsuoka, M., Takeuchi, M. et al. (2014). Understanding TiO_2 photocatalysis: mechanisms and materials. *Chem. Rev.* 114: 9919–9986.

10 Peng, Y., He, J., Liu, Q. et al. (2011). Impurity concentration dependence of optical absorption for phosphorus-doped anatase TiO_2. *J. Phys. Chem. C* 115: 8184–8188.

11 Wu, Y. and Ceder, G. (2013). First principles study on Ta_3N_5:$Ti_3O_3N_2$ solid solution as a water-splitting photocatalyst. *J. Phys. Chem. C* 117: 24710–24715.

12 Chan, M.K. and Ceder, G. (2010). Efficient band gap prediction for solids. *Phys. Rev. Lett.* 105: 196403.

13 Ju, M.G., Sun, G., Wang, J. et al. (2014). Origin of high photocatalytic properties in the mixed-phase TiO_2: a first-principles theoretical study. *ACS Appl. Mater. Interfaces* 6: 12885–12892.

14 Di Liberto, G., Tosoni, S., and Pacchioni, G. (2019). Role of heterojunction in charge carrier separation in coexposed anatase (001)–(101) surfaces. *J. Phys. Chem. Lett.* 10: 2372–2377.

15 Chen, X., Liu, L., Yu, P.Y., and Mao, S.S. (2011). Increasing solar absorption for photocatalysis with black hydrogenated titanium dioxide nanocrystals. *Science* 331: 746–750.

16 Ma, X., Dai, Y., Guo, M., and Huang, B. (2013). Insights into the role of surface distortion in promoting the separation and transfer of photogenerated carriers in anatase TiO_2. *J. Phys. Chem. C* 117: 24496–24502.

17 Wang, D., Liu, Z.P., and Yang, W.M. (2018). Revealing the size effect of platinum cocatalyst for photocatalytic hydrogen evolution on TiO_2 support: a DFT study. *ACS Catal.* 8: 7270–7278.

18 Wang, X., Maeda, K., Thomas, A. et al. (2009). A metal-free polymeric photocatalyst for hydrogen production from water under visible light. *Nat. Mater.* 8: 76–80.

19 Huang, C., Chen, C., Zhang, M. et al. (2015). Carbon-doped BN nanosheets for metal-free photoredox catalysis. *Nat. Commun.* 6: 7698.

20 Wan, Q., Wei, F., Ma, Z. et al. (2019). Novel porous boron nitride nanosheet with carbon doping: potential metal-free photocatalyst for visible-light-driven overall water splitting. *Adv. Theor. Simul.* 2: 1800174.

21 Xu, J., Wan, Q., Anpo, M., and Lin, S. (2020). Bandgap opening of graphdiyne monolayer via B, N-codoping for photocatalytic overall water splitting: design strategy from DFT studies. *J. Phys. Chem. C* 124: 6624–6633.

22 Wang, Z., Zhao, J., Wan, Q., and Lin, S. (2021). Halogen-driven bandgap opening in graphdiyne for overall photocatalytic water splitting. *Chin. J. Chem. Phys.* 34: 805–813.

23 Zhuo, Z., Wu, X., and Yang, J. (2016). Two-dimensional phosphorus porous polymorphs with tunable band gaps. *J. Am. Chem. Soc.* 138: 7091–7098.

24 Qiao, M., Chen, Y., Wang, Y., and Li, Y. (2018). The germanium telluride monolayer: a two dimensional semiconductor with high carrier mobility for photocatalytic water splitting. *J. Mater. Chem. A* 6: 4119–4125.

25 Zhang, G., Lan, Z.A., Lin, L. et al. (2016). Overall water splitting by Pt/g-C_3N_4 photocatalysts without using sacrificial agents. *Chem. Sci.* 7: 3062–3066.

26 Fang, Y., Merenkov, I.S., Li, X. et al. (2020). Vertically aligned 2D carbon doped boron nitride nanofilms for photoelectrochemical water oxidation. *J. Mater. Chem. A* 8: 13059–13064.

27 Lin, S., Ye, X., Gao, X., and Huang, J. (2015). Mechanistic insight into the water photooxidation on pure and sulfur-doped g-C_3N_4 photocatalysts from DFT calculations with dispersion corrections. *J. Mol. Catal. A Chem.* 406: 137–144.

28 Sun, J., Li, X., and Yang, J. (2019). Significantly enhanced charge separation in rippled monolayer graphitic C_3N_4. *ChemCatChem* 11: 6252–6257.

29 He, J., Wang, C., Zhou, B. et al. (2020). 2D van der Waals heterostructures: processing, optical properties and applications in ultrafast photonics. *Mater. Horizons* 7: 2903–2921.

30 Zhang, X., Meng, Z., Rao, D. et al. (2016). Efficient band structure tuning, charge separation, and visible-light response in ZrS_2-based van der Waals heterostructures. *Energy Environ. Sci.* 9: 841–849.

Part III

Photocatalytic Reduction of CO_2 and Fixation of N_2

13

Progress in Development of Cocatalysts for the Photocatalytic Conversion of CO_2 Using H_2O as an Electron Donor

Xuanwen Xu, Tsunehiro Tanaka, and Kentaro Teramura

Kyoto University, Graduate School of Engineering, Department of Molecular Engineering, Kyoto Daigaku Katsura, Kyoto 6158510, Japan

13.1 Background

13.1.1 Photocatalysis

Since the first industrial revolution, huge amounts of carbon dioxide (CO_2) have been emitted into the atmosphere as a result of the consumption of fossil fuels. Although various energy-generation technologies such as wind power have been developed to reduce our dependence on fossil fuels, global warming has continued to increase. Of the non-fossil-fuel-based energy sources, solar energy is the best choice for achieving green and sustainable power. As a result, there has been an upsurge in interest in photocatalysis as a means of storing and converting solar energy in recent years.

As shown in Scheme 13.1a, on light irradiation, the electrons in a typical semiconductor photocatalyst are excited from the valence band (VB) to the conduction band (CB) across the band gap. The photogenerated electrons and holes can be consumed by two pathways: (i) recombination or (ii) reaction with acceptors and donors. The reduction and oxidation potentials of the photogenerated electrons and holes depend significantly on the characteristics of the CB and VB of each photocatalyst. To enable the use of solar energy in the visible region of the spectrum ($\lambda > 420$ nm), Z-scheme photocatalysts are promising. Similar to natural photosynthesis, these systems comprise two photosystems (PSs), PS I and PS II, and an electron mediator that enables charge-carrier transport between the two PS, as shown in Scheme 13.1b.

13.1.2 Photocatalytic Conversion of CO_2 using H_2O as an Electron Donor

Two criteria are crucial for the photocatalytic conversion of CO_2 using H_2O as the electron donor. One is the balance between the electrons provided by H_2O (O_2 evolution) and those consumed by CO_2 and protons (CH_xO_y and H_2 evolution),

UV-Visible Photocatalysis for Clean Energy Production and Pollution Remediation: Materials, Reaction Mechanisms, and Applications, First Edition. Edited by Xinchen Wang, Masakazu Anpo, and Xianzhi Fu.
© 2023 WILEY-VCH GmbH. Published 2023 by WILEY-VCH GmbH.

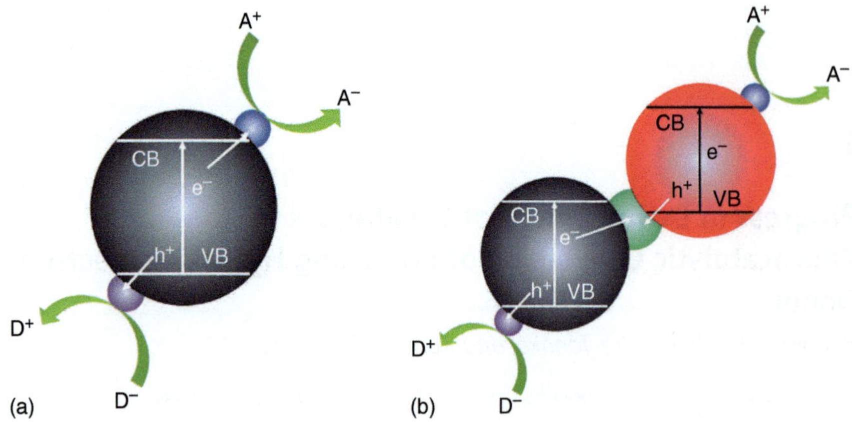

Scheme 13.1 Schematic of (a) a single photocatalyst and (b) Z-scheme photocatalyst showing PS I and PS II connected by the electron mediator. Here, CB and VB are the conduction and valence bands, respectively, e^- and h^+ are electrons and holes, respectively, and A and D are electron acceptor and donor species, respectively.

as shown in Eq. (13.1). The second is the selectivity of the products created by the photoreduction of CO_2 by the H_2 produced by water splitting, as stated in Eq. (13.2).

$$e^-/h^+ = (2\boldsymbol{R}_{\mathrm{HCOOH}} + 2\boldsymbol{R}_{\mathrm{CO}} + 4\boldsymbol{R}_{\mathrm{HCHO}} + 6\boldsymbol{R}_{\mathrm{CH_3OH}} + 8\boldsymbol{R}_{\mathrm{CH_4}} + 2\boldsymbol{R}_{\mathrm{H_2}})/4\boldsymbol{R}_{\mathrm{O_2}} \tag{13.1}$$

$$\text{Selectivity} = (2\boldsymbol{R}_{\mathrm{HCOOH}} + 2\boldsymbol{R}_{\mathrm{CO}} + 4\boldsymbol{R}_{\mathrm{HCHO}} + 6\boldsymbol{R}_{\mathrm{CH_3OH}} + 8\boldsymbol{R}_{\mathrm{CH_4}})/ (2\boldsymbol{R}_{\mathrm{HCOOH}} + 2\boldsymbol{R}_{\mathrm{CO}} + 4\boldsymbol{R}_{\mathrm{HCHO}} + 6\boldsymbol{R}_{\mathrm{CH_3OH}} + 8\boldsymbol{R}_{\mathrm{CH_4}} + 2\boldsymbol{R}_{\mathrm{H_2}}) \times 100\% \tag{13.2}$$

Here, ***R*** represents the rate of formation of each product formed by the photocatalytic conversion of CO_2 using H_2O as the electron donor.

However, achieving a higher selectivity toward carbon products, such as CO, CH_3OH, and CH_4, is difficult because water splitting is a strongly competitive reaction. As listed in Table 13.1 [1], the reduction potential of protons in water is much lower than that of HCOOH, CO, and HCHO; thus, the evolution of H_2 in the photocatalytic conversion of CO_2 using H_2O as the electron donor dominates the consumption of the photogenerated electrons. Although the required potentials for the formation of CH_3OH and CH_4 are lower than that of H_2, these reactions require a high concentration of protons, which will inevitably suppress the dissolution of CO_2. Furthermore, the poor solubility of CO_2 makes photocatalytic conversion of CO_2 challenging, as demonstrated by the significance of employing $NaHCO_3$ as an additive, which improves the dissolution of CO_2 in the reactions [2].

photocatalysts, were developed as an effective cocatalyst for the photocatalytic conversion of CO_2 by H_2O. The performances of Ag nanoparticles were influenced by many factors, including additives, surface properties of photocatalysts, morphologies, and dual cocatalysts. In addition to Ag nanoparticles, nonmetal cocatalysts such as LDHs, $Zn(OH)_2$, and the *in situ* added chromate ions were also found to enhance the evolution of CO during the photocatalytic conversion of CO_2 by H_2O.

Although various cocatalysts mentioned above were developed for the photocatalytic conversion of CO_2 by H_2O, they just exhibited photocatalytic performances only on the photocatalysts with wide bandgaps. It is rarely reported that proper cocatalysts, which are capable of proceeding with the photocatalytic conversion of CO_2 by H_2O on visible-light-driven photocatalysts, have been developed. This situation blocks the development of advanced photocatalysts for the photocatalytic conversion of CO_2 by H_2O harvesting visible light, which accounts for a large part of solar energy. First, it is an urgent task to make clear the mechanism of how cocatalysts interact with the main photocatalysts. Second, the bottleneck of cocatalysts used on the visible-light-driven photocatalysts for the photocatalytic conversion of CO_2 by H_2O should be determined. A clear guideline to design effective cocatalysts used on the visible-light-driven photocatalysts for the selective photocatalytic conversion of CO_2 by H_2O should be established.

References

1 White, J.L., Baruch, M.F., Pander, J.E. III, et al. (2015). Light-driven heterogeneous reduction of carbon dioxide: photocatalysts and photoelectrodes. *Chem. Rev.* 115: 12888–12935.

2 Pang, R., Teramura, K., Asakura, H. et al. (2017). Highly selective photocatalytic conversion of CO_2 by water over Ag-loaded $SrNb_2O_6$ nanorods. *Appl. Catal,. B* 218: 770–778.

3 Iguchi, S., Teramura, K., Hosokawa, S., and Tanaka, T. (2016). A $ZnTa_2O_6$ photocatalyst synthesized via solid state reaction for conversion of CO_2 into CO in water. *Catal. Sci. Technol.* 6: 4978–4985.

4 Pang, R., Teramura, K., Asakura, H. et al. (2019). Effect of thickness of chromium hydroxide layer on Ag cocatalyst surface for highly selective photocatalytic conversion of CO_2 by H_2O. *ACS Sustainable Chem. Eng.* 7: 2083–2090.

5 Iizuka, K., Wato, T., Miseki, Y. et al. (2011). Photocatalytic reduction of carbon dioxide over Ag cocatalyst-loaded $ALa_4Ti_4O_{15}$ (A = Ca, Sr, and Ba) using water as a reducing reagent. *J. Am. Chem. Soc.* 133: 20863–20868.

6 Wang, Z., Teramura, K., Hosokawa, S., and Tanaka, T. (2015). Photocatalytic conversion of CO_2 in water over Ag-modified $La_2Ti_2O_7$. *Appl. Catal., B* 163: 241–247.

7 Zhu, X., Anzai, A., Yamamoto, A., and Yoshida, H. (2019). Silver-loaded sodium titanate photocatalysts for selective reduction of carbon dioxide to carbon monoxide with water. *Appl. Catal., B* 243: 47–56.

8 Anzai, A., Fukuo, N., Yamamoto, A., and Yoshida, H. (2017). Highly selective photocatalytic reduction of carbon dioxide with water over silver-loaded calcium titanate. *Catal. Commun.* 100: 134–138.

9 Nakanishi, H., Iizuka, K., Takayama, T. et al. (2017). Highly active $NaTaO_3$-based photocatalysts for CO_2 reduction to form CO using water as the electron donor. *ChemSusChem* 10: 112–118.

10 Wang, S., Teramura, K., Asakura, H. et al. (2021). Effect of Zn in Ag-loaded Zn-modified $ZnTa_2O_6$ for photocatalytic conversion of CO_2 by H_2O. *J. Phys. Chem. C* 125: 1304–1312.

11 Teramura, K., Tatsumi, H., Wang, Z. et al. (2015). Photocatalytic conversion of CO_2 by H_2O over Ag-loaded SrO-modified Ta_2O_5. *Bull. Chem. Soc. Jpn.* 88: 431–437.

12 Yoshizawa, S., Huang, Z., Teramura, K. et al. (2019). Important role of strontium atom on the surface of $Sr_2KTa_5O_{15}$ with a tetragonal tungsten bronze structure to improve adsorption of CO_2 for photocatalytic conversion of CO_2 by H_2O. *ACS Appl. Mater. Interfaces* 11: 37875–37884.

13 Iguchi, S., Hasegawa, Y., Teramura, K. et al. (2017). Drastic improvement in the photocatalytic activity of Ga_2O_3 modified with Mg-Al layered double hydroxide for the conversion of CO_2 in water. *Sustain. Energy Fuels* 1: 1740–1747.

14 Wang, Z., Teramura, K., Hosokawa, S., and Tanaka, T. (2015). Highly efficient photocatalytic conversion of CO_2 into solid CO using H_2O as a reductant over Ag-modified $ZnGa_2O_4$. *J. Mater. Chem. A* 3: 11313–11319.

15 Pang, R., Teramura, K., Asakura, H. et al. (2019). Role of bicarbonate ions in aqueous solution as a carbon source for photocatalytic conversion of CO_2 into CO. *ACS Appl. Energy Mater.* 2: 5397–5405.

16 Teramura, K., Wang, Z., Hosokawa, S. et al. (2014). A doping technique that suppresses undesirable H_2 evolution derived from overall water splitting in the highly selective photocatalytic conversion of CO_2 in and by water. *Chem. Eur. J.* 20: 9906–9909.

17 Tatsumi, H., Teramura, K., Huang, Z. et al. (2017). Enhancement of CO evolution by modification of Ga_2O_3 with rare-earth elements for the photocatalytic conversion of CO_2 by H_2O. *Langmuir* 33: 13929–13935.

18 Huang, Z., Teramura, K., Asakura, H. et al. (2018). Flux method fabrication of potassium rare-earth tantalates for CO_2 photoreduction using H_2O as an electron donor. *Catal. Today* 300: 173–182.

19 Wang, S., Teramura, K., Hisatomi, T. et al. (2020). Effective driving of Ag-loaded and Al-doped $SrTiO_3$ under irradiation at $\lambda > 300$ nm for the photocatalytic conversion of CO_2 by H_2O. *ACS Appl. Energy Mater.* 3: 1468–1475.

20 Wang, S., Teramura, K., Hisatomi, T. et al. (2020). Optimized synthesis of Ag-modified Al-doped $SrTiO_3$ photocatalyst for the conversion of CO_2 using H_2O as an electron donor. *ChemistrySelect* 5: 8779–8786.

21 Ishida, M., Kikkawa, S., Hori, K. et al. (2020). Effect of surface reforming via O_3 treatment on the electrochemical CO_2 reduction activity of a Ag cathode. *ACS Appl. Energy Mater.* 3: 6552–6560.

22 Pang, R., Teramura, K., Tasumi, H. et al. (2018). Modification of Ga_2O_3 by an Ag-Cr core–shell cocatalyst enhances photocatalytic CO evolution for the conversion of CO_2 by H_2O. *Chem. Commun.* 54: 1053–1056.

23 Wang, S., Teramura, K., Hisatomi, T. et al. (2021). Dual Ag/Co cocatalyst synergism for the highly effective photocatalytic conversion of CO_2 by H_2O over Al-$SrTiO_3$. *Chem. Sci.* 12: 4940–4948.

24 Wang, S., Teramura, K., Hisatomi, T. et al. (2021). Highly selectivephotocatalytic conversion of carbon dioxide by water over Al-$SrTiO_3$ photocatalyst modified with silver–metal dual cocatalysts. *ACS Sustainable Chem. Eng.* 9: 9327–9335.

25 Iguchi, S., Teramura, K., Hosokawa, S., and Tanaka, T. (2015). Photocatalytic conversion of CO_2 in an aqueous solution using various kinds of layered double hydroxides. *Catal. Today* 251: 140–144.

26 Takemoto, M., Tokudome, Y., Kikkawa, S. et al. (2020). Imparting CO_2 reduction selectivity to $ZnGa_2O_4$ photocatalysts by crystallization from hetero nano assembly of amorphous-like metal hydroxides. *RSC Adv.* 10: 8066–8073.

27 Xu, X., Teramura, K., Asakura, H. et al. (2021). Development of zinc hydroxide as an abundant and universal cocatalyst for the selective photocatalytic conversion of CO_2 by H_2O. *ChemCatChem* 13: 4313–4317.

28 Xu, X., Teramura, K., Asakura, H. et al. (2021). Shift of active sites via in-situ photodeposition of chromate achieving highly selective photocatalytic conversion of CO_2 by H_2O over $ZnTa_2O_6$. *Appl. Catal., B* 298: 120508.

29 Xu, X., Asakura, H., Hosokawa, S. et al. (2022). Effect of the in situ addition of chromate ions on H_2 evolution during the photocatalytic conversion of CO_2 using H_2O as the electron donor. *Catal. Today* https://doi.org/10.1016/J.CATTOD.2022.05.045.

14

Preparation, Characterization, and Photocatalysts' Application of Silicas/Silicates with Nanospaces Containing Single-site Ti-oxo Species

Masashi Morita[1] *and Makoto Ogawa*[2]

[1] *Tokyo University of Agriculture and Technology, Graduate School of Engineering, Department of Applied Chemistry, 2-24-16 Naka-cho, Koganei, Tokyo 184-8588, Japan*
[2] *Vidyasirimedhi Institute of Science and Technology (VISTEC), School of Energy Science and Engineering, 555 Moo 1 Tumbol Payupnai, Amphoe Wangchan, Rayong 21210, Thailand*

14.1 Introduction

Precise design of nanostructures, or nanoarchitectonics, is a key concept for materials chemistry through important structure-property relationships [1]. The nanostructure design has been done to control functions as well as to find novel phenomena that are not available in bulk materials. Nanostructure design of titanium dioxide (TiO_2) has been done for catalysts and other applications. At a nanometer level, size reduction of crystalline TiO_2 (anatase and rutile) has resulted in the colloidal and optical properties being different from their bulk counterparts, so various methods to prepare TiO_2 nanoparticles for (photo)catalysts, electrochemical devices, and so on have been reported [2, 3]. Aggregation of nanoparticles has commonly been seen as a drawback in order to achieve reliable and reproducible results coming from the nanosize effects. The immobilization of TiO_2 in various nanospace materials (mesoporous silicas, silicates, etc.) has been investigated to suppress the aggregation. In addition, these silica/silicate-based nanospace materials have been used as templates to control the shape and size of TiO_2 nanoparticles [4].

Molecular Ti-oxo species are known as critical nano-objects useful as catalytically active sites. The (photo)catalytically active Ti with tetrahedrally coordinated oxygen has been shown to be (photo)catalytically active, and the immobilization of the molecular Ti-oxo species in/on various nanospace materials has been examined [5]. The shape and size, and the surface properties of the pores in nanospace materials may affect (photo)catalytic reactions (activity, selectivity, and catalyst lifetime). Such active species at atomic/molecular levels are designated as single-site catalysts [6] and also as single-atom catalysts, and they have been fixed in zeolites and mesoporous silicas [7]. Single-site Ti-oxo species have been prepared in graphene and the Ti-oxo cluster has been used for metal-organic frameworks (MOFs) or porous coordination polymers (PCPs) [8, 9]. Here, the preparation and characterization of

UV-Visible Photocatalysis for Clean Energy Production and Pollution Remediation: Materials, Reaction Mechanisms, and Applications, First Edition. Edited by Xinchen Wang, Masakazu Anpo, and Xianzhi Fu.
© 2023 WILEY-VCH GmbH. Published 2023 by WILEY-VCH GmbH.

single-site Ti-oxo species in various nanospace materials and their photocatalytic applications are summarized to clarify the possibilities and limitations.

14.2 Materials Variation of Single-site Ti-oxo Species in Nanospace Materials

14.2.1 Characterization of Ti-oxo Species

In the case of semiconducting bulk TiO_2 photocatalysts, photo-generated electrons and holes upon light irradiation are spatially separated in the conduction band and valence band, resulting in photocatalytic reactions on the surface. In contrast, Ti fixed in/on the silica-based nanospace materials as the isolated tetrahedrally coordinated species forms a charge transfer excited state ($[Ti^{4+}-O^{2-}] \rightarrow [Ti^{3+}-O^{-}]^*$) upon UV irradiation, and the electron–hole pairs are localized quite near to each other, which exhibits unique (photo)catalytic properties, photoinduced superhydrophilicity and associating self-cleaning properties. The Ti-oxo species fixed in/on nanospace materials have been characterized by several spectroscopic characterizations, including UV–Vis absorption, photoluminescence, ESR, and X-ray absorption (XANES and EXAFS) spectroscopies [10, 11]. As to the Ti K-edge XANES and the Fourier transformed EXAFS (FT-EXAFS) spectra (Figure 14.1), bulk TiO_2 shows several well-defined pre-edge peaks attributable to the Ti atom in octahedral coordination (Figure 14.1a). The tetrahedrally coordinated Ti-oxo species on the Y-zeolites prepared by ion exchange, on the other hand, show a very sharp single pre-edge peak at around 4970 eV, which is due to the allowed transitions from 1s to 3d level of an isolated Ti atom surrounded by four oxygen (Figure 14.1d,e). The Ti K-edge XANES spectra of Ti-oxo species on the Y-zeolites prepared by the impregnation of the molecular precursors presented a weak pre-edge peak with the increase in the Ti content, indicating the presence of an isolated tetrahedrally coordinated Ti-oxo species together with octahedrally coordinated Ti-oxo species (Figure 14.1b,c). Ti-oxo species fixed in/on silicas/silicates showed the absorption bands with the maxima at 220–280 nm, which were assignable to the charge transfer excited state ($[Ti^{4+}-O^{2-}] \rightarrow [Ti^{3+}-O^{-}]^*$). In addition, these Ti-oxo species exhibited photoluminescence at around 450–550 nm upon excitation at around 220–280 nm, which is assignable to the charge transfer process on the tetrahedrally coordinated Ti-oxo species. When it comes to controlling the state of Ti-oxo species, not only the preparation methods but also the amount and type of Ti sources used play an important role in the preparation of Ti-oxo species in/on silicas/silicates.

14.2.2 Ti-Containing Zeolites and Mesoporous Silicas/Silicates

Ti-containing zeolites and mesoporous silicas/silicates have been extensively reported as (photo)catalysts. The examples are summarized in Table 14.1. Titanium silicalite-1 (TS-1) exhibited selective oxidation of various hydrocarbons with H_2O_2 as a zeolite with isomorphous substitution of Si atoms with Ti atoms in the

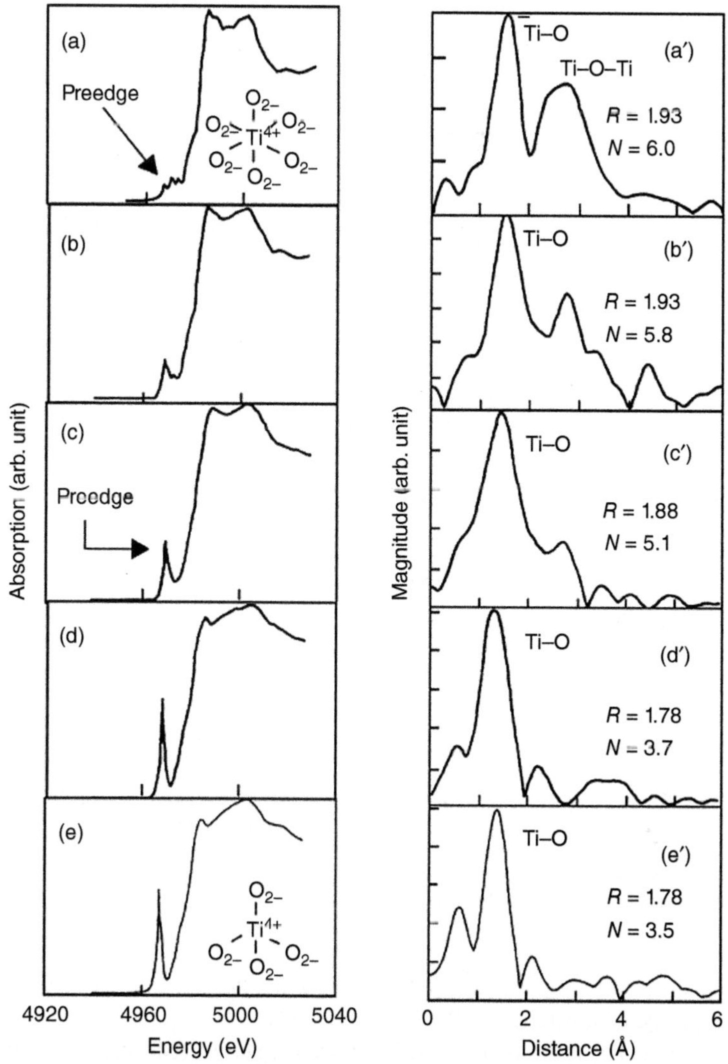

Figure 14.1 XANES (left) and FT-EXAFS (right) spectra: (a, a′) anatase TiO_2 powder, (b, b′) Ti-oxide/Y-zeolite (10 wt% as TiO_2) prepared by impregnating, (c, c′) Ti-oxide/Y-zeolite (1.0 wt% as TiO_2) prepared by impregnating, (d, d′) Ti-oxide/Y-zeolite prepared by ion-exchange, and (e, e′) Pt-loaded Ti-oxide/Y-zeolite catalysts prepared by ion-exchange. Source: Reproduced with permission from Yamashita and Anpo [11]/Elsevier.

MFI zeolite framework [12]. TS-1 is prepared by hydrothermal reaction of SiO_2 sources (tetraethoxysilane, fumed silica, etc.), Ti sources (tetraethyl orthotitanate [TEOT], tetrabutyl orthotitanate [TBOT], etc.), templates (tetrapropylammonium [TPA^+], tetrabutylammonium [TBA^+], etc.) and mineralizing agents (hexanediamine, methylamine, etc.) [18]. Incorporation of Ti into the silica frameworks of periodic mesoporous silicas such as MCM-41 (one-dimensional pore), MCM-48 (three-dimensional pore), and SBA-15 (one-dimensional pore) has been examined

Table 14.1 Ti contents for Ti-containing zeolites and mesoporous silicas/silicates.

Hosts	Ti contents	Ti sources	References
Silicalite-1	Si/Ti = 50	Tetraethyl orthotitanate (TEOT)	[12]
Hexagonal mesoporous silica (HMS)	Si/Ti = 100	Tetraisopropyl orthotitanate (TPOT)	[13]
MCM-41	Si/Ti = 71	Tetrabutyl orthotitanate (TBOT)	[14]
MCM-48	Si/Ti = 80–83	Tetrabutyl orthotitanate (TBOT)	[14]
Layered silicate HUS-2 ($Si_{20}O_{40}(OH)_4 \cdot 4[C_5H_{14}NO]$)	4.0–6.5 wt%[a] (e.g. TS-1: 2.2 wt%, Si/Ti = 35)	Ti(IV) acetylacetonate ($Ti(acac)_4$)	[15]
Layered silicate magadiite ($Na_2Si_{14}O_{29}$)	Si/Ti = 50–100	Titanium tetrachloride ($TiCl_4$)	[16]
Layered silicate octosilicate ($Na_2Si_8O_{17}$)	Si/Ti = 50–200	Titanium tetrachloride ($TiCl_4$)	[16]
Mesoporous silica films	Si/Ti = 50	Tetraisopropyl orthotitanate (TPOT)	[17]

a) Grafting of Ti(IV) acetylacetonate in the interlayer surface of HUS-2.

by the sol–gel process or the hydrothermal reactions in the presence of Ti precursor (TBOT, tetraisopropyl orthotitanate [TPOT], etc.) [13, 14].

Layered silicates such as kanemite ($NaHSi_2O_5$), magadiite ($Na_2Si_{14}O_{29}$), and octosilicate ($Na_2Si_8O_{17}$) are composed of alternately stacked silicate layers and exchangeable interlayer cations, and they are regarded as nanospace materials with two-dimensionally expandable nanospace (interlayer space). Some of them are naturally occurring and some of them are synthesized by hydrothermal reactions in the laboratory. The silanol groups are periodically arranged on the layer surface, and they have been used for cation exchange and grafting. The applications of the layered alkali silicates include adsorbents, ion exchangers, catalysts/catalyst supports, polymer additives, and photofunctional materials [19]. Taking advantage of the reactivity of silanol groups on the layer surface, Ti(IV) acetylacetonate ($Ti(acac)_4$) was immobilized in the interlayer surfaces of a layered silicate HUS-2 ($Si_{20}O_{40}(OH)_4 \cdot 4[C_5H_{14}NO]$) [15]. These Ti-oxo species on the silicate surface existed as an isolated tetrahedrally coordinated species as proposed by UV–Vis absorption, ^{13}C CP, and ^{29}Si NMR measurements. Ti(IV) acetylacetonate-immobilized HUS-2 exhibited higher photocatalytic activity than TS-1 and a high selectivity for partial

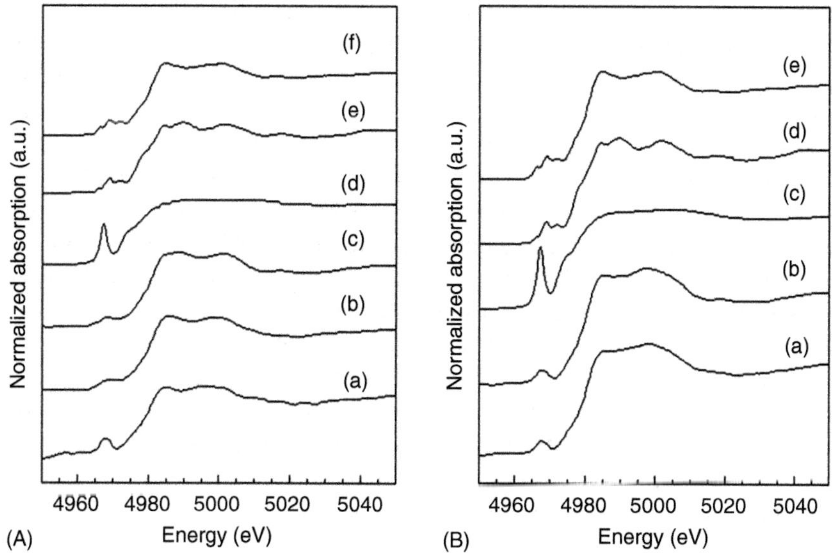

Figure 14.2 (A) XANES spectra of the Ti-containing octosilicates ((a) Si/Ti = 200, (b) 100, and (c) 50) and reference ((d) tetraisopropyl orthotitanate (TPOT), (e) TiO_2 rutile and (f) anatase). (B) XANES spectra of the Ti-containing magadiites ((a) Si/Ti = 100, (b) 50) and reference ((c) titanium tetraisopropoxide, (d) TiO_2 rutile and (e) anatase). Source: Reproduced with permission from Morita et al. [16]/American Chemical Society.

cyclohexane oxidation. During the crystallization of the silicate, titanium tetrachloride ($TiCl_4$), silica gel, and sodium hydroxide were used in hydrothermal processes to incorporate Ti into the frameworks of magadiite and octosilicate [16]. The UV–Vis absorption and XAFS spectra suggested that Ti-oxo species in the silicate frameworks existed as tetrahedrally coordinated species as a result of the isomorphous substitution of the Si atom in the silicate frameworks. In the XANES spectra of the Ti-containing silicates (Figure 14.2), the Ti K-edge XANES spectra of the Ti-containing magadiites (Si/Ti = 100 and 50) exhibited a weak pre-edge peak (Figure 14.2B(a),(b)), indicating that the Ti-oxo species existed mainly in tetrahedral coordination in the framework of magadiite. The XANES spectrum of the Ti-containing octosilicate (Si/Ti = 200) exhibited a weak but single pre-edge peak (Figure 14.2A(a)), indicating that a part of the Ti-oxo species was in tetrahedral coordination in the framework of tectosilicate. As the content of Ti increased (Si/Ti = 100 and 50), XANES spectra (Figure 14.2A(b) and A(c)) showed spectra similar to those of rutile and anatase (Figure 14.2A(e) and A(f)), indicating the presence of an isolated tetrahedrally coordinated Ti-oxo species in the framework of octosilicate together with extraframework octahedrally coordinated Ti species. These results suggest that the capacity of Ti incorporation is limited up to Si/Ti = 50 and 100 for magadiite and octosilicate, respectively. Further understanding of the arrangement of silanol groups in layered silicates may enable us to design dense immobilization of single-site metal-oxo species in/onto the silicate sheet of layered silicates.

The film has an ideal morphology for many applications and quantitative characterization. The transparent self-standing films of mesoporous silicas were synthesized by the solvent evaporation method from tetramethoxysilane and alkyltrimethylammonium salts [20]. Taking advantage of this simple process, Ti-containing mesoporous silica films (Si/Ti = 50) were successfully synthesized from tetramethoxysilane, vinyltrimethoxysilane, TPOT, and octadecyltrimethyl ammonium salts [17]. After the surfactant removal by the calcination at 550 °C, macroscopic morphology and highly ordered mesostructures were retained. In the mesostructures of the Ti-containing mesoporous silica films, periodic pore arrangements with hexagonal and cubic symmetries were controlled by the composition of the starting mixtures (water/Si ratio = 1/3 and 4/3). The Ti-oxo species in the silica frameworks of these films existed as an isolated tetrahedrally coordinated species, which was confirmed by UV–Vis absorption, photoluminescence, and XAFS measurements. Furthermore, these films exhibited high photocatalytic activity for the reduction of CO_2 with H_2O under UV irradiation, which led to the evolution of CH_4 and CH_3OH. The relationships between the structures and morphology of mesoporous silicas and photocatalytic activity (yields of products and selectivity) will be introduced in detail in Section 14.3.1.

14.2.3 Molecular Cluster of Ti Single-Site in Silica-Based Materials

As mentioned in Section 14.2.1, single-site Ti-oxo species in nanospace materials absorb only UV wavelengths (220–280 nm), so materials design is required to efficiently operate under both UV and visible irradiation. As one of the methods for visible light-responsive photocatalysts' design, the heteroelement (V, Cr, Fe, etc.) doping into porous silica/silicate frameworks containing single-site Ti-oxo species has been reported [21]. The incorporation of V ions into the frameworks of mesoporous silicas (Ti/HMS (hexagonal mesoporous silica) and Ti/MCM-41) containing single-site Ti-oxo species has been reported [21]. These materials exhibited useful catalytic abilities for the decomposition of NO into N_2 and O_2 under visible light irradiation ($\lambda > 420$ nm). In the diffuse reflectance UV–Vis absorption spectra of Ti/HMS and Ti/MCM-41 (Figure 14.3), absorption bands were observed at around 200–260 nm, which were assigned to the absorption of charge transfer excited state ($[Ti^{4+}–O^{2-}] \rightarrow [Ti^{3+}–O^{-}]^*$) of the isolated tetrahedrally coordinated Ti-oxo species. On the other hand, the incorporation of V ions into the framework of Ti/HMS and Ti/MCM-41 led to a large shift in their absorption spectra toward visible light regions, depending on the amount of the incorporated amount of V ions (0–2.0 μmol/g-cat). These results indicated that the interactions between the incorporated V ions and tetrahedrally coordinated Ti-oxo species affected the electronic transition of single-site Ti-oxo species in mesoporous silicas, resulting in the absorption shift. Mesoporous silica (SBA-15) with single-site Ti-oxo species and Fe oxide nanoparticles immobilized on the pore surface was found to have high activity for the partial oxidation of cyclohexane using O_2 under simulated solar light irradiation [22]. This material was synthesized by the post-synthetic incorporation by the reaction of pre-synthesized SBA-15 with Ti(IV) acetylacetonate and Fe(III) acetylacetonate successively (as schematically in Figure 14.4). Spectroscopic studies suggested that the immobilized tetrahedrally coordinated Ti-oxo species were

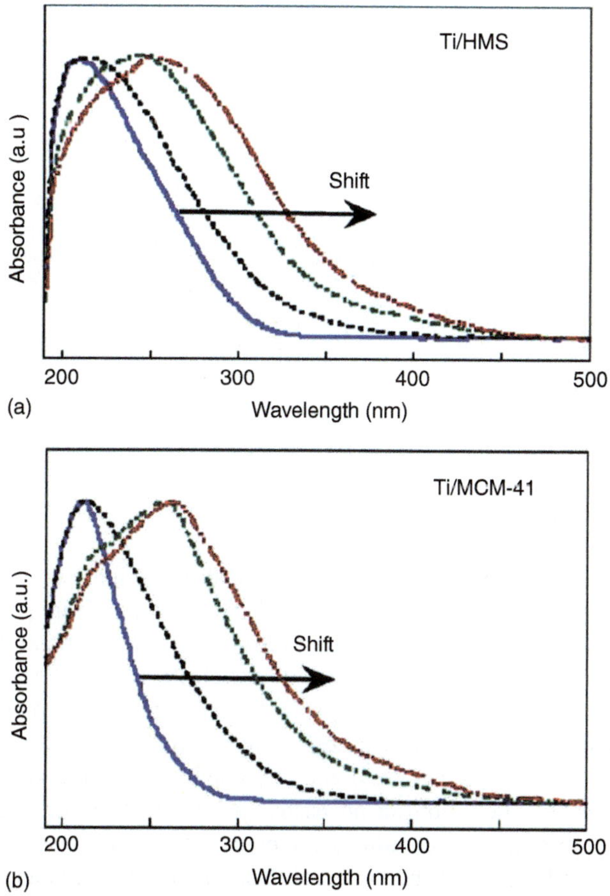

Figure 14.3 The diffuse reflectance UV–Vis absorption spectra of V ions incorporated (a) Ti/HMS and (b) Ti/MCM-41. The amount of incorporation with V ions (μmol/g-cat) are 0 (blue), 0.66 (black), 1.3 (green), and 2.0 (red), respectively. Source: Reproduced with permission from Anpo and Takeuchi [21]/Elsevier.

connected with Fe oxide nanoparticles via Ti—O—Fe bonds, resulting in the lengthened lifetime of the active Ti species by the electron delocalization, and visible light-response ($\lambda > 420$ nm) induced by Fe oxide nanoparticles coupled to the Ti-oxo species. Due to the formation of a molecular cluster, an induced new electronic transition, and electron delocalization via heterojunction, the incorporation/immobilization of multiple heteroelements in/on silica-based materials with a single Ti site is a material design for visible light-responsive photocatalysts (Ti—O—M bonds).

14.2.4 Other Ti-Containing Nanospace Materials

In addition to the isolated tetrahedrally coordinated Ti-oxo species, Ti-oxo clusters in MOFs or PCPs are regarded as single-site photocatalysts [23]. MOFs/PCPs are

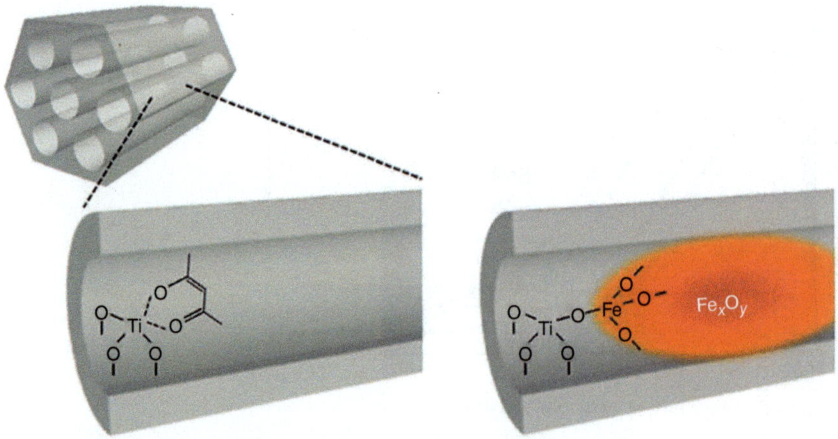

Figure 14.4 Schematic structures of $Ti_{0.2}$–SBA and $Fe_{0.2}Ti_{0.02}$–SBA. Source: Reproduced with permission from Ide et al. [22]/The Royal Society of Chemistry.

crystalline microporous materials consisting of metal and metal cluster ions (nodes) and organic ligands (linkers) by coordination bonds and have been applied for gas storage/separation, (photo)catalysts, and so on, taking advantage of their structural design freedom [9, 23]. The most common example is Ti-oxo clusters, $Ti_8O_8(OH)_4$, which have been used as metal nodes in MOFs. For the following purposes, MOF photocatalysts have been prepared: (i) *in situ* synthesis using polynuclear metal-oxo clusters/open metal sites (catalytic active sites) and organometallic ligands (e.g. metalloporphyrin), (ii) post-synthetic ion/ligand exchange, and (iii) grafting of active metal species on the pore surfaces coordinated to metal ions/organic ligands. MIL-125(Ti) ($[Ti_8O_8(OH)_4(bdc)_6]$, where bdc = terephthalate) has been used as a photocatalyst for H_2 production from water, CO_2 reduction, and so on [24]. Under visible light irradiation ($\lambda > 420$ nm), NH_2-MIL-125(Ti) with 2-aminoterephthalate as the organic linker proved effective for the photocatalytic conversion of CO_2 to formate ($HCOO^-$). The amine groups were explained to play a key role in (i) visible light absorption at around 550 nm attributed to ligand-to-metal charge transfer (LMCT, Figure 14.5A), and (ii) affinity to promote CO_2 adsorption. A photoexcited electron was transferred to the titanium oxo-clusters upon visible light irradiation, resulting in the reduction of Ti^{4+} and the formation of Ti^{3+}. The mechanism of the photocatalytic CO_2 reduction is proposed as CO_2 adsorption by NH_2-MIL-125(Ti) and the subsequent reduction to $HCOO^-$ at the Ti^{3+} site in the presence of triethanolamine (TEOA) as the electron donor (Figure 14.5B). The activity is still low, and the mechanism of the photocatalytic reactions on MOF photocatalysts has not been sufficiently clarified yet. Through materials design with systematic structural variation, it is worthwhile to investigate the relationship between the state/amount of catalytic active sites in/on MOFs and photocatalytic activity.

Graphenes are 2D materials composed of a single layer of sp^2-hybridized carbons arranged in a hexagonal lattice and are used as catalyst supports for single atom,

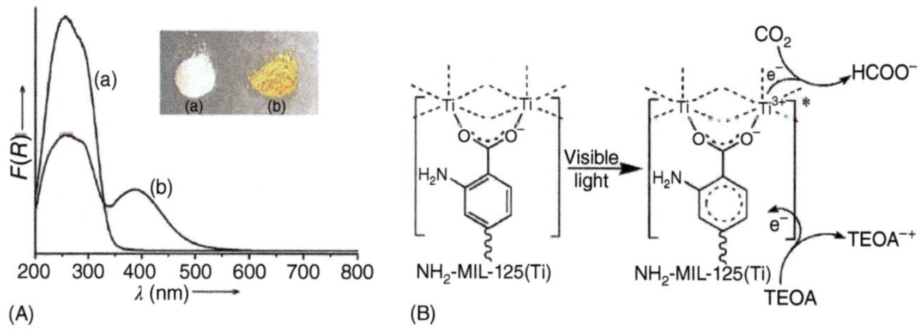

Figure 14.5 (A) The diffuse reflectance UV–Vis absorption spectra of (a) MIL-125(Ti) and (b) NH_2-MIL-125(Ti). The inset shows each sample. (B) Proposed reaction mechanism for photocatalytic CO_2 reduction to formate using NH_2-MIL-125(Ti). Source: Reproduced with permission from Fu et al. [24]/John Wiley & Sons.

nanoparticles, and clusters [8]. Fixation of single-site Ti catalysts (Ti coverage: 4.0×10^{13} sites/cm^2) on graphene was done to exhibit hydrogen spillover (H_2 molecules into H atoms) and then transfer them to the layer of graphene, resulting in the formation of C—H bonds on graphene as a way of hydrogen storage [25]. As the content of Ti increased (Ti coverage: above 1.3×10^{14} sites/cm^2), the Ti atoms aggregated to form cluster catalysts, which presented a larger hydrogen storage capacity (1.11 wt% at 1 mbar and room temperature) than the previously reported hydrogenation by graphene. Accordingly, the state of single-site metal species on graphene changes depending on the loading of the metal source, resulting in the difficulty of precise control of the coordination structure. The amount of doped metal in the graphene lattice is limited.

14.3 Applications

14.3.1 Photocatalytic Reduction of CO_2 with H_2O

Photocatalytic reduction of CO_2 with H_2O under UV irradiation is known as a unique property of single-site Ti-oxo species in silica-based nanospace materials [5–7]. In photocatalytic reduction, CH_4 and CH_3OH are the main products, and the photocatalytic activity (yields of products and selectivity) varies depending on the structures, pore size/surface properties, and morphology of the nanospace materials [17, 26–28]. Two types of Ti-β zeolites (abbreviated as Ti-β(OH) and Ti-β(F)), which were synthesized using OH^- and F^- as the structure-directing agents (SDAs) were examined [26]. Ti-β(OH) with hydrophilic properties showed higher photocatalytic activity than Ti-β(F), which was attributed to the easy access of H_2O molecules to tetrahedrally coordinated Ti-oxo species and the formation of the charge-transfer excited state ($[Ti^{4+}{-}O^{2-}] \rightarrow [Ti^{3+}{-}O^{-}]^*$) upon UV irradiation (Figure 14.6). On the other hand, Ti-β(F) with hydrophobic properties showed high selectivity for the formation of CH_3OH, indicating the suitable reactive field for the formation of CH_3OH due to the lower concentration of H_2O than Ti-β(OH).

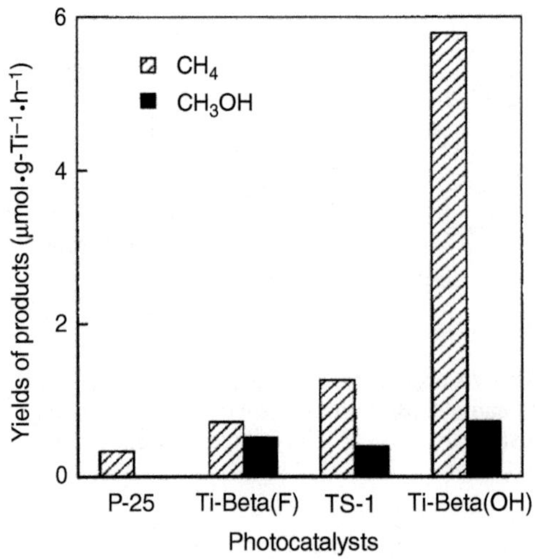

Figure 14.6 Yields of CH_4 and CH_3OH in the photocatalytic reduction of CO_2 with H_2O at 323 K on various photocatalysts. Reaction time is six hour and intensity of light is 265 μW cm^{-2}. Source: Reproduced with permission from Ikeue et al. [26]/American Chemical Society.

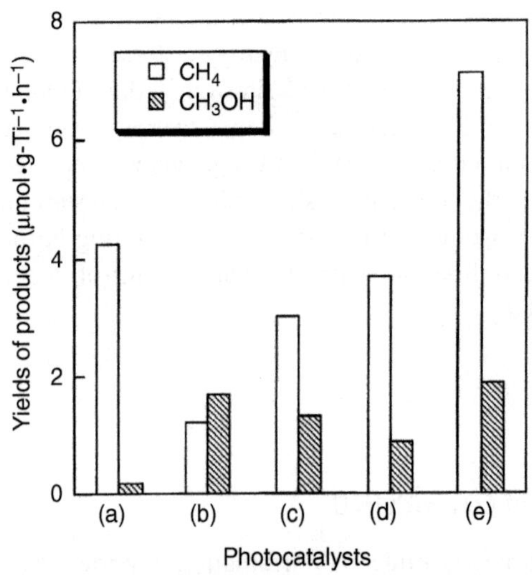

Figure 14.7 Yields of CH_4 and CH_3OH in the photocatalytic reduction of CO_2 with H_2O on various Ti-containing mesoporous silica films (Ti-MPS) and the powdered Ti-MCM-41; (a) Ti-MPS (hexagonal structure, Si/Ti = 25), (b) Ti-MPS (cubic structure, Si/Ti = 50), (c) Ti-MCM-41, (d) the powdered form of Ti-MPS (hexagonal structure, Si/Ti = 50) and (e) Ti-MPS (hexagonal structure, Si/Ti = 50). The condition of the catalytic reaction is the same in Figure 14.6. Source: Reproduced with permission from Ikeue et al. [27]/Elsevier.

In the photocatalytic reduction of CO_2 with H_2O, the Ti-containing mesoporous silica film (Si/Ti = 50) with a hexagonal structure exhibited higher activity than the powdered samples even with the same pore structure and Ti content (Figure 14.7(c)–(e)) [17, 27, 28]. Furthermore, the Ti-containing mesoporous silica film with a cubic structure showed lower photocatalytic activity for CH_4 and CH_3OH than that with a hexagonal structure even with the same Ti content (Figure 14.7(b)). Tetrahedrally coordinated Ti-oxo species and an aggregated octahedrally coordinated Ti-oxo species existed in the film with hexagonal structure as the Ti content increased (Si/Ti = 25) (Figure 14.7(a)), which showed the low photocatalytic activity. These Ti-containing mesoporous silica films had different concentrations of surface OH groups and the difference in the H_2O adsorption properties on the pore surface. Accordingly, H_2O adsorption properties attributed to the hydrophilicity/hydrophobicity of this single-site Ti-oxo species photocatalysts can be controlled by the activity and selectivity for the photocatalytic CO_2 reduction. The effect of light scattering by the particle surface is not negligible for powder samples, so effective light absorption and measurement are not straightforward. As a result of the advantages of transparent Ti-containing porous silica film photocatalysts for the quantitative evaluation, the quantum yield of the photocatalytic reaction was successfully and accurately evaluated [27, 28].

14.3.2 Other Application

Ti-containing mesoporous silica thin film (MSTF) shows not only photocatalytic activity but also photoinduced surface superhydrophilicity. MSTFs containing single-site photocatalysts (Ti, V, Cr, Mo, and W-oxo species), which were simply prepared on quartz plates using a sol–gel process and spin-coating method, exhibited strong hydrophilic surface properties compared to the MSTF without transition metal oxo species even before UV-light irradiation [29]. After UV-light irradiation, these thin films showed photoinduced superhydrophilic properties. Among them, the W-MSTF showed the highest hydrophilicity, and the water contact angle on that was 3° and 1° before and after UV-light irradiation. Furthermore, Ti-MSTF

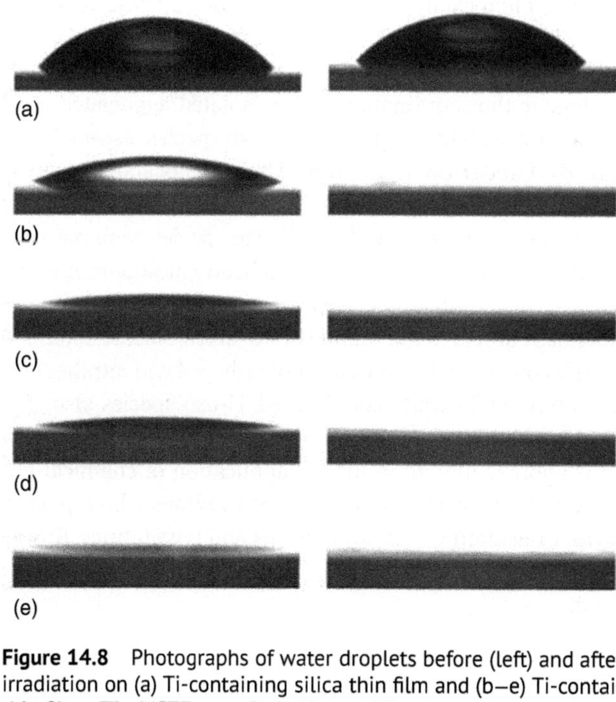

Figure 14.8 Photographs of water droplets before (left) and after (right) UV-light irradiation on (a) Ti-containing silica thin film and (b–e) Ti-containing mesoporous silica thin films (Ti-*n*MSTFs, $n = 2, 4, 10$, and 20) using various structure-directing agents (SDAs), (b) $C_{18}H_{35}(OCH_2CH_2)_2OH$ ($n = 2$), (c) $C_{12}H_{25}(OCH_2CH_2)_4OH$ ($n = 4$), (d) $C_{16}H_{33}(OCH_2CH_2)_{10}OH$ ($n = 10$), and (e) $C_{18}H_{37}(OCH_2CH_2)_{20}OH$ ($n = 20$). Source: Reproduced with permission from Horiuchi et al. [30]/American Chemical Society.

prepared by the same methods using various SDAs with different chain lengths has been investigated for their surface wetting properties [30]. The hydrophilic properties of Ti-*n*MSTFs changed depending on the length of hydrophilic domains of SDAs (R–$(OCH_2CH_2)_nOH$; $n = 2$, 4, 10, and 20 (Figure 14.8(b)–(e)). Among the Ti-*n*MSTFs series, Ti-20MSTF showed the highest hydrophilicity, and the water contact angle was 3° even before the UV irradiation (Figure 14.8(e)), SDA: $C_{18}H_{37}(OCH_2CH_2)_{20}OH$). On the contrary, nonporous silica thin films containing Ti-oxo species showed lower hydrophilicity than the Ti-*n*MSTFs series (Figure 14.8(a)). These results indicate that the longer hydrophilic domain of SDAs played a key role in the formation of silanol groups in the silica framework. In addition, the Ti-*n*MSTFs series exhibited photocatalytic activity for the degradation of methylene blue. These photoinduced surface superhydrophilic and associating self-cleaning properties are useful for coating ceramic tiles, glass, and so on.

14.4 Conclusions and Future Perspectives

Preparation and characterization (the local structures of Ti-oxo species were characterized by several key characterizations including UV–Vis absorption and XAFS spectroscopies) of single-site Ti-oxo species in nanospace materials (1D, 2D, and 3D structures) and their photocatalytic properties (photocatalytic reduction of CO_2 with H_2O, superhydrophilic surfaces, and self-cleaning properties) were summarized. The photocatalytic activity of the single-site Ti-oxo species is greatly influenced by a subtle change in the coordination states (isolated/aggregated tetrahedral and octahedral) and tetrahedrally coordinated Ti-oxo species, especially for the reduction of CO_2 with H_2O under UV irradiation. Their photocatalytic activity and selectivity for CH_4 and CH_3OH varied depending on the pore size/surface properties where the Ti-oxo species were fixed. Such Ti-oxo species were prepared by the isomorphous substitution of Si atoms in silica and silicate (mesoporous silicas and zeolites) with Ti atoms. To maintain the isolated tetrahedrally coordinated Ti-oxo species, the number of Ti atoms in the framework of silica/silicate should be as low as $Ti/Si > 1/50$. As the content of Ti increases, not only isolated tetrahedrally coordinated Ti-oxo species but octahedrally coordinated Ti-oxo species also exist in/on silica and silicate. In addition to the low population of the active site, a narrow absorption band is another limitation for the application of chemical fuel production. On the other hand, photocatalyst film has possibilities where powder samples do not have access. Considering these limitations and possibilities, further study on the preparation and modification of Ti-oxo species in various nanospace materials will be examined.

References

1 Ariga, K., Vinu, A., Yamauchi, Y. et al. (2012). Nanoarchitectonics for mesoporous materials. *Bull. Chem. Soc. Jpn.* 85: 1–32.

2 Chen, X. and Mao, S.S. (2007). Titanium dioxide nanomaterials: synthesis, properties, modifications, and applications. *Chem. Rev.* 107: 2891–2959.

3 Park, S., Keum, Y., and Park, J. (2022). Ti-based porous materials for reactive oxygen species-mediated photocatalytic reactions. *Chem. Commun.* 58: 607–618.

4 Vibulyaseak, K.G., Deepracha, S.B., and Ogawa, M. (2019). Immobilization of titanium dioxide in mesoporous silicas: structural design and characterization. *J. Solid State Chem.* 270: 162–172.

5 Matsuoka, M. and Anpo, M. (2003). Local structures, excited states, and photocatalytic reactivities of highly dispersed catalysts constructed within zeolites. *J. Photochem. Photobiol., C* 3: 225–252.

6 Anpo, M. (2013). Photocatalytic reduction of CO_2 with H_2O on highly dispersed Ti-oxide catalysts as a model of artificial photosynthesis. *J. CO_2 Util.* 1: 8–17.

7 Yamashita, H., Mori, K., Kuwahara, Y. et al. (2018). Single-site and nano-confined photocatalysts designed in porous materials for environmental uses and solar fuels. *Chem. Soc. Rev.* 47: 8072–8096.

8 Fei, H., Dong, J., Chen, D. et al. (2019). Single atom electrocatalysts supported on graphene or graphene-like carbons. *Chem. Soc. Rev.* 48: 5207–5241.

9 Kitagawa, S., Kitaura, R., and Noro, S. (2004). Functional porous coordination polymers. *Angew. Chem. Int. Ed.* 43: 2334–2375.

10 Blasco, T., Corma, A., Navarro, M.T., and Pariente, J.P. (1995). Synthesis, characterization, and catalytic activity of Ti-MCM-41 structures. *J. Catal.* 156: 65–74.

11 Yamashita, H. and Anpo, M. (2003). Local structures and photocatalytic reactivities of the titanium oxide and chromium oxide species incorporated within micro- and mesoporous zeolite materials: XAFS and photoluminescence studies. *Curr. Opin. Solid State Mater. Sci.* 7: 471–481.

12 Notari, B. (1991). Titanium silicalite: a new selective oxidation catalyst. *Stud. Surf. Sci. Catal.* 60: 343–352.

13 Tanev, P.T., Chibwe, M., and Pinnavaia, T.J. (1994). Titanium-containing mesoporous molecular sieves for catalytic oxidation of aromatic compounds. *Nature* 368: 321–323.

14 Koyano, K. and Tatsumi, T. (1996). Synthesis of titanium-containing mesoporous molecular sieves with a cubic structure. *Chem. Commun.* 145–146.

15 Tsunoji, N., Ide, Y., Yagenji, Y. et al. (2014). Design of layered silicate by grafting with metal acetylacetonate for high activity and chemoselectivity in photooxidation of cyclohexane. *ACS Appl. Mater. Interfaces* 6: 4616–4621.

16 Morita, M., Horiuchi, Y., Matsuoka, M., and Ogawa, M. (2022). Preparation of titanium-containing layered alkali silicates. *Cryst. Growth Des.* 22: 1638–1644.

17 Ogawa, M., Ikeue, K., and Anpo, M. (2001). Transparent self-standing films of titanium-containing nanoporous silica. *Chem. Mater.* 13: 2900–2904.

18 Perego, C., Carati, A., Ingallina, P. et al. (2001). Production of titanium containing molecular sieves and their application in catalysis. *Appl. Catal., A* 221: 63–72.

19 Ogawa, M. and Kuroda, K. (1995). Photofunctions of intercalation compounds. *Chem. Rev.* 95: 399–438.

20 Ogawa, M. (1996). A simple sol-gel route for the preparation of silica-surfactant mesostructured materials. *Chem. Commun.* 1149–1150.

21 Anpo, M. and Takeuchi, M. (2003). The design and development of highly reactive titanium oxide photocatalysts operating under visible light irradiation. *J. Catal.* 216: 505–516.

22 Ide, Y., Iwata, M., Yagenji, Y. et al. (2016). Fe oxide nanoparticles/Ti-modified mesoporous silica as a photo-catalyst for efficient and selective cyclohexane conversion with O_2 and solar light. *J. Mater. Chem. A* 4: 15829–15835.

23 Wen, M., Mori, K., Kuwahara, Y. et al. (2018). Design of single-site photocatalysts by using metal–organic frameworks as a matrix. *Chem. Asian J.* 13: 1767–1779.

24 Fu, Y., Sun, D., Chen, Y. et al. (2012). An amine-functionalized titanium metal–organic framework photocatalyst with visible-light-induced activity for CO_2 reduction. *Angew. Chem. Int. Ed.* 51: 3364–3367.

25 Chen, J.W., Hsieh, S.H., Wong, S.S. et al. (2022). Hydrogen spillover and storage on graphene with single-site Ti catalysts. *ACS Energy Lett.* 7: 2297–2303.

26 Ikeue, K., Yamashita, H., Anpo, M., and Takewaki, T. (2001). Photocatalytic reduction of CO_2 with H_2O on Ti-β zeolite photocatalysts: effect of the hydrophobic and hydrophilic properties. *J. Phys. Chem. B* 105: 8350–8355.

27 Ikeue, K., Nozaki, S., Ogawa, M., and Anpo, M. (2002). Characterization of self-standing Ti-containing porous silica thin films and their reactivity for the photocatalytic reduction of CO_2 with H_2O. *Catal. Today* 74: 241–248.

28 Shioya, Y., Ikeue, K., Ogawa, M., and Anpo, M. (2003). Synthesis of transparent Ti-containing mesoporous silica thin film materials and their unique photocatalytic activity for the reduction of CO_2 with H_2O. *Appl. Catal., A* 254: 251–259.

29 Horiuchi, Y., Mori, K., Nishiyama, N., and Yamashita, H. (2008). Preparation of superhydrophilic mesoporous silica thin films containing single-site photocatalyst (Ti, V, Cr, Mo, and W oxide moieties). *Chem. Lett.* 37: 748–749.

30 Horiuchi, Y., Ura, H., Kamegawa, T. et al. (2011). Controlled synthesis and surface hydrophilic properties of Ti-containing mesoporous silica thin films using various structure-directing agents. *J. Phys. Chem. C* 115: 15410–15415.

15

Surface Coordination Improved Photocatalytic Fixation of CO_2 over 2D Oxide Nanosheets

Zhiwen Wang and Yujie Song

Hainan Provincial Key Laboratory of Fine Chemicals, Hainan University, No. 58 Renmin Avenue, Haikou, Hainan 570228, P.R. China

15.1 Introduction

Fossil fuel consumption and CO_2 emissions have caused a great crisis in climate and energy [1–3]. There is an increased demand for innovative technology to reuse and recycle CO_2. Photocatalysis as a green and sustainable technology can utilize solar energy to produce clean fuels, which provides a promising approach for CO_2 conversion [4–6]. Since the breakthrough in photocatalytic CO_2 reduction over the GaP photocatalyst [7], enormous endeavors have been made to explore and improve the photocatalysts for promoting the conversion of CO_2, such as inorganic semiconductors [8], C_3N_4 [9], metal-organic frameworks (MOFs) [10], and covalent-organic frameworks (COFs) [11]. However, CO_2 molecules with high dissociation energy of 750 kJ mol^{-1} have a super stable thermodynamic structure, leading to a high energy barrier for conversion [12, 13]. Moreover, the reduction of CO_2 would undergo a complex multielectron process, forming multiple products, such as CO, HCOOH, CH_3OH, CH_4, and some C_2 products (C_2H_4, C_2H_6, and CH_3CH_2OH) [14]. Therefore, despite obvious progress, it has been a great challenge to achieve highly efficient conversion of CO_2 to a single product, depending on an efficient photocatalyst.

It is reported that many factors can influence the performance of CO_2 reduction on a photocatalyst, such as band structure, separation efficiency of electrons holes, light irradiation, reaction temperature, and time [15, 16]. Most researchers focus on regulating these factors to improve photocatalytic activity and selectivity. The reactants would, however, go through an adsorption and activation process on the catalyst's surface in heterogeneous catalysis, which is one of the crucial paths for defining the catalytic activity and selectivity[17]. Therefore, the capture and activation of CO_2 on photocatalysts' surfaces can be considered one of the crucial factors for the photocatalytic conversion of CO_2. Generally, CO_2 molecules chemisorb on the surface via a coordination bond. The coordination between CO_2 and a catalyst would lower the overall activation energy. The formed coordination bond would act as a bridge to transform photoelectrons into CO_2. In particular,

UV-Visible Photocatalysis for Clean Energy Production and Pollution Remediation: Materials, Reaction Mechanisms, and Applications, First Edition. Edited by Xinchen Wang, Masakazu Anpo, and Xianzhi Fu.
© 2023 WILEY-VCH GmbH. Published 2023 by WILEY-VCH GmbH.

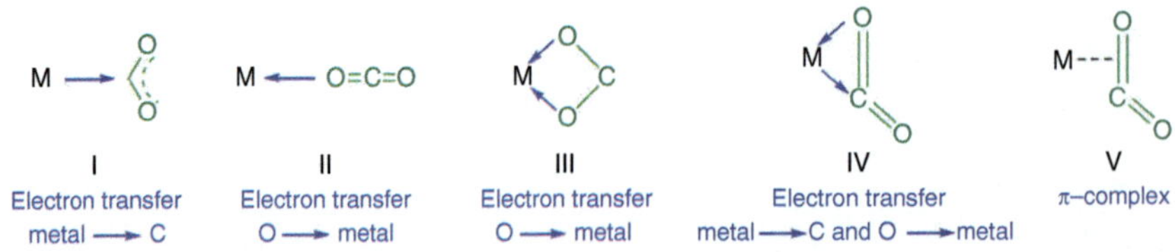

Figure 15.1 Coordination modes of CO_2 with transition metals. Source: Sun et al. [18]/Springer Nature.

the different coordination modes may induce the reaction to proceed in different directions, further controlling the selectivity of products.

As shown in Figure 15.1, there are five possible coordination modes of CO_2 with transition metal sites on the photocatalysts [18–20]. The C atoms of CO_2 are an electrophilic Lewis acid center that is likely to coordinate with an electron-rich metal via electron transfer from metal (M) to C (mode I). O atoms with the lone pair serve as nucleophilic Lewis bases to bond to M via electron transfer from O to M (mode I). In complex III, CO_2 coordinates via two M—O bonds. It means that the electron-deficient metal sites are favored for the formation of this mode via the electron transfer from O to M. Mode IV is generated by the combination of the two different electron transfers. A π-complex V is generated via transferring the π-electron of C=O to M. Moreover, the coordination modes of CO_2 are not just limited to these five. Based on the coordination modes of CO_2, precisely designing the catalyst with special active sites may promote the directional conversion of CO_2.

Ultrathin 2D nanosheets (NSTs) as photocatalysts have widely attracted attention due to their larger surface area, faster electron–hole separation, and more exposed surface sites. In the past decade, multiple 2D photocatalysts have been developed, such as metal oxides, layered double hydroxides, transition metal dichalcogenides, and graphitic carbon nitride [21–23]. In addition, MOFs and COFs with a nanosheet morphology are also successfully constructed as 2D photocatalysts [24, 25]. These 2D photocatalysts are employed for many catalytic reactions, such as the degradation of pollutants [26], fixation of N_2 [27], production of H_2 and O_2 from water [28], organic synthesis, and reduction of CO_2 [29]. 2D transition metal oxide nanosheets provide abundantly exposed transition metal atoms (M), which are beneficial for coordinating CO_2 molecules. In particular, M atoms are flexibly regulated to meet different coordination modes, thus tailoring the product's selectivity. Therefore, 2D transition metal oxide nanosheets may be an ideal material for exploring the role of CO_2 coordination in the photocatalytic process. Moreover, revealing the coordination between CO_2 molecules and the catalysts promotes the understanding of a pathway of photocatalytic CO_2 reduction at the molecular scale.

In the present chapter, how to precisely construct a photocatalyst that achieves a directional and high-efficiency conversion of CO_2 by the coordination of CO_2 is discussed. Moreover, the mechanism of the interface–surface interactions between the 2D metal oxide nanosheets and CO_2 molecules is also explored. Combining with our previous work, it is revealed that surface coordination greatly enhances the activity and tailors the product selectivity for photocatalytic reduction of CO_2.

15.2 Design of the Catalyst

How to create an efficient catalyst is a key issue for the directional conversion of CO_2. The different coordination modes would result in the generation of different products. The coordination modes are closely related to the properties of active sites. Based on the theoretical modes, the required sites, and further construction, the catalyst with the special active sites is predicted. Taking CH_4 and CO as samples, it is necessary to understand which mode benefits the formation of CH_4 or CO. Even though CO_2 activation has been extensively researched, precise coordination-based photocatalyst construction is still only sometimes documented [30, 31].Xie et al. reported that the metal oxide or sulfide photocatalysts with single-metal sites tend to coordinate CO_2 molecules via a weakly bonded M–C or M–O, which facilitates the formation of CO (Figure 15.2) [32]. However, if the catalyst possesses dual-metal sites, it would promote the generation of a highly stable configuration of M–C–O–M mode through the strong hybridization between the 2p orbitals of the C or O atom and the 3d orbitals of the M atom. This intermediate is beneficial to the further protonation of the C atoms, thus facilitating the production of CH_4. Based on

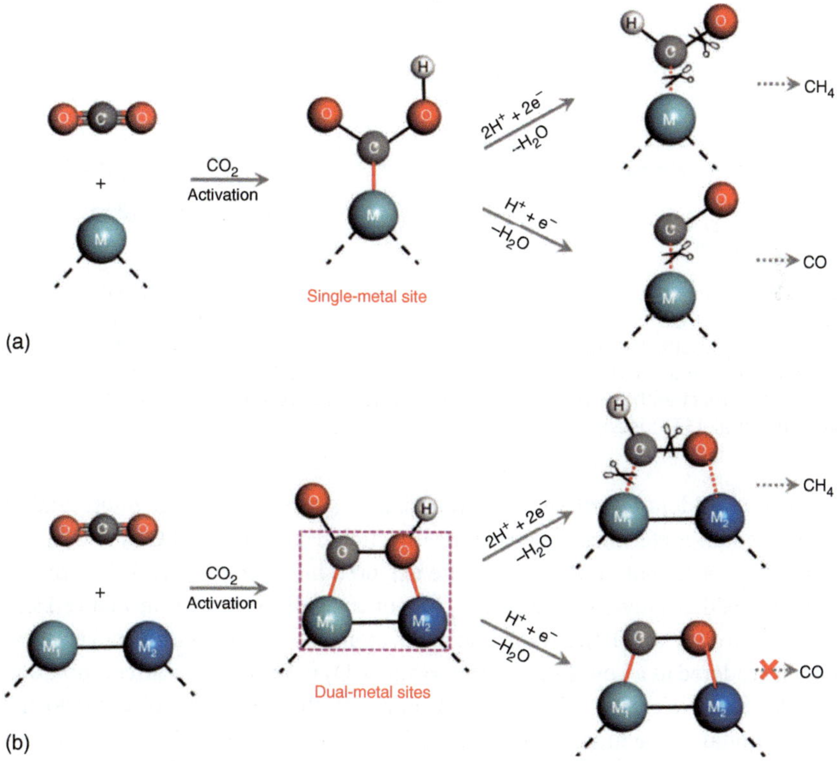

Figure 15.2 Manipulating reactivity and selectivity by modulating the reaction pathways. CO_2 photoreduction into fuels such as CH_4 and CO using single-metal site (a) and dual-metal site (b) catalytic systems (M represents the metal site). Source: Reproduced with permission from Li et al. [32]/Springer Nature.

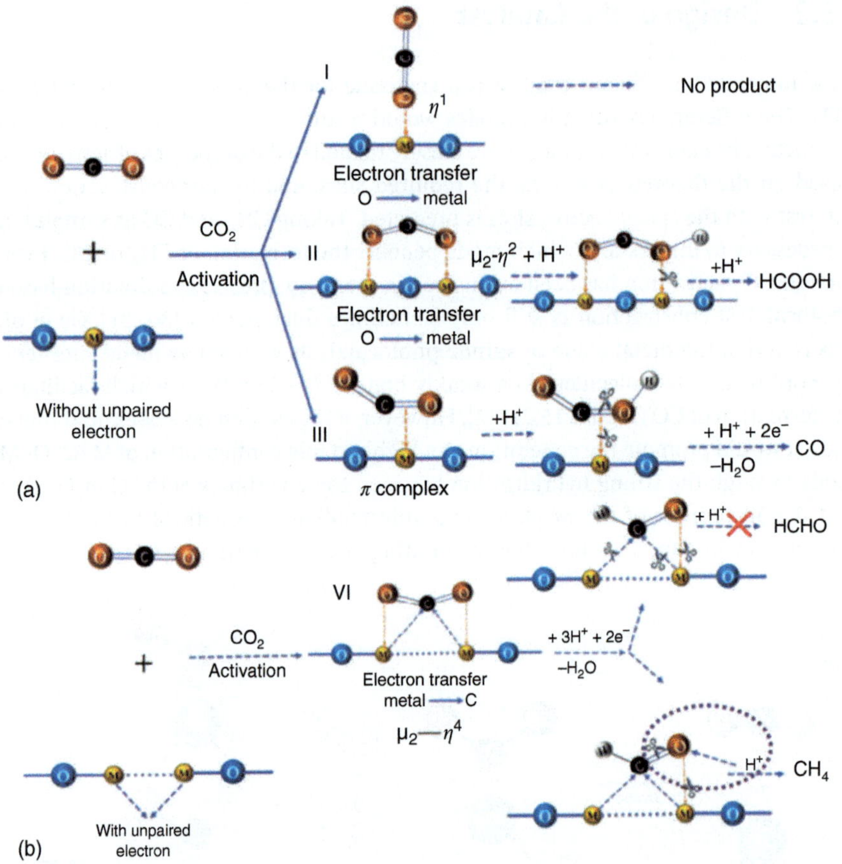

Figure 15.3 Manipulating reactivity and selectivity by modulating the coordination mode. Photoreduction of CO_2 on the catalyst: (a) Without unpaired electron and (b) With unpaired electron (M represents the metal site, "$H^+ + e^-$" refers to the proton-coupled electron transfer process, and "$-H_2O$" means the desorption of H_2O molecules after the intermediates react with the proton-electron pair). Source: Reproduced with permission from Wang et al. [33]/Elsevier.

the above consideration of CO_2 coordination, a single-component photocatalyst containing dual-metal sites may meet the directional conversion of CO_2 to CH_4.

Four CO_2 coordination possibilities are explored in our earlier work in a transition metal oxide catalyst with a single component [33]. As shown in Figure 15.3a (intermediate-I, II, and III), a transition metal atom M without an unpaired electron is only considered as a Lewis acid center to bind CO_2 molecules via the coordination of the lone pair electron of the oxygen atom (or the π-electron of C–O) to the empty d orbital of the metal. For route I, a low-hapticity (η^1) mode is formed via a weak interaction M—O bond, inferring that the CO_2 molecule is hardly activated. Intermediate-II is a μ_2-η^1 coordination mode with two M···O—C bonds, facilitating the formation of HCOOH. The intermediate III (π-coordination) is formed via transferring the π-electron to an empty d orbital, dominating CO_2 into CO by one activated C—O bond dissociation. In this case (Figure 15.3b), both the C and O atoms of the CO_2 molecule would be concurrently bonded with two metal sites to form the higher hapticity (μ_2-η^4) coordination mode (M$\cdots CO_2^- \cdots$M species), which is beneficial for forming the M$\cdots$CHO$\cdots$M intermediate by protonation. Considering the strong hybridization between the 2p orbital of the C or O atom and the d orbital of the M atom, the M—C, and M—O bonds are more stable than the activated C=O bond in the M$\cdots$CHO$\cdots$M intermediate. Therefore, the C=O bond is easy to dissociate via protonation for C—H bond generation, resulting in CH_4 formation by successive protonation. In other words, HCHO is more difficult to produce by simultaneously breaking the M—C and M—O bonds because it needs much more energy. According to the above discussion, it can be concluded that creating a suitable catalyst containing abundant surface M atoms with an unpaired electron would induce the formation of CH_4.

Another task is to convert CO_2 to CO over a photocatalyst. A key issue is to create another model which determines the generation of CO [34]. A material with two metal sites (M1 and M2) is created as the target photocatalyst. Based on Lewis's theory of acids (LA) and bases (LB), four combination patterns for M1 and M2 may appear on the catalyst: I. M1 and M2 (both of LA); II. M1 and M2 (both of LB); III. M1 and M2 (both of LA and LB); IV. M1 (LA) and M2 (LB). In this case, Figure 15.4

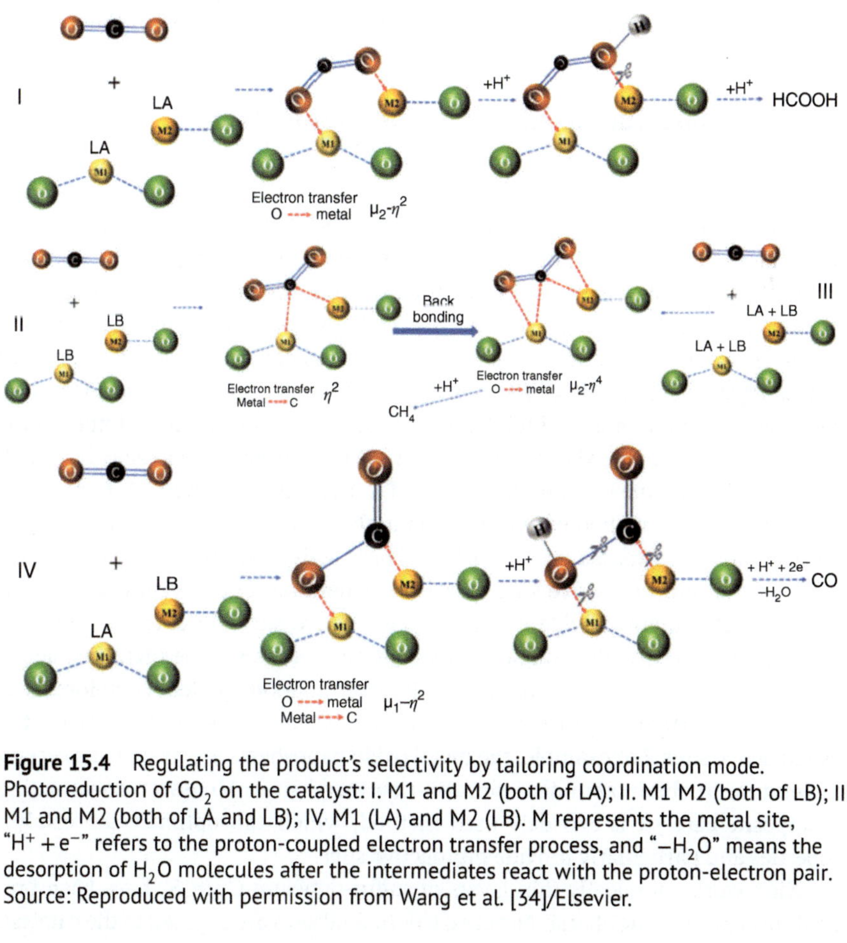

Figure 15.4 Regulating the product's selectivity by tailoring coordination mode. Photoreduction of CO_2 on the catalyst: I. M1 and M2 (both of LA); II. M1 M2 (both of LB); III. M1 and M2 (both of LA and LB); IV. M1 (LA) and M2 (LB). M represents the metal site, "$H^+ + e^-$" refers to the proton-coupled electron transfer process, and "$-H_2O$" means the desorption of H_2O molecules after the intermediates react with the proton-electron pair. Source: Reproduced with permission from Wang et al. [34]/Elsevier.

shows four possible coordination modes. For route I, both M1 and M2 serve as Lewi acid centers to bond with O via the coordination of the lone pair electron of an oxygen atom to empty the d orbital of the metal. Intermediate-I is a μ_2-η^2 coordination mode with two O → M bonds via transferring the lone pair electron of an oxygen atom to the empty d orbital of the metal, facilitating the formation of HCOOH. Route II exhibits that both M1 and M2 are considered Lewi base centers to bind with C through transferring unpaired electron coordination to the lowest-unoccupied molecular orbital (2πu LUMO) of CO_2. But this mode is unreasonable due to the formation of back bonding, which would be translated into a higher hapticity (μ_2-η^4) mode (Intermediate-III). Intermediate-III is a bridging M1···CO_2^-···M2 species, which promotes CH_4 production as shown in our previous work. Moreover, path IV (intermediate-IV) depicts that M1 and M2 play Lewi base and Lewi base, respectively, which bond CO_2 via forming the M1→C and O→M2 bonds. In this μ_1-η^2 species, the electron of M1 is transferred to C and then the electron of O (CO_2) is transferred to M2 via back bonding. It implies that the activated CO_2 molecule may obtain little charge due to the two different directions of electron transfer, which would impede the C atom's protonation. So, only one coordinated C=O bond is easy to dissociate to produce CO. According to the above discussion, it is one of the valid steps for regulating the selectivity of CO that the bimetal bridging M1···CO_2···M2 modes are tailored via creating the catalyst with M1 (LA) and M2 (LB) sites.

15.3 Preparation of 2D Transition Metal Oxide Nanosheets

It is reported that there are various methods for the synthetic of nanosheets, such as chemical vapor deposition with electron beam (CVD), top-down chemical swelling–exfoliation, mechanical exfoliation, bottom-up wet chemical exfoliation, and sonication exfoliation in solvents method. By these methods, many transition metal oxides with ultrathin nanosheet morphology have been developed as photocatalysts, such as TiO_2, WO_3, Fe_2O_3, CuO, $H_{1.4}Ti_{1.65}O_4 \cdot H_2O$, and HNb_3O_8. These nanosheet materials exhibit better photocatalytic performance. Importantly, abundant metal atoms (M) can be exposed on the surface of nanosheets, which would greatly promote the coordination of CO_2 molecules. Taking TiO_2 nanosheets as an example, Ti^{4+} with low coordination numbers, electron-rich Ti^{3+}, and oxygen deficiency may be exposed on the surface. These active sites bond CO_2 molecules. However, the multiple active sites cause the generation of different coordination modes, which cannot provide a single desired active site. Moreover, a transition metal oxide catalyst with a single component only possesses the identified active site, which restricts the modulation of the coordination model. Therefore, it is necessary to further improve the pristine nanosheets to accurately create the desired active sites. For example, the metal oxide nanosheets are heated at a certain temperature to obtain abundant electron-rich metal sites. Moreover, constructing a composite catalyst is a widely used method, which can optimize the catalyst properties and particularly provide multiactive sites.

In Xie's work, atomically thin layers of sulfur-deficient $CuIn_5S_8$ have been prepared successfully. The charge-enriched Cu–In dual sites are exposed to the catalyst, which meets the presupposed dual-metal sites (M–M). This catalyst achieves near 100% selectivity for visible-light-driven CO_2 reduction to form CH_4. In our work, an ultrathin TiO_2 nanosheet with abundant Ti^{3+} sites has been developed as a photocatalyst. Ti^{3+} sites have been considered as the abundant surface M atoms with the unpaired electron. The catalyst exhibits a high yield (147.2 μmol g^{-1} h^{-1}) and selectivity (96.8%) for the reduction of CO_2 to form CH_4. Moreover, in another work, $Cu^{1-\delta}$ and $Ti^{4+\delta}$ have been generated on Cu_2O/TiO_2 nanosheet via a Cu^+–Ti^{3+} interface interaction. These metal sites meet with M1 (LA) and M2 (LB) sites. CO_2 can be converted to CO with a selectivity of 98% and a yield of 162.6 μmol g^{-1} h^{-1} on 5 wt% Cu_2O/TiO_2 nanosheet with $Cu^{1-\delta}$ and $Ti^{4+\delta}$ sites.

15.4 Coordination of CO_2

CO_2 molecules are bonded to the active sites of a catalyst via forming coordinated bonds along with the surface–interface charge transfer. Different coordination modes are generated in this process. Therefore, it is a key pathway for revealing coordinated modes to explore the formed coordinated bonds and the electron transfers between CO_2 molecules and the catalysts.

In our previous study, the temperature-programmed desorption of CO_2 (CO_2-TPD) revealed that CO_2 is coordinated on Lewis base sites via the weak chemical bond in an ultrathin TiO_2 nanosheet (Figure 15.5a). The result of *in situ* Fourier Transform Infrared Spectrometer (FTIR) suggested that the strength of the C=O bond (CO_2) has received a weakening due to the interaction of CO_2 with TiO_2 nanosheet (NST). In particular, the bridging $Ti\cdots CO_2^-\cdots Ti$ species are formed on the surface of an ultrathin TiO_2 nanosheet (Figure 15.5b). *In situ* Electron Paramagnetic Resonance Spectrometer (EPR) results reveal the inexistence of the electron transfer between CO_2 and the catalyst (Figure 15.5c). Furthermore, Density functional theory (DFT) calculation results (Figure 15.6) demonstrate that the significant charge (−0.90 |e|) transfer from the TiO_2 to the adsorbed CO_2, is inconsistent with our EPR observation. In particular, the DFT configurations of CO_2 exhibit a binding structure (bridging $Ti\cdots CO_2^-\cdots Ti$ species), which is similar to the CO_2 coordination mode deduced from the experiment discussion. Therefore, it can be concluded that the bridging $Ti\cdots CO_2^-\cdots Ti$ species is formed on NST via transferring about one electron from the defective Ti to the C atom. This coordination mode is consistent with our presupposed mode, which determines the generation of CH_4 from the photocatalytic conversion of CO_2. Therefore, CO_2 can be photocatalytically converted to CH_4 with a high yield (147.2 μmol g^{-1} h^{-1}) and selectivity (96.8%) over an ultrathin TiO_2 nanosheet with abundant Ti^{3+} sites.

In another study, the coordination of CO_2 is explored over Cu_2O/TiO_2 nanosheet with $Cu^{1-\delta}$ and $Ti^{4+\delta}$ sites. The XPS results reveal that $Ti^{4+\delta}$ sites obtain electrons from CO_2 and the electrons of Cu sites are transferred into CO_2 (Figure 15.7). *In situ* FT-IR results suggest that only one C=O of CO_2 is involved in coordination (Figure 15.8). It could be concluded that the coordination mode on Cu_2O/TiO_2 nanosheet possesses C=O and C—O bonds, and the bimetal bridging $Cu\cdots CO_2\cdots Ti$ mode is generated via the electron transfer between CO_2 and active sites ($O \rightarrow Ti^{4+\delta}$ and $Cu^{1-\delta} \rightarrow C$). This coordination mode also meets our presupposed mode, which

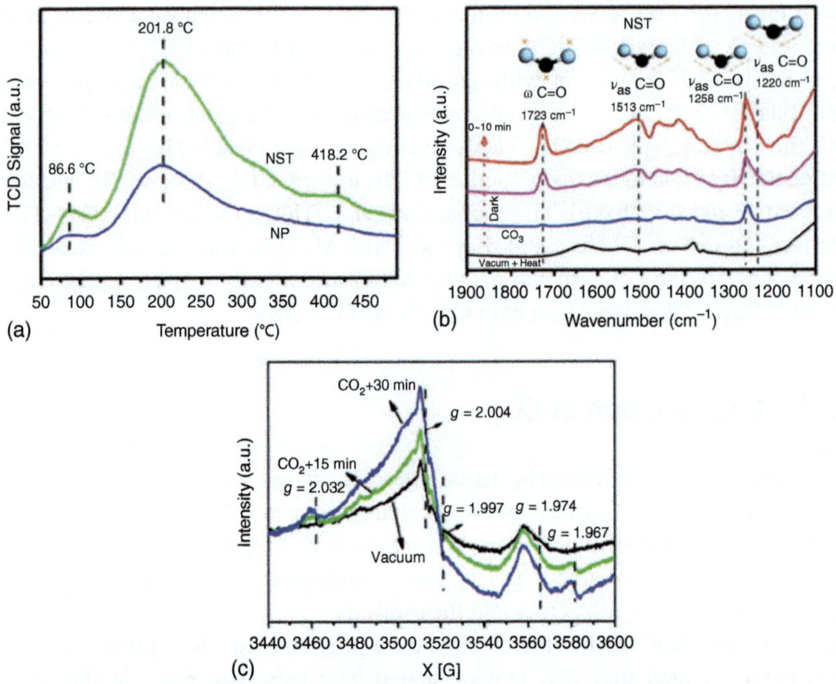

Figure 15.5 (a) CO_2-TPD for TiO_2 nanoparticle (NP) and nanosheets (NST); *In situ* FT-IR spectra of the absorption for CO_2 over NST (b) under the dark, respectively; (c) *In situ* EPR spectra of NST in CO_2 atmosphere under the dark. Source: Reproduced with permission from Wang et al. [33]/Elsevier.

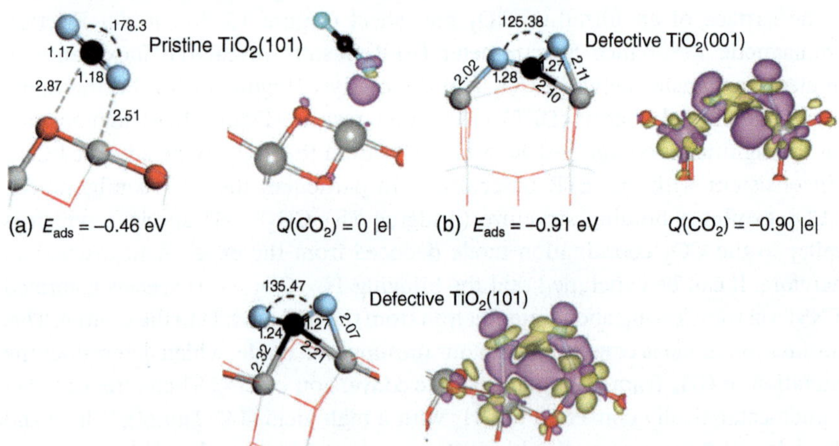

Figure 15.6 DFT optimized CO_2 adsorption configurations on (a) pristine TiO_2(101), (b) defective TiO_2(001), and (c) defective TiO_2(101) with the corresponding plots of charge density difference. The bond distance (Å), bond angle (°), adsorption energy (Eads) of CO_2 as well as the Bader charge (Q) of CO_2 are also provided. The purple area represents the accumulation of the charge while the yellow region represents the depletion of the charge. The iso-surface level is 0.003 e/bohr^3. Color scheme: Ti, gray; O in TiO_2, red; O in CO_2, blue; C, black. Source: Reproduced with permission from Wang et al. [33]/Elsevier.

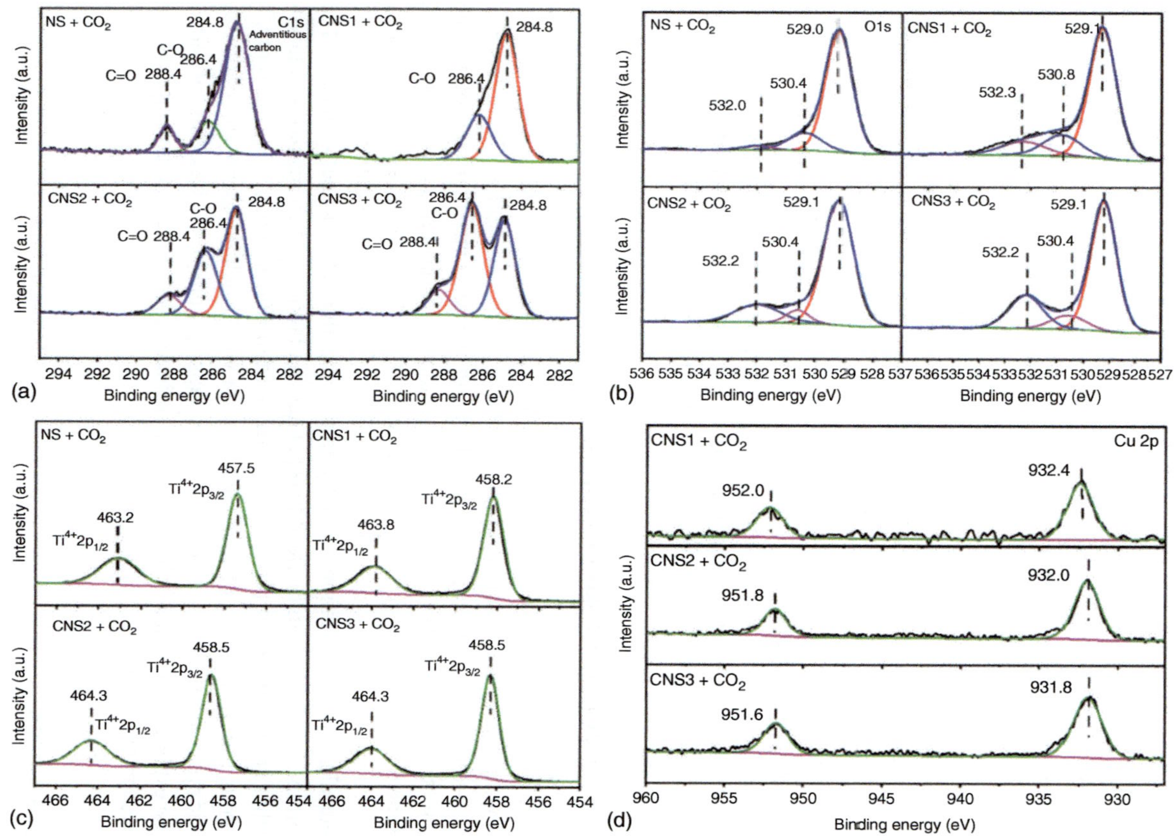

Figure 15.7 X-ray photoelectron spectroscopy (XPS) spectra of the samples after adsorption of CO_2. (a) C 1s; (b) O 1s; (c) Ti 2p; (d) Cu 2p. The samples labels as follows: TiO_2 nanosheets (NSTs), 1 wt% Cu_2O/TiO_2 (CNS1), 3 wt% Cu_2O/TiO_2 (CNS2), and 5 wt% Cu_2O/TiO_2 (CNS3). Source: Reproduced with permission from Wang et al. [34]/Elsevier.

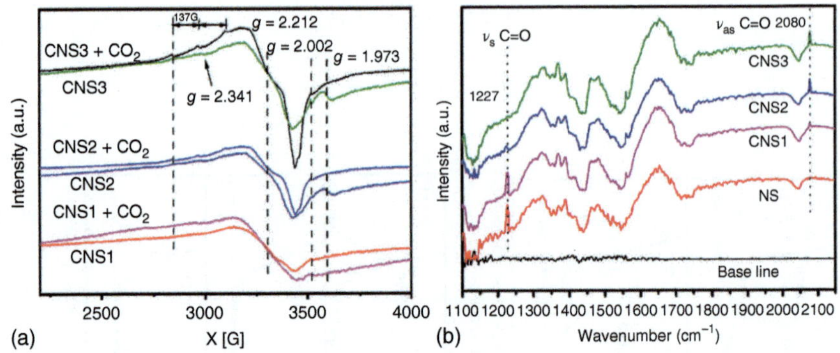

Figure 15.8 (a) *In situ* EPR spectra of the samples in CO_2 atmosphere under the dark; (b) *In situ* FTIR spectra of the absorption of CO_2 over the samples. The samples labels as follows: 1 wt% Cu_2O/TiO_2 (CNS1), 3 wt% Cu_2O/TiO_2 (CNS2), and 5 wt% Cu_2O/TiO_2 (CNS3). Source: Reproduced with permission from Wang et al. [34]/Elsevier.

facilitates the formation of CO. Thus, the catalyst achieves 98% selectivity for the formation of CO (162.6 $\mu mol\,g^{-1}\,h^{-1}$).

These studies have thus demonstrated that surface coordination may offer a special guideline for designing an effective photocatalyst and understanding the photocatalytic process for CO_2 conversion at a molecule level.

15.5 Conclusion and Prospects

Based on the surface coordination of CO_2, we can precisely construct a photocatalyst with special transition metal atoms (M), which can coordinate CO_2 molecules via the desired mode, thus greatly enhancing the reactivity and controlling the product's selectivity. Besides the quick separation of photogenerated electron hole, 2D metal oxide nanosheets with a thickness of several atoms possess abundantly exposed metal atoms, and surface defects can effectively capture and activate CO_2 molecules via coordination, signally improving the catalytic performance. The coordination modes can be flexibly regulated via tailoring M atoms of the catalyst, further controlling each product's selectivity from CO_2 reduction. Furthermore, a thorough understanding of CO_2 coordination would help to clarify the photocatalytic process at the molecular level. This promising strategy still has many novelties to be explored. We believe that surface coordination would effectively improve the development of photocatalytic CO_2 reduction.

References

1 Chang, X., Wang, T., Yang, P. et al. (2019). The development of cocatalysts for photoelectrochemical CO_2 reduction. *Adv. Mater.* 31: 1–13.

2 Birdja, Y., Pérez-Gallent, E., Figueiredo, M. et al. (2019). Advances and challenges in understanding the electrocatalytic conversion of carbon dioxide to fuels. *Nat. Energy* 4: 732–745.

3 Li, X., Yu, J., Jaroniec, M., and Chen, X. (2019). Cocatalysts for selective photoreduction of CO_2 into solar fuels. *Chem. Rev.* 119: 3962–4179.

4 Dong, Y., Duchesne, P., Mohan, A. et al. (2020). Shining light on CO_2: from materials discovery to photocatalyst, photorcactor and process engineering. *Chem. Soc. Rev.* 49: 5641–6114.

5 Modak, A., Bhanja, P., Dutta, S. et al. (2020). Catalytic reduction of CO_2 into fuels and fine chemicals. *Green Chem.* 22: 4002–4033.

6 Gong, E., Ali, S., Hiragond, C. et al. (2011). Solar fuels: research and development strategies to accelerate photocatalytic CO_2 conversion into hydrocarbon fuels. *Energy Environ. Sci.* 15: 880–937.

7 Halmann, M. (1978). Photoelectrochemical reduction of aqueous carbon dioxide on p-type gallium phosphide in liquid junction solar cells. Characterisation of organic acids trapped in coals. *Nature* 275: 115–116.

8 Habisreutinger, S., Schmidt-Mende, L., and Stolarczyk, J. (2013). Photocatalytic reduction of CO_2 on TiO_2 and other semiconductors. *Angew. Chem. Int. Ed.* 52: 7372–7408.

9 Sun, Z., Wang, H., Wu, Z., and Wang, L. (2018). g-C_3N_4 based composite photocatalysts for photocatalytic CO_2 reduction. *Catal. Today* 300: 160–172.

10 Chen, X., Peng, C., Dan, W. et al. (2022). Bromo- and iodo-bridged building units in metal-organic frameworks for enhanced carrier transport and CO_2 photoreduction by water vapor. *Nat. Commun.* 13: 4592.

11 Nguyen, H.L. and Alzamly, A. (2021). Covalent organic frameworks as emerging platforms for CO_2 photoreduction. *ACS Catal.* 11: 9809–9824.

12 Ulmer, U., Dingle, T., Duchesne, P. et al. (2019). Fundamentals and applications of photocatalytic CO_2 methanation. *Nat. Commun.* 10: 3169.

13 Corma, A. and Garcia, H. (2013). Photocatalytic reduction of CO_2 for fuel production: possibilities and challenges. *J. Catal.* 308: 168–175.

14 Albero, J., Peng, Y., and García, H. (2020). Photocatalytic CO_2 reduction to C_{2+} products. *ACS Catal.* 10: 5734–5749.

15 Wang, J., Lin, S., Tian, N. et al. (2021). Nanostructured metal sulfides: classification, modification strategy and solar-driven CO_2 reduction application. *Adv. Funct. Mater.* 31: 1–39.

16 Fu, J., Jiang, K., Qiu, X. et al. (2020). Product selectivity of photocatalytic CO_2 reduction reactions. *Mater. Today* 32: 222–243.

17 Chang, X., Wang, T., and Gong, J. (2016). CO_2 photo-reduction: insights into CO_2 activation and reaction on surfaces of photocatalysts. *Energy Environ. Sci.* 9: 2177–2196.

18 Sun, Z., Talreja, N., Tao, H. et al. (2015). Using carbon dioxide as a building block in organic synthesis. *Nat. Commun.* 6: 5933.

19 Vasileff, A., Xu, C., Jiao, Y. et al. (2018). Surface and interface engineering in copper-based bimetallic materials for selective CO_2 electroreduction. *Chem* 4: 1809–1831.

20 Huang, K., Sun, C., and Shi, Z. (2011). Transition metal-catalyzed C—C bond formation through the fixation of carbon dioxide. *Chem. Soc. Rev.* 40: 2435–2452.

21 Liang, S., Wen, L., Lin, S. et al. (2014). Monolayer HNb_3O_8 for selective photocatalytic oxidation of benzylic alcohols with visible light response. *Angew. Chem. Int. Ed.* 53: 2951–2955.

22 Wang, H., Luo, S., Song, Y. et al. (2020). Enhanced photocatalytic hydrogen evolution over monolayer HTi_2NbO_7 nanosheets with highly dispersed Pt nanoclusters. *Appl. Surf. Sci.* 511: 145501.

23 Song, Y., Wang, H., Wang, Z. et al. (2018). Selective photocatalytic synthesis of haloanilines from halonitrobenzenes over multifunctional AuPt/monolayer titanate nanosheet. *ACS Catal.* 8: 9656–9664.

24 Zhou, H. and Kitagawa, S. (2014). Metal-organic frameworks (MOFs). *Chem. Soc. Rev.* 43: 5415–5418.

25 Feng, X., Ding, X., and Jiang, D. (2012). Covalent organic frameworks. *Chem. Soc. Rev.* 41: 6010–6022.

26 Chen, Z., Jiang, X., Zhu, C., and Shi, C. (2016). Chromium-modified $Bi_4Ti_3O_{12}$ photocatalyst: application for hydrogen evolution and pollutant degradation. *Appl. Catal., B* 199: 241–251.

27 Guo, B., Cheng, X., Tang, Y. et al. (2022). Dehydrated UiO-66 $(SH)_2$: The Zr-O cluster and its photocatalytic role mimicking the biological nitrogen fixation. *Angew. Chem. Int. Ed.* 66: e202117244.

28 Luo, J., Lin, Z., Zhao, Y. et al. (2020). The embedded $CuInS_2$ into hollow-concave carbon nitride for photocatalytic H_2O splitting into H_2 with S-scheme principle. *Chin. J. Catal.* 41: 122–130.

29 Di, J., Chen, C., Yang, S. et al. (2019). Isolated single atom cobalt in Bi_3O_4Br atomic layers to trigger efficient CO_2 photoreduction. *Nat. Commun.* 10: 2840.

30 Meng, A., Zhang, L., Cheng, B., and Yu, J. (2019). TiO_2-MnO*x*-Pt hybrid multi-heterojunction film photocatalyst with enhanced photocatalytic CO_2-reduction activity. *ACS Appl. Mater. Interfaces* 11: 5581–5589.

31 Zhu, S., Chen, X., Li, Z. et al. (2020). Cooperation between inside and outside of TiO_2: lattice Cu^+ accelerates carrier migration to the surface of metal copper for photocatalytic CO_2 reduction. *Appl. Catal., B* 264: 118515.

32 Li, X., Sun, Y., Xu, J. et al. (2019). Selective visible-light-driven photocatalytic CO_2 reduction to CH_4 mediated by atomically thin $CuIn_5S_8$ layers. *Nat. Energy* 4: 690–699.

33 Wang, Z., Wan, Q., Shi, Y. et al. (2021). Selective photocatalytic reduction CO_2 to CH_4 on ultrathin TiO_2 nanosheet via coordination activation. *Appl. Catal., B* 288: 120000.

34 Wang, Z., Shi, Y., Liu, C. et al. (2022). Cu^+–Ti^{3+} interface interaction mediated CO_2 coordination model for controlling the selectivity of photocatalytic reduction CO_2. *Appl. Catal., B* 301: 120803.

16

Recent Progress on Layered Double Hydroxides-Based Nanomaterials for Solar Energy Conversion

Sha Bai, Chenjun Ning, Tianyang Shen, and Yu-Fei Song

Beijing University of Chemical Technology, State Key Laboratory of Chemical Resource Engineering, No.15 Beisanhuan East Road, Chaoyang District, Beijing 100029 P.R. China

16.1 Introduction

With the growth of the population and the rapid development of science and technology, the consumption of fossil energy has put tremendous pressure on the ecological environment [1]. The development of clean energies is in great demand. Solar energy has attracted widespread attention due to its sustainable properties as one type of clean energy [2]. In this regard, photocatalysis provides an environmentally friendly and sustainable strategy to harvest and store solar energy in the form of molecular bonds of high-value-added chemicals, in which thermodynamic energy barriers can be efficiently overcome [2, 3].

In 1968, Boddy first reported the photoelectrochemical catalysis of water splitting reaction using TiO_2 [4]. Pioneering work using a TiO_2 electrode for the electrochemical photolysis of water was reported by Honda and Fujishima in 1972 [5]. To date, a large number of materials, such as TiO_2, ZnO, GO, CdS, MOF, MXene, Metal Sulfide, BiOX, g-C_3N_4, and Polyoxometalates (POMs), have been widely explored and applied for photocatalytic reactions, and remarkable progress has been made. These photocatalysts provide a boost platform for solar conversion of N_2, H_2O, CO_2, etc. into high-value-added chemicals, and exhibit excellent performance in organic conversion, pollutant degradation, or adsorption. In order to facilitate solar energy absorption ability, increase the separation of photoinduced carriers' efficiency and improve the activity of photocatalysts, the rational design of photocatalysts needs to meet the following requirements: (i) an appropriate bandgap to promote the absorption of light; (ii) the high photogenerated electron–hole separation efficiency; (iii) the appropriate adsorption sites to optimize the adsorption and activation of substrates; (iv) fast transfer efficiency of interfacial charge to improve the reactivity; (v) weak adsorption of product, which is conducive to their release.

Layered double hydroxides (LDHs), a typical 2D layered materials with the formula $[M^{2+}{}_{1-x}M^{3+}{}_x(OH)_2](A^{n-}{}_{x/n}\cdot mH_2O)$ ($M^{2+} = Mg^{2+}$, Zn^{2+}, Ni^{2+}, Cu^{2+}, etc. $M^{3+} = Cr^{3+}$, Al^{3+}, Ga^{3+}, In^{3+}, Ti^{3+}, etc; $x = M^{3+}/(M^{2+} + M^{3+})$, A^{n-} is the interlayer

UV-Visible Photocatalysis for Clean Energy Production and Pollution Remediation: Materials, Reaction Mechanisms, and Applications, First Edition. Edited by Xinchen Wang, Masakazu Anpo, and Xianzhi Fu.
© 2023 WILEY-VCH GmbH. Published 2023 by WILEY-VCH GmbH.

anions), are made up of $M(OH)_6$ octahedra in the laminate and the intercalated anions for charge balance [6]. The M^{2+}, M^{3+}, and interlayer anions in LDHs can be easily exchanged, which enables multiple tunable factors, including the composition, ratios, anion species, and interlayer distance. Based on this, by introducing photocatalytically active metal ions on the LDHs layer, we are able to adjust the bandgap of LDHs in the range of 2.1–3.4 eV [7]. Furthermore, various strategies, such as exfoliation strategies, plasma methods, etching methods, and doping have been developed to modulate the size, morphology, electronic structure, surface defects, and coordination structures of LDHs [8, 9]. As a result, the catalytic performance of LDHs can be effectively improved accordingly. In addition, the development of LDH-based composite materials is another method to improve its catalytic performance. Interestingly, after topological transformation, LDHs can be converted to the corresponding composite materials, mixed metal oxides (MMO), or alloys, which exhibit a wide range of applications for different catalytic reactions [10, 11]. On the basis of the unique layered structure and multiple tunable factors, LDHs show great potential as photocatalysts by making use of visible light and even infrared light.

This chapter reports the latest progress on LDHs-based photocatalysts (Figure 16.1). First of all, DFT calculation can be applied to guide rational design of LDHs, in which the analysis for physical/chemical properties of LDHs structures including bandgap, DOS, electronic structure, adsorption energy, and Gibbs free energy can be conducted to judge the catalytic activity of predesigned materials. Then, various strategies are summarized by tuning the elemental composition,

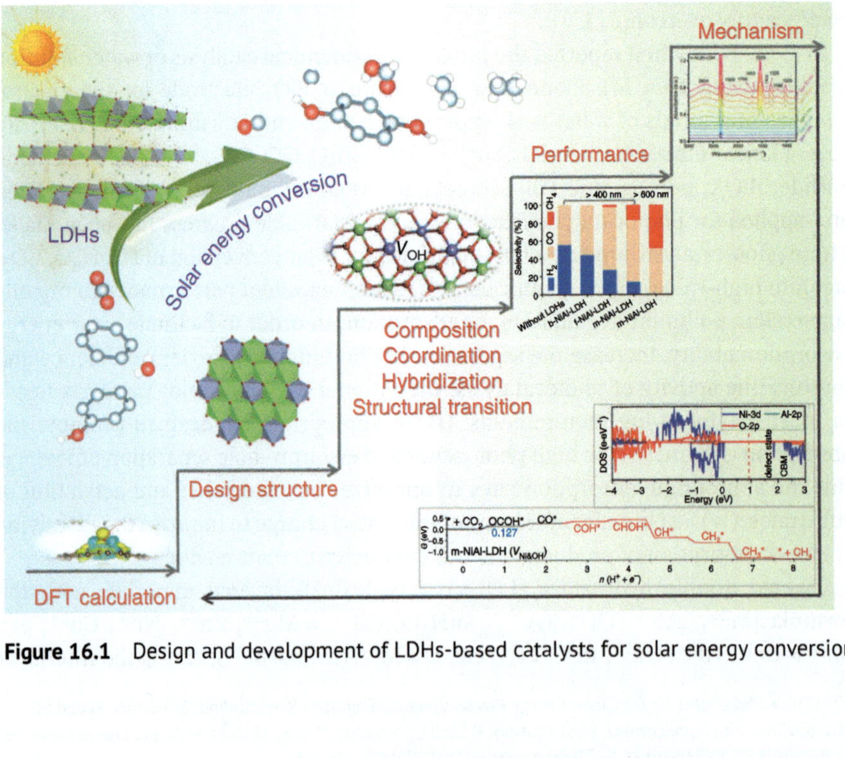

Figure 16.1 Design and development of LDHs-based catalysts for solar energy conversion.

coordination structure, and topological transformation of LDHs. Detailed characterizations, such as XPS, X-ray absorption fine structure (XAFS), and ESR are carried out to reveal the defect structure and electronic structure of LDHs-based catalysts. In addition, to enhance the development of LDH-based materials in photocatalytic reactions, in-depth structural understanding and *in situ* experiments play a very important role in revealing the structure-activity relationship. Finally, we wish that the review herein can draw researchers' attention to the photocatalysis of LDHs-based materials.

16.2 Prediction of the Reactivity via DFT Calculations

DFT can provide theoretical data for material design and reactivity. For example, Xie and coworkers [12] reported that the presence of oxygen vacancies (OVs) in Bi_2O_3 resulted in the localization of electrons near the OVs, suggesting that OVs were more conducive to electronic excitation. To go further, the calculation of CO_2 adsorption was performed. The Bi_2O_3 with OVs exhibited a promising CO_2 adsorption capacity of −0.30 eV, while Bi_2O_3 without OVs showed a very weak interaction with CO_2. As demonstrated, Bi_2O_3 with OVs displayed higher dimethyl carbonate (DMC) productivity in the presence of methanol with a selectivity of about 100%. Therefore, DFT calculation gave a prediction of the activity in CO_2 reaction, which can guide the design of the materials in experiments.

Similarly, the electronic properties and mechanisms toward photocatalytic oxygen evolution reaction (OER) of a series of LDHs were studied by DFT (Figure 16.2a) [7]. Among these LDHs, CoAl-LDH could get over the reaction barrier (0.653 eV) via hole (0.836 eV) oxidation under photoexcitation, resulting in high activity in OER. Song and coworkers [13] used DFT calculations at first to study the band structures and adsorption energies of CO_2 and H_2O for MAl LDH (M = Mg^{2+} ($t_{2g}{}^0e_g{}^0$), Co^{2+} ($t_{2g}{}^6e_g{}^1$), Ni^{2+} ($t_{2g}{}^6e_g{}^2$), and Zn^{2+} ($t_{2g}{}^6e_g{}^4$)) with different electrons in e_g orbitals of M, in order to reasonably design photocatalytic CO_2 reduction catalyst with excellent performance. They found that among these MAl-LDH, the bandgap of CoAl-LDH was the smallest (Figure 16.2b) and the CO_2 and H_2O adsorption energies were the strongest (Figure 16.2c), which may promote H_2O to generate active hydrogen in u-CoAl-LDH, thereby promoting the conversion of CO_2. As expected, among these LDHs, CoAl-LDH exhibited a very high CO_2 to CO with a conversion rate of 43.73 mmol(g h)$^{-1}$ under 600 nm irradiation.

To further improve the utilization of Co atoms, Co atoms were dispersed into MgAl-LDH [15]. XAFS and PDF showed that the introduction of the appropriate amount of Co atoms into the LDH laminate can promote the dispersion of Co. The resulting CoMgAl-LDH nanosheets exhibited excellent CO_2 to CO conversion activity with a time-of-flight (TOF) of 11.57 h^{-1}, three times that of CoAl-LDH at $\lambda > 400$ nm. Furthermore, in order to promote the separation of carriers and realize the regulation of the syngas ratio in photocatalytic CO_2 reduction, Song and coworkers predicted that the combination of LDH and TiC to regulate the ratio of syngas in photocatalytic CO_2 reduction through DFT calculation [14].

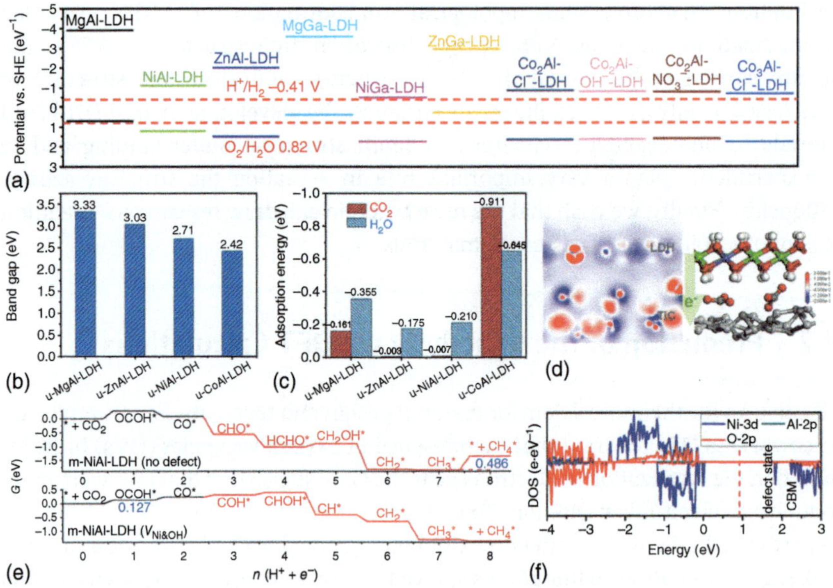

Figure 16.2 (a) Calculated band edge of M^{II}M^{III}-LDHs. Source: Reproduced with permission from Xu et al. [7]/American Chemical Society; (b) Calculated bandgaps and (c) adsorption energy of H_2O and CO_2 for u-MAl-LDH (M = Mg, Zn, Ni, Co) photocatalysts. Source: Reproduced with permission from Bai et al. [13]/American Chemical Society. (d) the charge density difference slices of LDH/TiC. Source: Reproduced with permission from Wang et al. [14]/The Royal Society of Chemistry. (e) The Gibbs free energy diagrams for photocatalytic CO_2 reduction to CH_4 on m-NiAl-LDH (no defect) and m-NiAl-LDH ($V_{Ni\&OH}$); (f) The projected density of states with the defect state and conduction band minimum (CBM) labeled. Source: Reproduced with permission from Tan et al. [8]/John Wiley & Sons.

Firstly, they found that the charge transfer took place from LDH to TiC, as revealed by the work function and electron density difference (Figure 16.2d). In addition, the adsorption energies of LDH and TiC for CO and H_2 were calculated, and it was found that TiC was beneficial to the desorption of H_2, while the adsorption energy of CO was stronger (−2.22 eV) than that of LDH (−0.14 eV), which facilitated the production of H_2 by HER in TiC and may control the distribution of products. Therefore, regulating the LDH/TiC heterojunction concentration was a potential strategy for regulating CO_2 reduction to different ratios of syngas. Based on this, the LDH/TiC materials with different ratios (LDH/TiC-x) were developed. As expected, the LDH/TiC-x can effectively modulate the syngas ratio from 1 : 1 to 1 : 3. Overall, DFT calculations can help us to reasonably predict the performance before the experiment and provide guidance for rational design of catalysts.

Besides, DFT calculation is beneficial to reveal reaction mechanisms and demonstrate the relationship between structure and performance. For example, DFT calculations showed that the Gibbs free energy barrier for CO_2 reduction to CH_4 decreased from 0.486 eV in monolayer NiAl-LDH (m-NiAl-LDH) (no defect) to 0.127 eV in m-NiAl-LDH with metal and hydroxyl defects (m-NiAl-LDH ($V_{Ni\&OH}$)) (Figure 16.2e), which enabled better selectivity of CH_4 over m-NiAl-LDH ($V_{Ni\&OH}$).

Moreover, a new defect state in m-NiAl-LDH ($V_{Ni\&OH}$) can overcome the barrier of CO_2 to CH_4 (0.127 eV) instead of generating H_2 (0.425 eV), which explained the complete suppression of H_2 production at $\lambda > 600$ nm (Figure 16.2f). In summary, DFT calculation can not only help us predict the feasibility of the reaction but also corroborate the experimental data in revealing the reaction pathway. As a result, a better understanding of structure-activity relationship and the reaction mechanism can be achieved.

16.3 Controllable Synthesis

16.3.1 Modulation of the Compositions

The pioneering work for CO_2 photoreduction was reported by Izumi and coworkers in 2011 [16], and they found that methanol was the main product formed when using ZnCuAl-LDH and ZnCuGa-LDH as photocatalysts. Subsequently, Tanaka and coworkers constructed a series of M^{2+}–M^{3+} LDHs (M^{2+} = Mg^{2+}, Zn^{2+}, Ni^{2+}; M^{3+} = Al^{3+}, Ga^{3+}, In^{3+}) for photocatalytic CO_2 reduction, through isotopic experiments, it was found that CO_2 in the gas phase was dissolved in water and reduced to CO under light irradiation [17]. In 2018, Wang et al. [18] reported that CoAl-LDH nanosheets exhibited efficient photocatalytic activity at atmospheric concentration for photoreduction of CO_2 to CH_4 with no deactivation of the reaction within 55 hours. This is closely related to the Co^{2+} in CoAl-LDH.

Recently, the introduction of Ni into CoFe-LDH (denoted as NiCoFe-LDH) as the photocatalyst can reduce CO_2 to CH_4 and CO with high selectivity (Figure 16.3a) [9]. More interestingly, at $\lambda > 500$ nm, the CH_4 selectivity increased from 0% in CoFe-LDH to 78.9% in NiCoFe-LDH, and the H_2 selectivity decreased from 30.5% in CoFe-LDH to 1.7% in NiCoFe-LDH (Figure 16.3b). The XAFS results indicated that the introduction of Ni increased the defect concentration of the LDH and promoted the separation of the carrier. This work demonstrated that the synergistic effect of Ni incorporation and defects in the LDH structure can be beneficial to enhance the conversion of CO_2 to CH_4 and suppress the generation of byproduct H_2. Notably, the electronic structure of divalent metals can be effectively modulated by tuning the trivalent metal species on the LDH layer.

A recent study showed that the selectivity of photocatalytic CO_2 to CH_4 by using monolayer m-Ni_3X-LDH (X = Cr, Mn, Fe, Co) exhibited a volcano-like trend with the highest point appearing at m-Ni_3Mn-LDH (Figure 16.3c) [19]. This was due to the fact that the electron-rich Ni—O bonds in m-Ni_3Mn-LDH can improve the CO_2 photoreduction activity by enhancing the adsorption for CO*, benefiting the hydrogenation of CO to CH_4 (Figure 16.3d). This can be observed directly from the *in situ* DRIFTS (Figure 16.3e,f), in which the absorption peaks at 1529 and 1196 cm^{-1} could be assigned to OCOH*, and the peaks at 1989 and 2664 cm^{-1} were ascribed to CO* intermediate. It is worth noting that the peak intensity of CO* was the strongest for m-Ni_3Mn-LDH among these samples, implying that m-Ni_3Mn-LDH was beneficial to promote CO hydrogenation to CH_4 (Figure 16.3f).

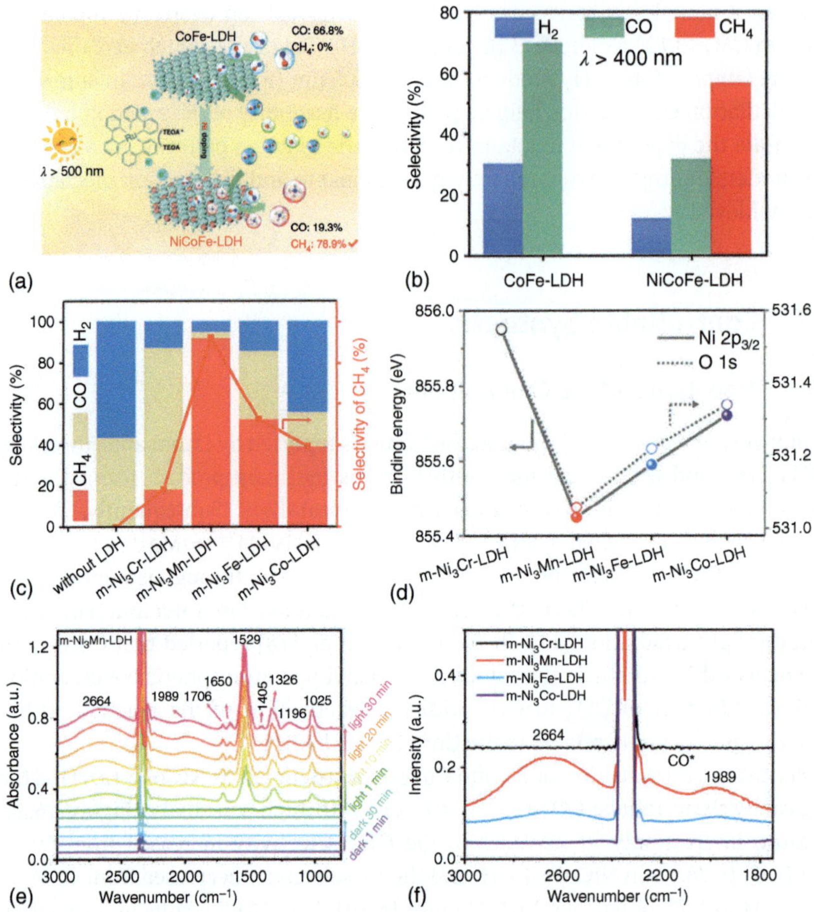

Figure 16.3 (a) Schematic illustration of photocatalytic CO_2 to CH_4 in CoFe-LDH and NiCoFe-LDH under $\lambda > 500$ nm; selectivity of CH_4, CO, and H_2 for (b) CoFe-LDH and NiCoFe-LDH and (c) m-Ni_3X-LDH under $\lambda > 400$ nm; (d) the binding energy of Ni 2p and O 1s for m-Ni_3X-LDH; *In situ* DRIFT of photocatalytic CO_2 reduction on (e) m-Ni_3Mn-LDH and (f) m-Ni_3X-LDH under $\lambda > 400$ nm. Source: (a and b) Reproduced with permission from Hao et al. [9]/American Chemical Society and (c, d, e, and f) Reproduced with permission from Tan et al. [19]/Elsevier.

The DFT calculation showed that the Ni_3Mn-LDH has the strongest adsorption energy for CO, which was consistent with the result of *in situ* DRIFTS. After optimization, m-Ni_3Mn-LDH can achieve the selectivity of CH_4 to nearly 99% under irradiation at $\lambda = 600$ nm, and completely inhibit the byproduct of H_2. Note that different interlayer anions in LDHs can affect photocatalytic performance. For example, the NO_3^- intercalated NiAl-LDH (NiAl-NO_3) exhibited higher CH_4 selectivity compared with the CO_3^{2-} intercalated NiAl-LDH (NiAl-CO_3) [20]. This was because the abundant defects in Ni_2Al-NO_3 as active sites facilitated electron transfer. From the above results, we can conclude that by modulating the composition of LDH, the electronic structure or defect structure can be further regulated, and thereby affect the photocatalytic activity. As such, tuning the composition of LDHs-based catalysts can have a significant effect on the active sites and mechanism of photocatalysis.

16.3.2 Modulation of the Coordination Environment

Defect engineering in photocatalytic materials has been proven as a versatile approach to affect their performance in solar-to-chemical energy conversion. Due to most active sites in LDHs are inhibited by surface hydroxyl groups and intercalated anions. Therefore, the introduction of unsaturated metal sites and ultrathin structures can effectively expose active sites and improve catalytic activity. As reported, oxygen and metal defects can be easily introduced into LDHs by modulating the doping, thickness, and morphology.

In general, the construction of the unsaturated coordination structure enables LDH-based materials to have unparalleled advantages in photocatalytic reactions. Song et al. reported a series of studies on defect engineering to enhance the photocatalytic activity of LDHs. As we can see in Figure 16.4a, the authors prepared several NiAl-LDHs with different thicknesses from bulk phase, few layers to monolayer [8]. The corresponding (00l) peaks were disappeared in XRD of monolayer NiAl-LDH (m-NiAl-LDH) (Figure 16.4b). The CO_2 photoreduction using m-NiAl-LDH as catalyst exhibited the highest CH_4 selectivity of 70.3% under $\lambda > 600$ nm irradiation, and byproduct H_2 was completely inhibited. The XAFS and positron annihilation spectrometry (PAS) demonstrated that the m-NiAl-LDH possessed the highest concentration of metal and hydroxyl defects (Figure 16.4c). Under irradiation of $\lambda > 600$ nm, the photogenerated electrons localized at defect states can overcome the Gibbs free energy barrier of CO_2 to CH_4, rather than HER. Thus, it promoted the photocatalytic reduction of CO_2 to CH_4 while suppressing the H_2 evolution. Similar results were also found in NiFe-LDH with different thicknesses, in which monolayer NiFe-LDH nanosheets with abundant defects exhibited excellent CO_2 photoreduction to CH_4 [24].

Besides, the strategy of constructing defects in the LDHs structure was also widely applicable to other photocatalytic reactions. For example, in the photo-driven Fenton hydroxylation reaction, defect-rich NiFe-NS (NS = nanosheets) can achieve a high selectivity of 99% from phenol to dihydroxybenzenes with H_2O_2 and a phenol conversion rate of 70% under visible light irradiation [21] (Figure 16.4d). This result was attributed to the fact that the unsaturated sites promoted the adsorption of H_2O_2 on the catalyst surface, and the generation of ·OH radicals to participate in the hydroxylation of phenol. Moreover, in the photocatalytic selective oxidation of benzene to phenol in water, Zn_2Ti-LDH with an abundance of OVs exhibited a selectivity of 87.1% (Figure 16.4e) [22]. Through ESR and free radical quenching experiments, it was found that superoxide radicals ($^{\cdot}O_2^-$) played an important role in the reaction. DFT calculation revealed that a large number of OVs can generate additional defect states and become electron-trapping sites, which promote the formation of superoxide radicals ($^{\cdot}O_2^-$) and in turn boost the photocatalytic performance. Similar results can be obtained in the photocatalytic conversion of olefins to alkanes [25].

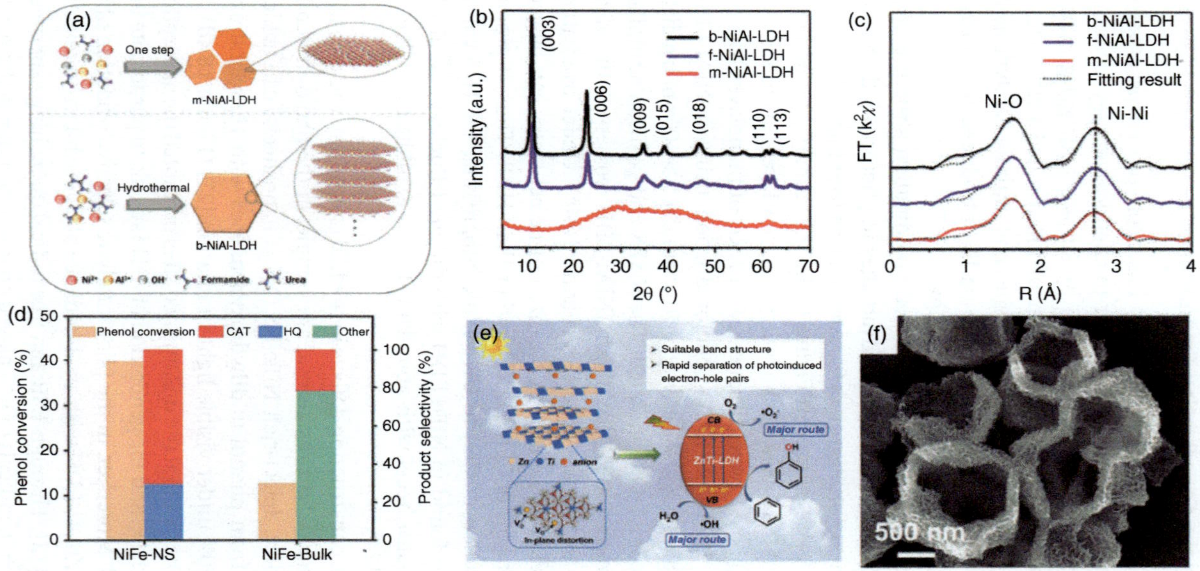

Figure 16.4 (a) Illustration for the synthesis of different thickness NiAl-LDH; (b) X-ray diffraction patterns and (c) magnitude of weighted FT of Ni K-edge extended X-ray absorption fine-structure spectra of b-NiAl-LDH, f-NiAl-LDH, and m-NiAl-LDH, respectively. Source: Reproduced with permission from Tan et al. [8]/John Wiley & Sons. (d) Phenol conversion and product selectivity over NiFe-NS and NiFe-Bulk under irradiation at >400 nm. Source: Reproduced with permission from Wang et al. [21]/The Royal Society of Chemistry. (e) Illustration of Zn_2Ti-LDH containing oxygen vacancies in photocatalytic selective oxidation of benzene to phenol in H_2O. Source: Reproduced with permission from Li et al. [22]/Elsevier. (f) SEM image of the HC-NiCo-LDH. Source: Reproduced with permission from An et al. [23]/The Royal Society of Chemistry.

Apart from adjusting the thickness of the LDHs to construct the defects in LDHs, the modulation of morphologies for LDHs can affect the formation of defects and thereby improve the catalytic performance. A recent study reported a self-sacrificing template strategy by using ZIF-67 as a template to construct NiCo-LDH with a hollow nanocage morphology (HC-NiCo-LDH) (Figure 16.4f). The as-prepared HC-NiCo-LDH showed increased CH_4 selectivity compared with bulk NiCo-LDH [25]. This was due to the rich metal and OVs in HC-NiCo-LDH, as certified by ESR and XAFS results, promoted the activity of photoreduction of CO_2. Therefore, the defect structure of LDHs can be modulated through morphology control.

16.3.3 Hybridization LDHs with Other Materials

Hybridization of LDHs with other materials is considered to be an effective strategy to optimize photocatalyst activity and selectivity by accelerating the separation of electron and hole, which stems from the fact that the large contact area between the layered structure and the substrate can promote the generation of photogenerated carriers, and heterojunction of hybrid materials was beneficial for the migration of charge carriers.

It is well-known that Pd has been regarded as cocatalysts that can promote the generation of photoinduced charge carriers and improve the performance of HER [26]. Song and coworkers designed and synthesized a series of palladium nanoparticles (NPs) supported on CoAl-LDH heterostructure photocatalysts by the coprecipitation method (Figure 16.5a) [27]. The photocatalytic reduction of CO_2 to syngas with different ratios (CO/H_2 = 1:0.74–1:3) can be achieved ($\lambda > 400$ nm) by adjusting the loading amount of Pd NPs (Figure 16.5b). Their findings demonstrated that the loading of Pd effectively enhanced the charge transfer efficiency, and a large number of H_2 were generated on Pd sites at the same time. As a result, the ratio of CO/H_2 can be finely tuned. To further understand the mechanism of photocatalytic CO_2 reduction in regulating the syngas ratio, a CoAl-LDH/MoS_2 heterojunction photocatalyst was successfully synthesized by electrostatic interaction.

By adjusting the catalyst concentration (0.2–1.5 mg ml^{-1}), the syngas ratio at 1.3 : 1–15 : 1 can be achieved under visible light [28]. The authors discovered that with the formation of the CoAl-LDH/MoS_2 heterojunction, a new Mo—O bond was formed (Figure 16.5c), which can accelerate the electron transfer from CoAl-LDH to MoS_2 and promote the coupling of H* on the MoS_2 site to generate H_2. Overall, the regulation of the syngas ratio by the composites depends on the modulation of the HER and CO_2 reactions through the heterostructure.

16.3.4 Topological Transformation of LDHs

The structure of LDHs is diversified through topological transformations, such as calcination or liquid phase transformation. The calcination of LDHs can be used to prepare MMO, bi/multi-metal alloys, etc., and they were widely used in many industrial reactions [29].

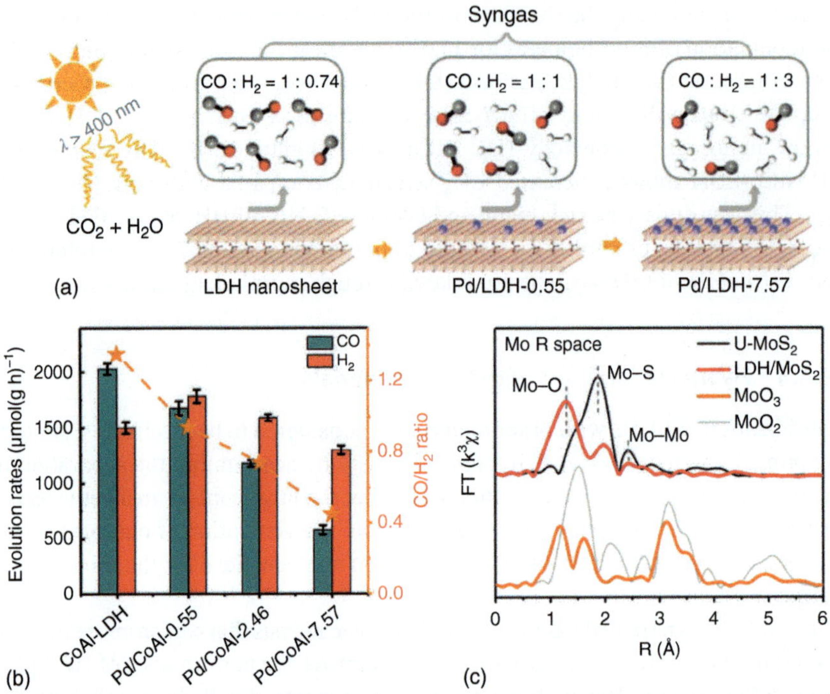

Figure 16.5 (a) The selectivity of CoAl-LDH and Pd/CoAl-x for CO_2 reduction under visible light irradiation; (b) The evolution rates and ratio of CO and H_2 on CoAl-LDH, Pd/CoAl-0.55, Pd/CoAl-2.46, and Pd/CoAl-7.57, respectively. Source: Reproduced with permission from Wang et al. [27]/Elsevier. (c) Mo magnitude of weighted FT of the K-edge extended X-ray absorption fine-structure spectra. Source: Reproduced with permission from Qiu et al. [28]/The Royal Society of Chemistry.

Song et al. found that the corresponding MMO with abundant interfaces and the exposure of specific crystal facets can be obtained by calcination of LDH at different temperatures, resulting in improved catalytic performance [29]. Recently, they reported the topotactic preparation of a series of NiO with different amounts of Ni and O vacancies from NiAl-LDH precursors in air from 200 to 800 °C (denote as NiAl-x) (Figure 16.6a) [10]. The Ni K-edge XAFS showed the lowest coordination number in Ni–O and Ni–Ni shells, indicating that NiAl-275 was rich in oxygen vacancy (V_O) and nickel vacancy (V_{Ni}). PAS and O K-edge XES data suggested that the concentration of oxygen vacancy in NiAl-x decreased with increasing calcination temperature, and NiAl-275 exhibited the highest defect concentration. Interestingly, for photocatalytic CO_2 reduction reaction, with the increase of Ni and O vacancies, the selectivity of CH_4 showed a similar regularity, reaching the highest selectivity of 22.8% in NiAl-275 (Figure 16.6b). More importantly, the CH_4 selectivity can be improved to 38.5%, and the competitive reaction HER can be completely inhibited under the irradiation of $\lambda > 600$ nm.

Recently, a heterostructured Co-doped MgO-based catalyst with highly dispersed active sites was successfully prepared through topological transformation of

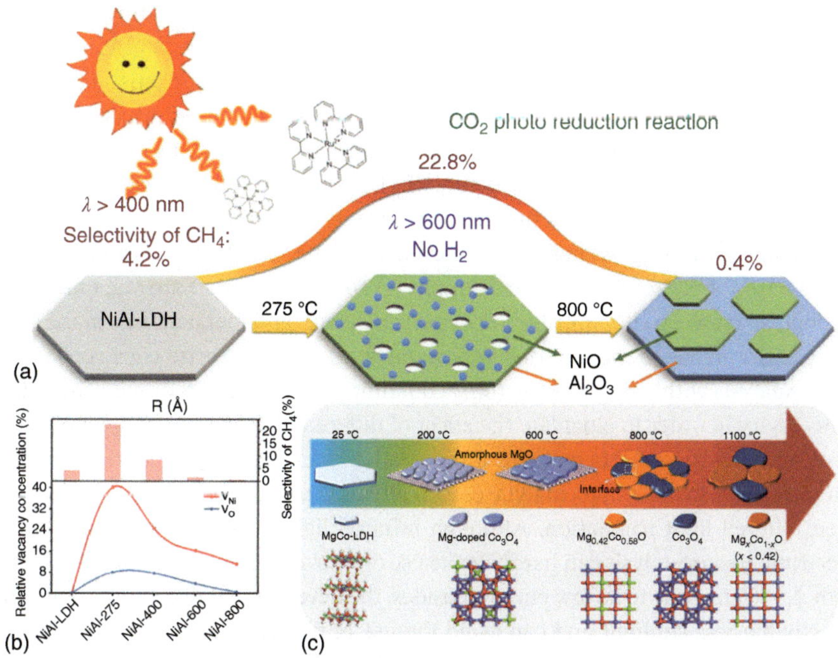

Figure 16.6 (a) Schematic illustration of NiO ($V_{Ni\&O}$) derived from the topological transformation of NiAl-LDH; (b) the relationship between selective CH_4 and relative vacancy concentration. Source: Reproduced with permission from Wang et al. [10]/Elsevier. (c) Schematic illustration of the topological transformation of MgCo-LDH at different temperatures. Source: Reproduced with permission from Xu et al. [30]/John Wiley & Sons.

MgCo-LDH by calcination from 200 to 1100 °C (Figure 16.6c) [30]. LDH calcined at 800 °C exhibited excellent activity for CH_4 coupling to generate C_2H_6 with a selectivity of 41.60%. X-ray total scattering experiments and high-energy XRD and TOF neutron diffraction confirmed that Mg only crystallized to form a MgO/CoO solid solution ($Mg_{0.42}Co_{0.58}O$) with no detectable Mg in the spinel phase, which was favorable for the activation of CH_4 to generate CH_3* and thus the coupling to C_2H_6. In addition to calcination of LDH, the liquid-phase reduction can achieve *in situ* partial transformation of LDH into a metallic phase. It was reported that highly dispersed metallic Ni catalysts supported on LDH nanosheets exhibited high performance for activating CH_4 to H_2 (388.28 μmol $(g\,min)^{-1}$) [11]. Based on the above facts, it can be seen that the topological transformation of LDH is a promising pathway for the generation of abundant defects, exposure of active sites, and promotion of photocatalytic reaction.

16.4 Summary and Perspectives

This chapter summarizes recent progress in the design and preparation of LDH-based catalysts and their applications in photocatalytic reactions, such as

CO_2 reduction, CH_4 conversion, benzene to phenol, and phenol to dihydroxybenzenes. Taking advantage of LDHs in composition, hybridization, and topological transformation, we are able to adjust the band structure, carrier separation, and adsorption of LDH-based photocatalysts. A number of advanced techniques, such as HRTEM, STEM, XPS, DRIFTS, EXAFS, ESR, and PAS have been used to elucidate the coordination and electronic structure of LDHs-based photocatalysts in order to have a better understanding of the structure-activity relationship.

Although great progress has been achieved, there are still challenges and difficulties in terms of photocatalysis. For example: (i) To date, scientists are still unable to distinguish between the contribution of oxygen defects and metal defects in the catalytic performance of the reported photocatalysts. As such, precise construction of photocatalysts with a single defect (only containing oxygen defects or metal defects) is necessary in order to elucidate the effect of different defect types and concentrations on catalytic performance. (ii) More strategies are required to fabricate novel photocatalysts so that photocatalytic reactions can take place under long-wavelength even infrared light irradiation. Although infrared light occupies 53% of the solar spectrum, photocatalysis can hardly make use of near-infrared or even infrared light with $\lambda > 800$ nm due to its low energy. Besides, the development of catalysts that can directly harness sunlight and can avoid the use of photosensitizers and sacrificial agents at the same time is highly desirable. (iii) Exploring more *in situ* techniques, such as *in situ* Ramen, XPS, FTIR, and XAFS and ultrafast transient absorption spectroscopy techniques, are of significance in order to have an in-depth understanding of the photogenerated carrier transfer and reaction mechanisms.

In conclusion, LDH-based materials exhibit great potential in photocatalytic reactions. We hope that the present chapter provides a better understanding of LDHs-based photocatalysis and thereby guides future design of highly efficient and selective photocatalysts.

Acknowledgments

This research was supported by the National Nature Science Foundation of China (22178019, U1707603, 21625101) and the Fundamental Research Funds for the Central Universities (XK1802-6, XK1803-05, XK1902, 12060093063).

References

1 Zhao, Y., Gao, W., Li, S. et al. (2019). Solar- versus thermal-driven catalysis for energy conversion. *Joule* 3: 920–937.

2 Ran, J., Zhang, J., Yu, J. et al. (2014). Earth-abundant cocatalysts for semiconductor-based photocatalytic water splitting. *Chem. Soc. Rev.* 43: 7787–7812.

3 Tan, L., Wang, Z., Zhao, Y., and Song, Y.-F. (2020). Recent progress on naon-structured layered double hydroxides for visible-light-induced photoreduction of CO_2. *Chem. Asian J.* 15: 3380–3389.

4 Boddy, P.J. (1968). Oxygen evolution on semiconducting TiO_2. *J. Electrochem. Soc.* 115: 199–203.

5 Fujishima, A. and Honda, K. (1972). Electrochemical photolysis of water at a semiconductor electrode. *Nature* 238: 37–38.

6 Fan, G., Li, F., Evans, D.G., and Duan, X. (2014). Catalytic applications of layered double hydroxides: recent advances and perspectives. *Chem. Soc. Rev.* 43: 7040–7066.

7 Xu, S.-M., Pan, T., Dou, Y.-B. et al. (2015). Theoretical and experimental study on $M^{II}M^{III}$-layered double hydroxides as efficient photocatalysts toward oxygen evolution from water. *J. Phys. Chem. C* 119: 18823–18834.

8 Tan, L., Xu, S.-M., Wang, Z. et al. (2019). Highly selective photoreduction of CO_2 with suppressing H_2 evolution over monolayer layered double hydroxide under irradiation above 600 nm. *Angew. Chem. Int. Ed.* 58: 11860–11867.

9 Hao, X., Tan, L., Xu, Y. et al. (2020). Engineering active Ni sites in ternary layered double hydroxide nanosheets for a highly selective photoreduction of CO_2 to CH_4 under irradiation above 500 nm. *Ind. Eng. Chem. Res.* 59: 3008–3015.

10 Wang, Z., Xu, S.-M., Tan, L. et al. (2020). 600 nm-driven photoreduction of CO_2 through the topological transformation of layered double hydroxides nanosheets. *Appl. Catal., B* 270: 118884.

11 Li, T., Tan, L., Zhao, Y., and Song, Y.-F. (2021). Solar-driven hydrogen production from steam methane reforming using highly dispersed metallic Ni catalysts supported on layered double hydroxide nanosheets. *Chem. Eng. Sci.* 245: 116839.

12 Chen, S., Wang, H., Kang, Z. et al. (2019). Oxygen vacancy associated single-electron transfer for photofixation of CO_2 to long-chain chemicals. *Nat. Commun.* 10: 788.

13 Bai, S., Wang, Z., Tan, L. et al. (2020). 600 nm irradiation-induced efficient photocatalytic CO_2 reduction by ultrathin layered double hydroxide nanosheets. *Ind. Eng. Chem. Res.* 59: 5848–5857.

14 Wang, H., Bai, S., Zhao, P. et al. (2021). Green light (550 nm) driven tunable syngas synthesis from CO_2 photoreduction using heterostructured layered double hydroxide/TiC photocatalysts. *Catal. Sci. Technol.* 11: 7091–7097.

15 Ning, C., Wang, Z., Bai, S. et al. (2021). 650 nm-driven syngas evolution from photocatalytic CO_2 reduction over Co-containing ternary layered double hydroxide nanosheets. *Chem. Eng. J.* 412: 128362.

16 Ahmed, N., Shibata, Y., Taniguchi, T., and Izumi, Y. (2011). Photocatalytic conversion of carbon dioxide into methanol using zinc–copper–M(III) (M = aluminum, gallium) layered double hydroxides. *J. Catal.* 279: 123–135.

17 Teramura, K., Iguchi, S., Mizuno, Y. et al. (2012). Photocatalytic conversion of CO_2 in water over layered double hydroxides. *Angew. Chem. Int. Ed.* 51: 8008–8011.

18 Wang, K., Zhang, L., Su, Y. et al. (2018). Photoreduction of carbon dioxide of atmospheric concentration to methane with water over CoAl-layered double hydroxide nanosheets. *J. Mater. Chem. A* 6: 8366–8373.

19 Tan, L., Xu, S.-M., Wang, Z. et al. (2021). 600 nm induced nearly 99% selectivity of CH_4 from CO_2 photoreduction using defect-rich monolayer structures. *Cell Rep. Phys. Sci.* 2: 100322.

20 Kipkorir, P., Tan, L., Ren, J. et al. (2020). Intercalation effect in NiAl-layered double hydroxide nanosheets for CO_2 reduction under visible light. *Chem. Res. Chin. Univ.* 36: 127–133.

21 Wang, J., Xu, Y., Li, J. et al. (2020). Highly selective photo-hydroxylation of phenol using ultrathin NiFe-layered double hydroxide nanosheets under visible-light up to 550 nm. *Green Chem.* 22: 8604–8613.

22 Li, J., Xu, Y., Ding, Z. et al. (2020). Photocatalytic selective oxidation of benzene to phenol in water over layered double hydroxide: a thermodynamic and kinetic perspective. *Chem. Eng. J.* 388: 124248.

23 An, J., Shen, T., Chang, W. et al. (2021). Defect engineering of NiCo-layered double hydroxide hollow nanocages for highly selective photoreduction of CO_2 to CH_4 with suppressing H_2 evolution. *Inorg. Chem. Front.* 8: 996–1004.

24 Bai, S., Li, T., Wang, H. et al. (2021). Scale-up synthesis of monolayer layered double hydroxide nanosheets via separate nucleation and aging steps method for efficient CO_2 photoreduction. *Chem. Eng. J.* 419: 129390.

25 Ma, X., Xu, Y., Tan, L. et al. (2020). Visible-light-induced hydrogenation of C=C bonds by hydrazine over ultrathin layered double hydroxide nanosheets. *Ind. Eng. Chem. Res.* 59: 14315–14322.

26 Wang, J., Tan, H., Jiang, D., and Zhou, K. (2017). Enhancing H_2 evolution by optimizing H adatom combination and desorption over Pd nanocatalyst. *Nano Energy* 33: 410–417.

27 Wang, X., Wang, Z., Bai, Y. et al. (2020). Tuning the selectivity of photoreduction of CO_2 to syngas over Pd/layered double hydroxide nanosheets under visible light up to 600 nm. *J. Energy Chem.* 46: 1–7.

28 Qiu, C., Hao, X., Tan, L. et al. (2020). 500 nm induced tunable syngas synthesis from CO_2 photoreduction by controlling heterojunction concentration. *Chem. Commun.* 56: 5354–5357.

29 Xu, Y., Wang, Z., Tan, L. et al. (2018). Fine tuning the heterostructured interfaces by topological transformation of layered double hydroxide nanosheets. *Ind. Eng. Chem. Res.* 57: 10411–10420.

30 Xu, Y., Sun, X., Wang, X. et al. (2021). Topological transformation of Mg-containing layered double hydroxide nanosheets for efficient photodriven CH_4 coupling. *Chem. Eur. J.* 27: 13211–13220.

17

The Significance and Current Status of Photocatalytic N_2 Fixation Study

Tingting Dong and Guohui Dong

Shaanxi University of Science and Technology, School of Environmental Science and Engineering, Xi'an 710021, China

17.1 Introduction

Nitrogen (N) is an essential element for the composition of biological proteins, and thus it is one of the essential elements for biological growth [1]. Meanwhile, N participates in a variety of life activities. Although the content of N_2 in the atmosphere accounts for more than 70%, it is combined with nitrogen–nitrogen triple bonds (N≡N), which makes its structure very stable and not easily absorbed by organisms (Figure 17.1) [2]. Therefore, it is very important to convert N_2 into a nitrogen substance that can be easily absorbed by plants. This process is called "nitrogen fixation." The synthesis of NH_3 from N_2 is a common nitrogen fixation reaction in nature, and it is also one of the most important reactions in social life [3]. With the continuous growth of the world population and the increasing demand for food, the production of NH_3 through N_2 fixation reaction has attracted more and more attention from scientists. In addition, 17.6 weight percent of liquid ammonia is hydrogen; as a result, NH3 is also regarded as an energy storage medium and may be crucial to the development of the hydrogen storage sector in the future [4].

In nature, N_2 fixation methods include biological N_2 fixation and high-energy N_2 fixation. However, the efficiency of this N_2 fixation is low and cannot meet the needs of crop growth and social production. Until the 20th century, Haber–Bosch synthesized ammonia with iron-based catalysts for the first time under the action of high temperature and high pressure. However, this synthetic method needs complicated equipment, complicated operation, high energy consumption, and serious pollution. In this way, the realization of the NH_3 preparation in mild conditions has become the focus of current researchers. At present, the methods reported in the world to synthesize NH_3 under mild conditions mainly include electrocatalytic NH_3 synthesis and photocatalytic NH_3 synthesis. Compared with thermocatalytic NH_3 synthesis and electrocatalytic NH_3 synthesis, photocatalytic NH_3 synthesis is more energy saving and environmentally friendly [5]. This is because solar energy is clean and renewable. As early as 1977, Schrauze and Guth used TiO_2 as the photocatalyst to convert

UV-Visible Photocatalysis for Clean Energy Production and Pollution Remediation: Materials, Reaction Mechanisms, and Applications, First Edition. Edited by Xinchen Wang, Masakazu Anpo, and Xianzhi Fu.
© 2023 WILEY-VCH GmbH. Published 2023 by WILEY-VCH GmbH.

Figure 17.1 The process of converting free nitrogen to compound nitrogen. Source: Tomasz/Adobe Stock.

N_2 into NH_3 under the irradiation of ultraviolet light, which opened up a precedent for the photocatalytic N_2 fixation reaction. In the 21st century, more researchers have joined the research area of photocatalytic nitrogen fixation [6].

17.2 The Mechanism of Photocatalytic N_2 Fixation

The photocatalytic reaction mainly includes three processes: (i) the adsorption of reactant molecules on the photocatalyst surface; (ii) the photocatalyst is excited by light to produce photogenerated carriers; and (iii) the photogenerated carriers are separated and migrate to the photocatalyst surface to trigger the redox reactions. To increase photoconversion efficiency, it is necessary to restrain the recombination of photogenerated carriers. As for photocatalytic N_2 fixation, the photogenerated electrons in the conduction band (CB) of the photocatalyst can reduce the adsorbed N_2 to produce NH_3, and the photogenerated holes in the valence band (VB) can oxidize H_2O to produce O_2. The reduction potential of typical hydrogenation reactions is related to the reduction of N_2 to NH_3 (Figure 17.2). It is well known that the photocatalytic redox ability is related to the energy band position of the photocatalyst and the redox potential of the reactions. Therefore, the CB potential of the photocatalyst

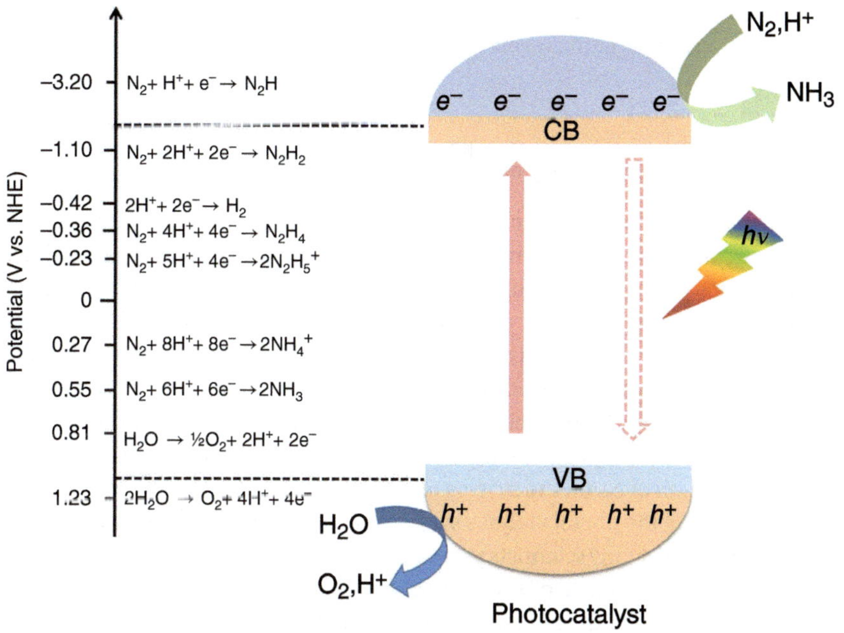

Figure 17.2 Reduction potential (compared to NHE at pH 0) and photocatalytic N_2 fixation for a typical hydrogenation reaction associated with the reduction of N_2 to NH_3.

should be higher than the reduction potential of N_2, so that the electrons in CB are able to reduce N_2 [1, 7].

At present, the mechanisms of photocatalytic N_2 fixation mainly have three types: (i) dissociative pathway; (ii) alternative associative pathway; (iii) distal associative pathway. In the dissociative pathway, the N_2 molecule first dissociates into two N atoms, and then each N atom combines with a proton until it becomes NH_3. In the associative mechanism, the two nitrogen atoms remain connected to each other when they are hydrogenated. There are two possible hydrogenation pathways in the associative mechanism: one is that one N atom of an N_2 molecule is adsorbed on the surface of the catalyst, and the hydrogenation process first occurs on another N atom until one NH_3 molecule is generated. Then, the hydrogenation process occurs on the N atom, which is connected to the catalyst surface until the second NH_3 molecule is generated [6, 8]. The second method involves adsorbing one N atom from a N_2 molecule onto the catalyst's surface, where the hydrogenation process alternately affects two N atoms until all hydrogenation has occurred on both N atoms [9, 10].

17.3 Influencing Factors of Photocatalytic N_2 Fixation Efficiency

There are many factors that could affect the efficiency of photocatalytic N_2 fixation reactions. These factors include the N_2 adsorption ability of the photocatalyst, the physical and chemical properties of the photocatalyst, and environmental factors such as reaction temperature, reaction pressure, reaction light intensity, and catalyst dosage.

17.3.1 N_2 Adsorption Ability of Photocatalyst

It is well known that the first step of the catalytic reaction is the adsorption of reactants on the catalyst surface. Therefore, the amount of N_2 adsorption is proportional to the efficiency of the N_2 fixation reaction. Many studies try to increase the specific surface area of the photocatalyst to improve the adsorption capacity of N_2 through morphology control or other methods. In addition to the adsorption amount, previous studies have shown that the adsorption strength of the photocatalyst to N_2 is also very important for the N_2 fixation reaction. The stronger the adsorption strength, the better the effect of N_2 activation, and the easier the N_2 reduction reaction. The adsorption strength of N_2 on the surface of the photocatalyst is closely related to the atomic structure of the photocatalyst's surface and the structure of the exposed crystal plane. Therefore, many scientists pursue efficient N_2 chemisorption by regulating the surface atomic structure of the photocatalyst, such as by constructing surface defects on the surface of the photocatalyst to create N_2 chemisorption sites [11].

17.3.2 Intrinsic Properties of Photocatalysts

The intrinsic properties of photocatalysts are the key factors affecting the photocatalytic N_2 fixation reaction. Intrinsic properties include band gaps, potentials of the energy bands, and the recombination rate of photogenerated carriers. The ideal band gap for photocatalytic N_2 fixation is around 2.0 eV. On this basis, the CB potential needs to be more negative than the redox potential of the N_2 reduction reaction. Only in this way do the electrons from the CB have the ability to reduce the N_2. Before the photocatalytic reaction, the photocatalyst needs to be excited by light to produce photogenerated carriers (electrons and holes). After the generation of photogenerated carriers, the recombination rate has a great influence on photocatalytic N_2 fixation. The lower the recombination rate, the more photogenerated electrons will participate in the N_2 reduction reaction. In recent years, many scientists have strived to prolong the lifetime of photogenerated carriers. Many strategies have been developed to prolong the lifetime of photogenerated carriers. These strategies [12, 13] include the construction of heterojunctions, the modification of functional groups, the loading of cocatalysts, and element doping.

17.3.3 Environmental Factors of Photocatalytic Reaction

In addition to the intrinsic properties of photocatalysts, the influence of environmental factors on the photocatalytic N_2 fixation reaction cannot be ignored. There are many environmental factors, mainly temperature, pressure, pH value, sacrificial agent, and so on.

The reaction temperature during the photocatalytic N_2 fixation has an important influence on catalytic efficiency. First, the increase in temperature could accelerate the collision between N_2 molecules, and the collision is conducive to the generation of ammonia. Second, the N_2 fixation process is an exothermic reaction, so the low temperature is beneficial to the forward progress of the N_2 fixation reaction. However, the low temperature will affect the reaction rate, thus prolonging the reaction time and decreasing the yield of the NH_3. Therefore, finding a suitable temperature in the photocatalytic N_2 fixation process is very important.

If the photocatalytic N_2 fixation reaction is carried out under the gas-solid reaction condition, the pressure of the reaction vessel has a great influence on the reaction process. It is well known that within a certain range, the higher the reaction pressure corresponds to the higher concentrations of reactive gas molecules. The higher concentrations of the reactants may induce more reaction products.

If the photocatalytic N_2 fixation reaction is carried out under the liquid–solid reaction condition, the pH value of the reaction liquid has a great influence on the reaction process. Since NH_3 itself is an alkaline substance and the N_2 fixation reaction is a reversible process, a strongly alkaline solution is not conducive to the photocatalytic N_2 fixation reaction. However, if the reaction solution is too acidic, the excess H^+ in the solution will combine with the hydroxyl groups on the surface of the photocatalyst. In this case, the utilization of holes will be reduced. As a result, the recombination rate of photogenerated electrons and holes may be increased.

Under the liquid–solid reaction condition, the sacrificial agent is also a crucial factor. This is because the sacrificial agent could consume the photogenerated holes and restrain the recombination of photogenerated holes and electrons. Therefore, the sacrificial agent could help the photocatalyst produce more electrons to reduce N_2. Commonly used sacrificial agents include triethanolamine ($C_6H_{15}NO_3$) [14], sodium sulfite (Na_2SO_3) [15], methanol (CH_3OH) [16, 17], ethylene glycol (C_3H_8O) [18], etc. However, the sacrificial agent may produce some substances, which may inhibit the photocatalytic N_2 fixation reaction. Therefore, scientists have paid attention to the exploitation of photocatalysts that could directly reduce N_2 without sacrificial agents.

17.4 Photocatalytic N_2 Fixation Materials

With the development of photocatalytic N_2 fixation, many scientists are exploring different materials for photocatalytic N_2 fixation. At present, the materials used in the field of photocatalytic N_2 fixation can be divided into the following types: metal oxides, hydrated metal oxides, metal sulfides, and other materials (Figure 17.3).

17.4.1 Metal oxide

Among many single metal oxides, TiO_2 is an outstanding photocatalyst due to its high physical and chemical stability, strong activity, low cost, and nontoxicity. In 1977, Schrauze and Guth first reported the use of TiO_2 as the N_2 fixation catalyst

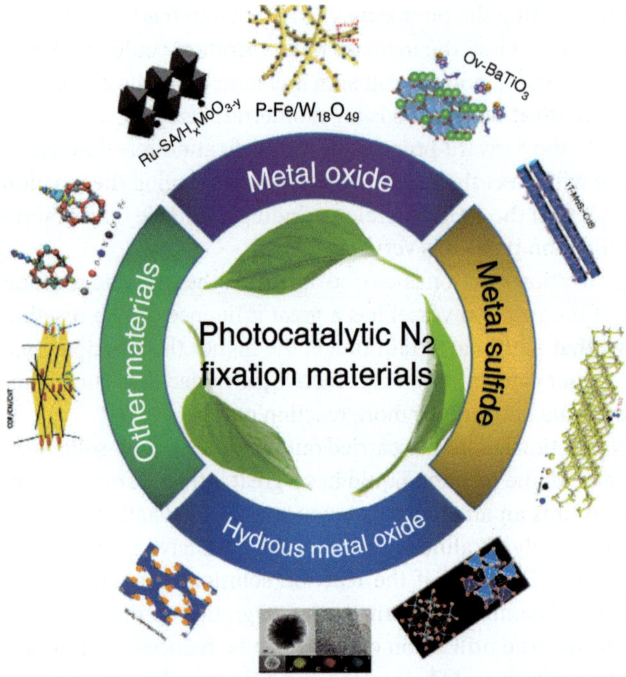

Figure 17.3 Classification of materials in the field of photocatalysis. Source: Li et al. [19]/Reproduced with permission from John Wiley & Sons.

to convert N_2 into NH_3 under ultraviolet light irradiation. Subsequently, to further improve the photocatalytic N_2 fixation performance of TiO_2, the scientists doped TiO_2 with different elements. For example, iron ion doping can improve the yield of NH_3. Meng et al. reported a high-efficiency Fe-mediated Bi_2MoO_6 nitrogen-fixing photocatalyst [20], Fe-BMO samples with different ratios of Fe:Bi were synthesized by the solvothermal method, and the morphology of pristine BMO shows a hollow microsphere structure. As the amount of Fe^{3+} increased, the yolk-shell structure was gradually formed in Fe-BMO samples. By analyzing the HADDF- STEM image (Figure 17.4a) and EDX mapping (Figure 17.4b) of the 0.5% Fe-BMO sample, it can be seen that Bi, Mo, O, and Fe are uniformly distributed in the 0.5% Fe-BMO, indicating that there is Fe element in Bi_2MoO_6. The authors compared the photocatalytic NH_3 production of BMO and different ratios of Fe : Bi, and Fe-modulated Bi_2MoO_6 exhibited excellent visible-light-driven nitrogen fixation photocatalytic activity (Figure 17.4c). The existence of oxygen vacancies can greatly improve the adsorption and activation capacity of N_2 [20], thereby improving the N_2 fixation effect of MoO_3.

Li et al. prepared [22] a nitrogen-fixing photocatalyst MoO_{3-x} nanobelt without a sacrificial agent and noble metal promoter at room temperature and pressure by hydrothermal method. As shown in the HAADF image (Figure 17.4d), the white arrows point to the ordered oxygen vacancies residing at the (001) and (100) surfaces of MoO_{3-x}. In addition, the red lines mark the dislocation formed in MoO_{3-x}. The

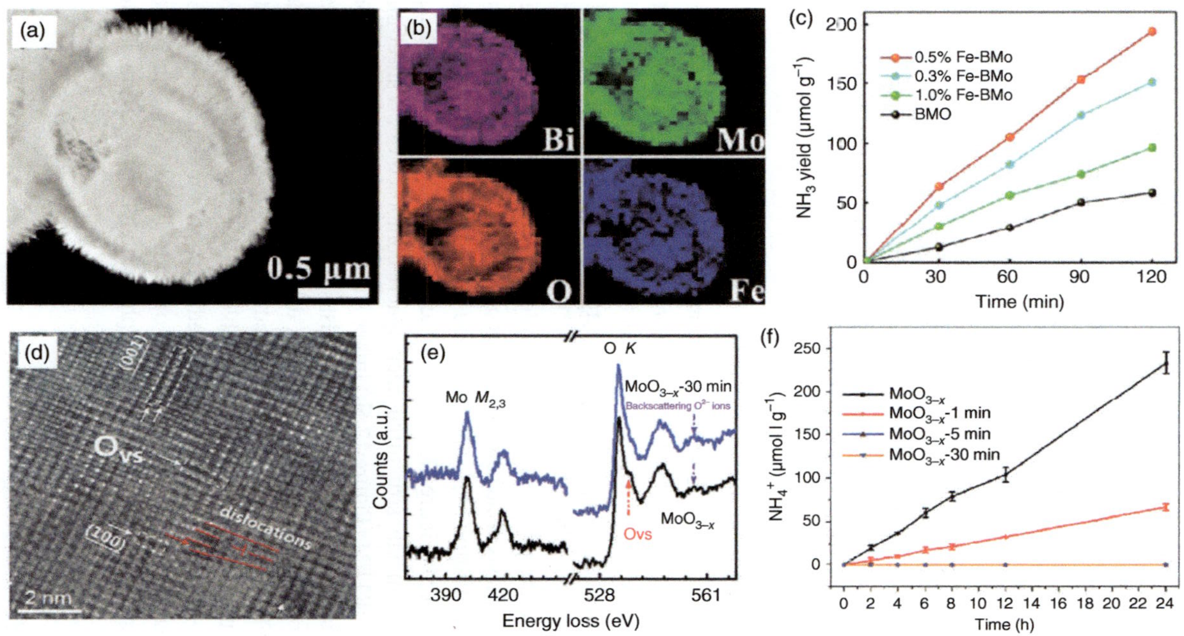

Figure 17.4 (a) HADDF STEM image and (b) EDX mapping were used to analyze the elemental composition of 0.5% Fe-BMo sample; (c) N_2 fixation efficiency of photocatalysts with Fe/Bi molar ratios of 0, 0.3, 0.5, and 1%. Source: Meng et al. [20]/Reproduced with permission from Elsevier. (d) HAADF image analysis of oxygen vacancies on the (001) and (100) planes of MoO_{3-x} samples; (e) EELS spectra of MoO_{3-x} and MoO_{3-x}-30 minutes (MoO_{3-x} was annealed in a tube furnace at 400 °C for 30 minutes in air); (f) N_2 fixation efficiency of MoO_{3-x}, MoO_{3-x}-1 minute, MoO_{3-x}- 5 minutes, MoO_{3-x}-30 minutes (experimental conditions: Xenon lamp 300 w to simulate sunlight, photocatalyst 50 mg, pure water 100 ml). Source: Li et al. [21]/Reproduced with permission from Elsevier.

dislocations formed in the MoO_{3-x} may partially release the internal strain induced by the formation of oxygen vacancies. In order to reflect the local electron density of MoO_{3-x}, the Mo and O of MoO_{3-x} and MoO_{3-x}-30 minutes were analyzed by EELS (Figure 17.4e), and the lattice distortion and reduced symmetry of MoO_{3-x} led to enhanced peaks of oxygen K-edges. By comparing the photocatalytic nitrogen fixation activity of samples with different annealing times (Figure 17.4f); it is found that the oxygen vacancies gradually disappear with the prolongation of annealing time, which leads to the weakening of the photocatalytic nitrogen fixation activity. Besides MoO_3, many other single-metal oxide catalysts have also been found to have photocatalytic N_2 fixation effects, such as Fe_2O_3, WO_3, Ga_2O_3, Bi_2O_3, and TiO_2. In addition, bimetallic oxides have also been studied in the field of photocatalytic N_2 fixation.For example, $SrTiO_3$ and $BaTiO_3$, which loaded RuO_2 and NiO_2 have good photocatalytic N_2 fixation effects. The hydrogen bond center and transition metal iron of the phosphoric acid-treated perovskite ($LaFeO_3$) as two active sites can enhance the adsorption and dissociation of N_2, thus exhibiting excellent photocatalytic N_2 fixation activity.

17.4.2 Hydrous Metal Oxide

In 1987, Tennakone and his co-workers found that hydrated iron oxide has better photocatalytic N_2 fixation activity than commercial TiO_2 photocatalysts under visible-light irradiation [23]. The reason for that is this is that hydrated iron oxide has a relatively negative flat band potential and a strong adsorption effect for N_2. In 1992, researchers found that the introduction of V^{3+} sites into hydrated iron oxide would further improve the photocatalytic N_2 fixation effect [24]. Li et al. utilized microwave-assisted synthesis of carbon-surface-decorated HWO/C-X photocatalysts for the synergistic optimization of N_2 activation and subsequent photoinduced protonation. As shown in TEM (Figure 17.5a), the surface of HWO/C modified with carbon appears as a hierarchically structured microsphere composed of 2D sheet-like building blocks. The HRTEM (Figure 17.5b) image shows that the carbon is amorphous on the HWO layer. Carbon is uniformly distributed in the surface layer of HWO in an amorphous form (Figures 17.5c and d). The photocatalytic nitrogen fixation activity on different materials was explored (Figure 17.5e), and the results showed that carbon modification was indispensable for the photoactive hydrated $WO_3{\cdot}H_2O$ to improve the efficiency of nitrogen reduction to ammonia [19]. Since then, hydrated cuprous oxide, hydrated samarium oxide, hydrated vanadium oxide, and other hydrated oxides have also been used as photocatalytic N_2 fixation photocatalysts.

17.4.3 Metal Sulfide

Compared with metal oxides, the band gap of metal sulfides is narrower than that of oxides, which is more conducive to the absorption of visible light. A large number of researchers continue to develop metal sulfides as catalysts for photocatalytic N_2

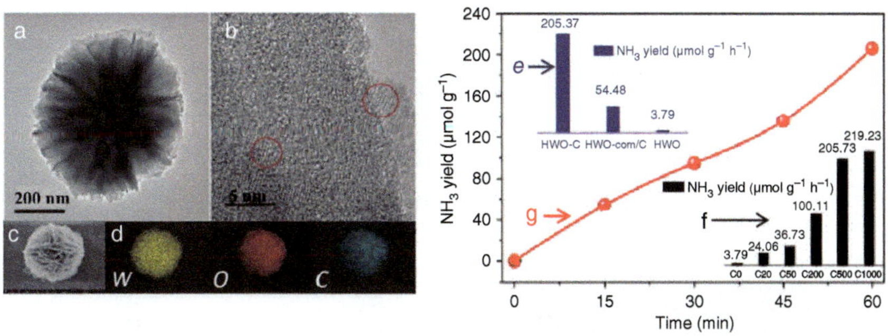

Figure 17.5 (a) TEM image; (b) HRTEM image; (c) SEM image of HWO/C samples; (d) EDS mapping of HWO/C samples; (e) N_2 fixation yield of HWO/C, HWO-com/C samples, and HWO; (f) Photocatalytic N_2 efficiency (HWO/C-*x*) of samples with different quality of glucose added for surface decoration; (g) Photocatalytic N_2 fixation rates of HWO/C500 with time under UV-Vis light. Source: Li et al. [19]/Reproduced with permission from John Wiley & Sons.

fixation. Miyama et al. applied CdS and Pt/CdS as photocatalysts in the N_2 reduction reaction [25]. After the loading of Pt nanoparticles, the photocatalytic N_2 fixation activity of CdS was significantly improved. Besides Pt, other noble metals such as Ag and Ru can also improve the photocatalytic N_2 fixation effect of metal sulfides.

17.4.4 Other Materials

In recent years, 2D layered semiconductor materials, such as bismuth oxyhalide (BiOX, X = Cl, Br, I) and graphitic carbon nitride ($g\text{-}C_3N_4$), have been found to have the photocatalytic N_2 fixation effect. Scientists found that oxygen vacancies can improve the photocatalytic N_2 fixation activity of BiOBr. The photocatalytic N_2 fixation reaction mechanisms of different crystal faces of bismuth oxyhalide are different. Lan et al. used a one-step thermal solvent process to synthesize Bi@BiOBr heterostructured microspheres (Figure 17.6a) [26]. As shown in Figure 17.6b, the TEM image can clearly observe the fringes of the BiOBr (110) crystal plane and the metallic Bi (200) crystal plane, respectively. As can be observed from Figure 17.6c, the pristine BiOBr possesses a slight OV signal at $g = 2.002$, but the generated Bi@BiOBr has a stronger OV signal with increasing glycerol addition, meaning the one-pot solvothermal strategy is a simple and effective method to obtain Bi@BiOBr with plentiful OVs. As shown in Figure 17.6d,e, the glycerol concentration significantly affects the mesoporous structure, specific surface area, and photocatalytic activity of the samples. Compared to Bi@BiOBr-1 and Bi@BiOBr-3, Bi@BiOBr-2 shows the largest specific surface area ($346.9\,m^2\,g^{-1}$) and better photocatalytic activity for nitrogen fixation. With the development of the field of photocatalytic NH_3 synthesis, nonmetallic semiconductor catalysts are also used by scientists as a research focus. Boron-doped diamond and metal-diamond heterojunctions were used for photocatalytic N_2 fixation. But more researchers are focusing on carbon-based nanomaterials, such as carbon nitride.

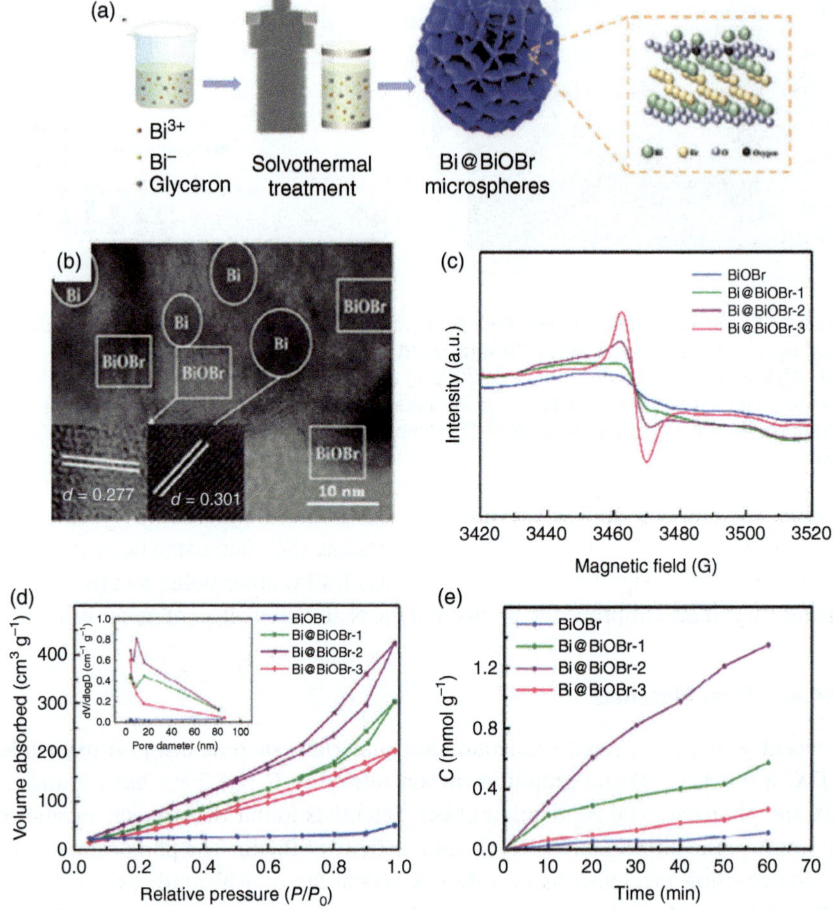

Figure 17.6 (a) Schematic diagram of the synthetic route of Bi@BiOBr microspheres; (b) TEM image of Bi@BiOBr-2; (c) EPR patterns of different samples; (d) N_2 adsorption – desorption curves of different samples (the inset is the pore size distribution of the samples); (e) Photocatalytic N_2 fixation activity of different samples; (0.5, 1.0, and 1.5 mmol of $Bi(NO_3)_3 \cdot 5H_2O$ precursors were dissolved in 0.5 mmol NaBr in 25 ml of glycerol, respectively, to obtain Bi@BiOBr-1, Bi@BiOBr-2, Bi@BiOBr-3. BiOBr was prepared by replacing the solvent in the above method with deionized water). Source: Lan et al. [26]/Reproduced with permission from Elsevier.

17.5 Challenges and Opportunities

The use of solar energy to drive N_2 fixation has attracted widespread attention due to its environmental friendliness and cost-effectiveness. However, as far as current research is concerned, there are still great challenges in realizing industrial applications. First of all, the photocatalytic N_2 fixation reaction is accompanied by the photocatalytic water splitting reaction. The H_2 evolution reaction in the photocatalytic water splitting reaction will compete with electrons with N_2. Generally, the

H_2 evolution reaction will occur at a more positive potential compared to the N_2 reduction, and thus it will limit the N_2 reduction reaction. Second, the difficulty of adsorption and activation of N_2 molecules on photocatalysts greatly limits the N_2 fixation ability. This is because solar energy cannot provide the huge energy to split the N≡N. In order to weaken the N≡N bond energy of N_2, N_2 molecules need to be adsorbed on the surface of the photocatalyst and thus be activated. Finally, how to achieve an efficient photocatalytic N_2 fixation without using hole sacrificial agents? The addition of hole sacrificial reagents is not an ideal choice for improving N_2 fixation efficiency. This is because the addition of sacrificial reagents can interfere with the detection of NH_3 and may cause the corrosion of photocatalyst. The photocatalytic N_2 fixation performance is poor and much below the minimal requirement for large-scale applications when the combined effects of the aforementioned conditions are present.

In order to overcome the above factors, future photocatalytic nitrogen fixation materials need to possess several characteristics: (i) photogenerated electron and hole can be efficiently separated without the presence of sacrificial agents; (ii) photocatalysts possess suitable properties to adsorb and active N_2; and (iii) photocatalyst is selective for the N_2 reduction reactions.

References

1 Chen, L.W., Hao, Y.C., Guo, Y. et al. (2021). Metal–organic framework membranes encapsulating gold nanoparticles for direct plasmonic photocatalytic nitrogen fixation. *J. Am. Chem. Soc.* 143: 5727–5736.

2 Bo, Y., Wang, H., Lin, Y. et al. (2021). Altering hydrogenation pathways in photocatalytic nitrogen fixation by tuning local electronic structure of oxygen vacancy with dopant. *Angew. Chem. Int. Ed.* 60: 16085–16092.

3 Canfield, D.E., Glazer, A.N., and Falkowski, P.G. (2010). The evolution and future of Earth's nitrogen cycle. *Science* 330: 192–196.

4 Wang, Y., Suzuki, H., Xie, J. et al. (2018). Mimicking natural photosynthesis: solar to renewable H_2 fuel synthesis by Z-scheme water splitting systems. *Chem. Rev.* 118: 5201–5241.

5 Zuo, S., Zhang, H., Li, X. et al. (2022). Dual active sites boosting photocatalytic nitrogen fixation over upconversion mineral nanocomposites under the full spectrum. *ACS Sustainable Chem. Eng.* 10: 1440–1450.

6 Wang, L., Wu, W., Liang, K., and Yu, X. (2022). Advanced strategies for improving the photocatalytic nitrogen fixation performance: a short review. *Energy Fuels* 36: 11278–11291.

7 Dong, G., Ho, W., and Wang, C. (2015). Selective photocatalytic N_2 fixation dependent on g-C_3N_4 induced by nitrogen vacancies. *J. Mater. Chem. A* 3: 23435–23441.

8 Shen, Z.K., Yuan, Y.J., Wang, P. et al. (2020). Few-layer black phosphorus nanosheets: a metal-free cocatalyst for photocatalytic nitrogen fixation. *ACS Appl. Mater. Interfaces* 12: 17343–17352.

9 Bian, S., Wen, M., Wang, J. et al. (2020). Edge-rich black phosphorus for photocatalytic nitrogen fixation. *J. Phys. Chem. Lett.* 11: 1052–1058.

10 Xu, C., Qiu, P., Li, L. et al. (2018). Bismuth subcarbonate with designer defects for broad-spectrum photocatalytic nitrogen fixation. *ACS Appl. Mater. Interfaces* 10: 25321–25328.

11 Gao, X., An, L., Qu, D. et al. (2019). Enhanced photocatalytic N_2 fixation by promoting N_2 adsorption with a co-catalyst. *Sci. Bull.* 64: 918–925.

12 Vu, M.H., Sakar, M., Hassanzadeh-Tabrizi, S.A., and Do, T.O. (2019). Photo (electro) catalytic nitrogen fixation: problems and possibilities. *Adv. Mater. Interfaces* 6: 1900091.

13 Li, H., Mao, C., Shang, H. et al. (2018). New opportunities for efficient N_2 fixation by nanosheet photocatalysts. *Nanoscale* 10: 15429–15435.

14 Fan, J., Zuo, M., Ding, Z. et al. (2020). A readily synthesis of oxygen vacancy-induced $In(OH)_3$/carbon nitride 0D/2D heterojunction for enhanced visible-light-driven nitrogen fixation. *Biochem. Eng. J* 396: 125263.

15 He, C., Li, X., Chen, X. et al. (2020). Palygorskite supported rare earth fluoride for photocatalytic nitrogen fixation under full spectrum. *Appl. Clay Sci.* 184: 105398.

16 Huang, T., Pan, S., Shi, L. et al. (2020). Hollow porous prismatic graphitic carbon nitride with nitrogen vacancies and oxygen doping: a high-performance visible-light-driven catalyst for nitrogen fixation. *Nanoscale* 12: 1833–1841.

17 Wang, H., Bu, Y., Wu, G., and Zou, X. (2019). The promotion of the photocatalytic nitrogen fixation ability of nitrogen vacancy-embedded graphitic carbon nitride by replacing the corner-site carbon atom with phosphorus. *Dalton Trans.* 12: 11724–11731.

18 Ge, J., Xu, J., Liu, Y. et al. (2019). Enhanced nitrogen photo fixation performance of transition metal-doped urchin-like $W_{18}O_{49}$ under visible-light irradiation. *Nano* 14: 1950143.

19 Li, X., Wang, W., Jiang, D. et al. (2016). Efficient solar – Driven nitrogen fixation over carbon–tungstic-acid hybrids. *Chem. Eur. J.* 22: 13819–13822.

20 Meng, Q., Lv, C., Sun, J. et al. (2019). High-efficiency Fe-Mediated Bi_2MoO_6 nitrogen-fixing photocatalyst: Reduced surface work function and ameliorated surface reaction. *Appl. Catal., B* 256: 117781.

21 Li, G., Yang, W., Gao, S. et al. (2021). Creation of rich oxygen vacancies in bismuth molybdate nanosheets to boost the photocatalytic nitrogen fixation performance under visible light illumination. *Biochem. Eng. J* 404: 127115.

22 Li, Y., Chen, X., Zhang, M. et al. (2019). Oxygen vacancy-rich MoO_{3-x} nanobelts for photocatalytic N_2 reduction to NH_3 in pure water. *Catal. Sci. Technol.* 9: 803–810.

23 Tennakone, K., Wickramanayake, S., Fernando, C.A.N. et al. (1987). Photocatalytic nitrogen reduction using visible light. *J. Chem. Soc., Chem. Commun.* 14: 1078–1080.

24 Tennakone, K., Thaminimulla, C.T.K., and Bandara, J.M.S. (1992). Nitrogen photoreduction by vanadium (III)-substituted hydrous ferric oxide. *J. Photochem. Photobiol., A* 68: 131–135.

25 Miyama, H., Fujii, N., and Nagae, Y. (1980). Heterogeneous photocatalytic synthesis of ammonia from water and nitrogen. *Chem. Phys. Lett.* 74: 523–524.
26 Lan, M., Zheng, N., Dong, X. et al. (2021). One-step in-situ synthesis of Bi-decorated BiOBr microspheres with abundant oxygen vacancies for enhanced photocatalytic nitrogen fixation properties. *Colloids Surf., A* 623: 126744.

18

Photocatalytic N_2 Fixation: A Step Closer to the Solar Farm

Yuanyi Zhou[1], Cailin Xiao[1], Songmei Sun[1], Xiaoman Li[1], and Wenzhong Wang[1,2]

[1] *Shanghai Institute of Ceramics, Chinese Academy of Sciences, State Key Laboratory of High Performance Ceramics and Superfine Microstructure, 1295 Dingxi Road, Shanghai, Shanghai 200050, China*
[2] *University of Chinese Academy of Sciences, Hangzhou Institute for Advanced Study, School of Chemistry and Materials Science, 1 Sub-lane Xiangshan, Hangzhou 310024, China*

18.1 Introduction

As the building block of proteins and nucleic acids, nitrogen is essential for life. Benefiting from the Haber–Bosch process for ammonia (NH_3) synthesis, the production and industrial use of artificial nitrogen fertilizers support the burgeoning human population and social prosperity. Prized for its considerable hydrogen content (17.6 wt%), high energy density (12.8 GJ m^{-3}), and facile liquefaction under mild conditions, NH_3 has emerged as a strong and increasingly compelling candidate as the renewable energy-sourced fuel of the future. More importantly, the only exhaust upon NH_3 combustion could be N_2 and H_2O, which would be safely released into the environment directly. As a result, the "ammonia economy" (Figure 18.1), a sustainable cycle created by the conversion of N_2 and NH_3, has caught the attention of both the academic community and business [1].

Currently, NH_3 is the second-highest produced chemical commodity after sulfuric acid (H_2SO_4), globally yielding about 175 million tons per year. Although the reaction is thermodynamically accessible (N_2 (g) + 3 H_2 (g) → 2 NH_3 (g), ΔH_{298K} = −92.2 kJ mol^{-1}), the Haber-Bosch process is conducted under high temperature (400–600 °C) and high pressure (200–300 atm) over multicomponent metal-based catalysts, attributing to the robust N−N triple bond (941.69 kJ mol^{-1}). This energy-intensive working condition is responsible for 1–2% of the global energy consumption (mainly via steam reforming for H_2 production and 30% for NH_3 synthesis) and 1.5% of global CO_2 emissions, thus causing complex energy and environmental concerns [2]. Under the context of expanding clean and low-carbon contents in the global energy structure, developing more sustainable and resilient approaches for N_2 fixation is craved, especially those that could be realized under ambient conditions like nitrogenase.

UV-Visible Photocatalysis for Clean Energy Production and Pollution Remediation: Materials, Reaction Mechanisms, and Applications, First Edition. Edited by Xinchen Wang, Masakazu Anpo, and Xianzhi Fu.
© 2023 WILEY-VCH GmbH. Published 2023 by WILEY-VCH GmbH.

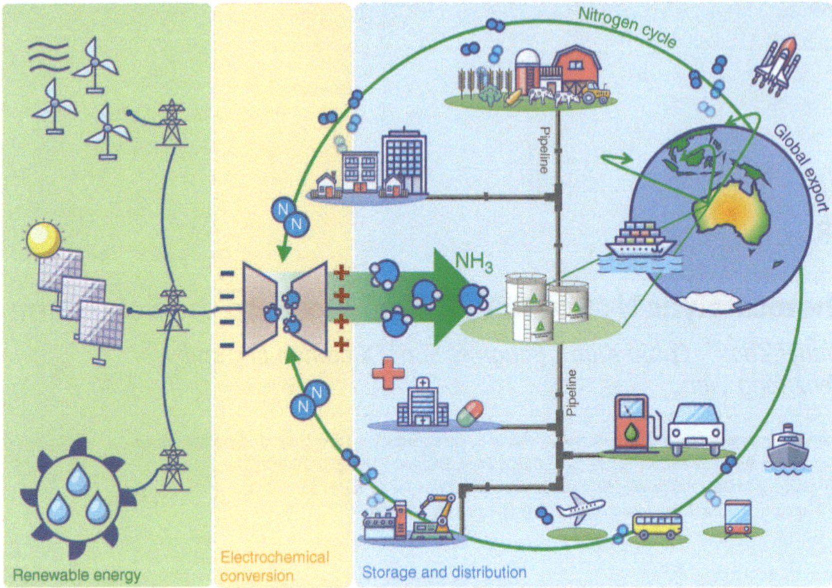

Figure 18.1 Vision of the "ammonia economy" in which the energy sources and uses are all based on ammonia. Source: Reproduced with permission from MacFarlane *et al.* [1]; © 2020 Elsevier Inc.

18.2 Photocatalytic N_2 Fixation

Photocatalysis allows promoting difficult reactions under mild conditions by photoexcitation instead of thermal activation, arising as an ideal, energy saving, and environmentally benign approach [3]. The application of this technique to N_2 fixation dates back to the 1940s, when Dhar *et al.* found that sunlight illumination enhanced the rate of N_2 fixation in sterilized soils [4]. In 1977, the phenomenon of N_2 fixation was rediscovered by Schrauzer *et al.* during the evaluation of acetylene photoreduction over TiO_2 catalyst [5]. When quantifying the products formed under different concentrations of Ar and N_2, NH_3 and hydrazine (N_2H_4) were detected. Even though much of these early works might be questionable, the results and main conclusions opened the discussion on photocatalytic N_2 fixation.

After decades of exploration, numerous semiconductor systems have been investigated as photocatalysts for N_2 fixation. Generally, this process includes several steps: first of all, the photocatalyst is excited under light irradiation, and the photogenerated electrons are stimulated to the conduction band (CB), leaving holes in the valence band (VB). Afterward, the holes that made it to the surface participate in H_2O oxidation, generating protons (H^+) and O_2. At the same time, the counterpart electrons react with the activated N_2 molecule with the assistance of H^+ to produce NH_3. Consequently, NH_3 comes from N_2 and H_2O under ambient conditions using sunlight as the energy source ($2\ N_2\ (g) + 6\ H_2O\ (l) + h\nu \rightarrow 4\ NH_3\ (l) + 3\ O_2\ (g)$). In other words, N_2 activation and the transfer of electrons and protons are critical for the reductive half reaction.

Since N_2 represents one of the most inert chemical species, weakening its triple bond is the prerequisite for photocatalytic N_2 fixation [6]. Metal elements with d-orbitals are competent to achieve the goal. As the electron-rich active center, the metal site binds to N_2 with its unoccupied d-orbital accepting the electron density from the occupied p-orbital of N, while donating the valence electron from its occupied d-orbital to the π^*-orbital of N (π backdonation). This interaction would maneuver the electron distribution in the N_2 molecule. The stronger the π backdonation is realized, the easier the N_2 molecule is activated. Hence, the key to promoting this step lies in constructing active sites with N_2 bonding capability and an electron-abundant nature.

The next step is the transfer of electrons and protons. The injection of electrons into the N_2 molecule would not only strengthen the π backdonation but also facilitate the following hydrogenation. Hence, no matter the heterogeneous Fe and Ru catalyst in the Haber–Bosch process or the FeMo-cofactor in nitrogenase, the "electron promoter" is essential (corresponding to the alkali/alkaline metals and the electron-donation Fe-protein, respectively). For photocatalysis, this could be reflected by the separation and transfer of the charge carriers. Efficient charge carrier dynamics would benefit both electron injection through the active sites and H_2O oxidation to ensure the supply of protons. As a result, numerous methods, such as crystal, defect, and surface/interface engineering, were created to regulate photoabsorption and speed up charge transfer on the surface.

In the following section, we present our group's efforts on photocatalytic N_2 fixation. It should be noted that, although our work might be introduced focusing on one of the essentials, the realization of photocatalytic N_2 fixation is the integrated embodiment of the factors above. Furthermore, we also combined the functional units to build composite materials, aiming to improve the performance. We end this chapter with some remarks that depict the challenges and opportunities in this research field.

18.3 Current Progress

The interaction between carbon materials and N_2 molecules provides a promising strategy for N_2 activation via carbon-based surface engineering [7, 8]. Taking tungsten acid ($WO_3{\cdot}H_2O$, and HWO) as a model photocatalyst, the decoration of carbon on the surface for effective N_2 adsorption has been explored [9]. Carbon modification was realized through a facile microwave treatment. Temperature programmed desorption (TPD) investigation (Figure 18.2a) was carried out to visualize N_2 activation over the dehydrated HWO (HWO-500). The results indicated N_2 chemisorption on both tungsten and carbon sites. The moderate interaction between carbon and N_2 favored the reduction. Transient current responses provided another convincing supplement for the decisive role of carbon (Figure 18.2b and c). A more pronounced current decay over HWO/C in the N_2-saturated electrolyte than that over HWO implied a higher degree of photo-induced electron consumption by N_2. Considerable improvements in the photocatalytic performance further verified

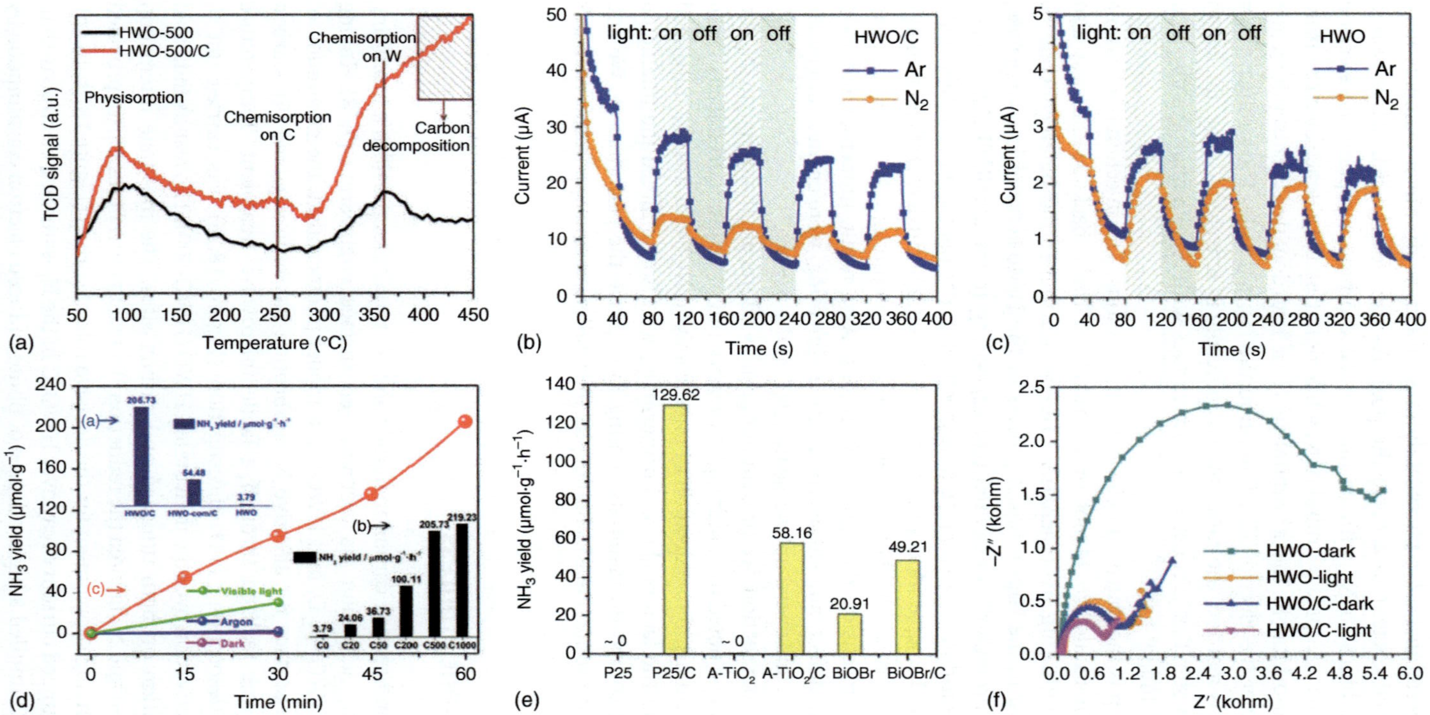

Figure 18.2 (a) N_2-TPD profiles of the as-prepared HWO-500 and HWO-500/C photocatalysts; Transient current-time curves over HWO/C (b) and HWO (c) electrodes in an Ar or N_2 atmosphere; (d) Photocatalytic N_2 fixation rate; (e) The photocatalytic N_2 fixation rates of P25, Anatase TiO_2 (A-TiO_2), BiOBr; the P25/C, A-TiO_2/C, and BiOBr/C prepared by the same method as the HWO/C; (f) Electrochemical impedance spectroscopy Nyquist plots with or without Xe lamp irradiation over HWO and HWO/C. Source: Adapted from Wang *et al.* [9].

the feasibility of this strategy. After decoration, the NH_3 evolution rate over HWO increased from 3.8 to 205.7 μmol g^{-1} h^{-1} (Figure 18.2d). Moreover, a similar promotion was also achieved over other common photocatalysts (Figure 18.2e), such as P25 TiO_2, anatase TiO_2, BiOBr, *etc.*, which confirmed the universality of carbon modification on photocatalytic N_2 fixation. Despite facilitating N_2 chemisorption, carbon decoration would also accelerate the charge carrier dynamics Figure 18.2f) that offered sufficient electrons for the photoreduction.

Considering that the conduction band of a metal oxide semiconductor is usually hybridized by the metal p, d orbitals while the photogenerated electrons are mainly located on the metal sites, coordination-unsaturated low-valent metal species may serve as both N_2 activation and hydrogenation centers that allow quick reaction kinetics for N_2 fixation [10]. To underline this opinion, bismuth monoxide (BiO) quantum dots were synthesized (Figure 18.3a), achieving an NH_3 evolution rate of up to 1226 μmol g^{-1} h^{-1} without the assistance of a sacrificial agent or cocatalyst (Figure 18.3b) [11]. Quantum chemical calculations based on density functional theory (DFT) indicated that Bi atoms underwent sp^3d hybridization, resulting in a trigonal bipyramid orbital distribution and leaving a lone pair of electrons on the Bi atom. When interacting with the catalyst, the N_2 molecule might be stretched and activated by three alternatively arranged Bi atoms through electron donation to the empty Bi 6d orbitals and π backdonation to the unoccupied anti-bonding orbitals, bringing about a $1N_2$-3Bi side-on bond structure (Figure 18.3c). This configuration not only strongly weakened the N–N triple bond but also contributed to a one-step three-electron N_2 reduction process instead of the traditional consecutive single-electron ones. The conjecture was proved in the electrochemical measurement. The electron transfer number involved in the N_2 reduction was calculated to be 2.93 in the cyclic voltammograms (Figure 18.3d), which was in good agreement with the one-step three-electron process.

Inspired by the potassium promoter in the industrial N_2 fixation catalyst that could facilitate electron injection, NH_3 desorption, and lower the reaction temperature [12, 13], KOH treatment was applied to enhance the photocatalytic performance of carbon nitride (C_3N_4) [14]. With methanol as a facile proton source, the evolution rate of NH_3 over KOH-etched C_3N_4 boosted to 3.632 mmol g^{-1} h^{-1}, which was four times superior to that of the pristine C_3N_4 (Figure 18.4a and b). Characterizations were conducted to study the internal reasons for the promotion. KOH at moderate concentrations caused a soft crack on C_3N_4 with K^+ bonded through van der Waals forces (Figure 18.4c). The *s*-triazine rings of C_3N_4 were broken, and the C—N=C coordination became more active, leading to better physical and chemical N_2 adsorption over the catalyst (Figure 18.4d). Interestingly, 0.39 atom% of ^{15}N was detected in the photocatalyst after NH_3 synthesis under the $^{15}N_2$ atmosphere and the subsequent calcination to remove the adsorbed nitrogen. In combination with the limited but not negligible performance under the Ar atmosphere, this result implied that after KOH treatment, nitrogen in the C_3N_4 structure could react with the proton in methanol directly, and the remaining vacancies could further adsorb and activate N_2 molecules. Transient current responses revealed the role of methanol (Figure 18.4e). On one hand, methanol could react with the photogenerated holes

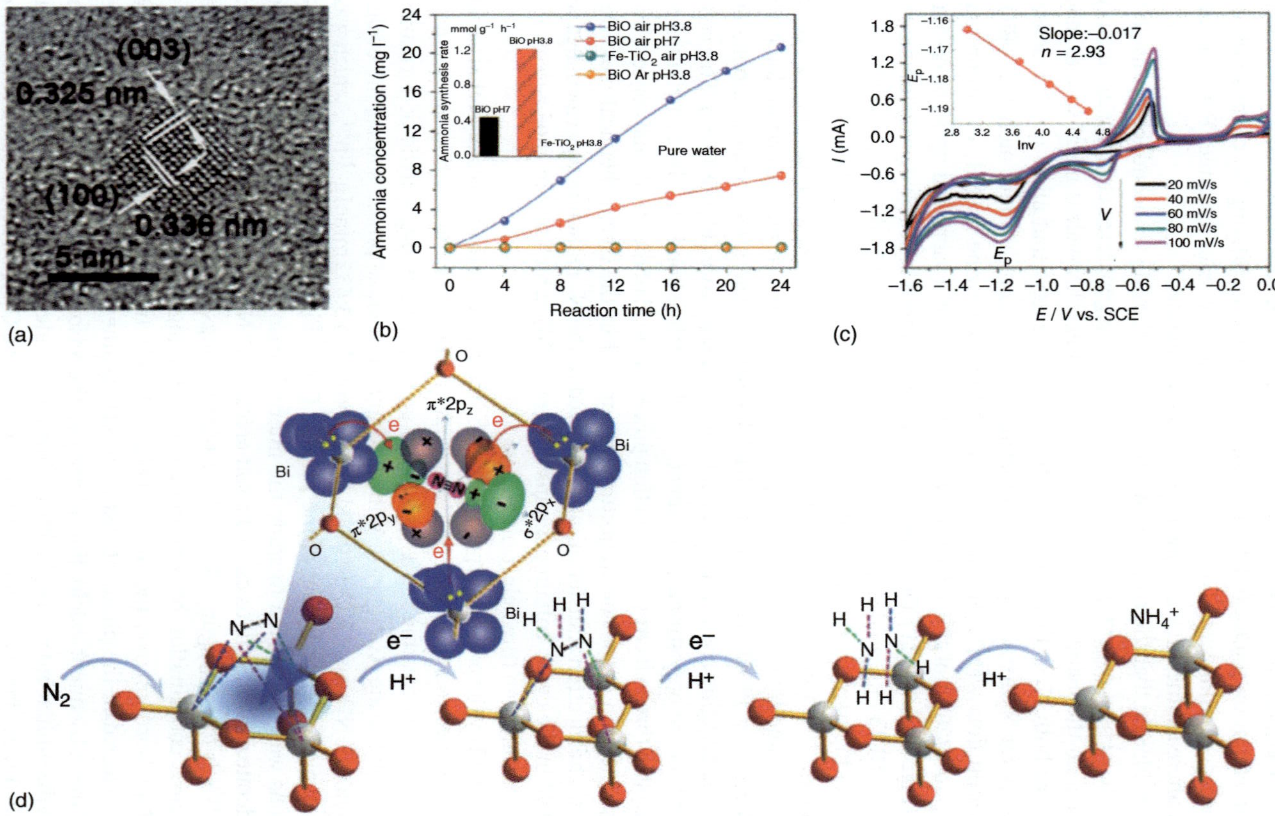

Figure 18.3 (a) High resolution TEM image of the BiO sample; (b) The generated ammonia concentrations at different irradiation times in pure water; (c) Cyclic voltammograms of the BiO electrode at pH = 3.8 in N_2 saturated 0.5 M Na_2SO_4 under room temperature and simulated solar light irradiation; (d) Possible pathway for N_2 activation and hydrogenation on the BiO catalyst. Source: Sun *et al.* [11] Reproduced with permission from Royal Society of Chemistry.

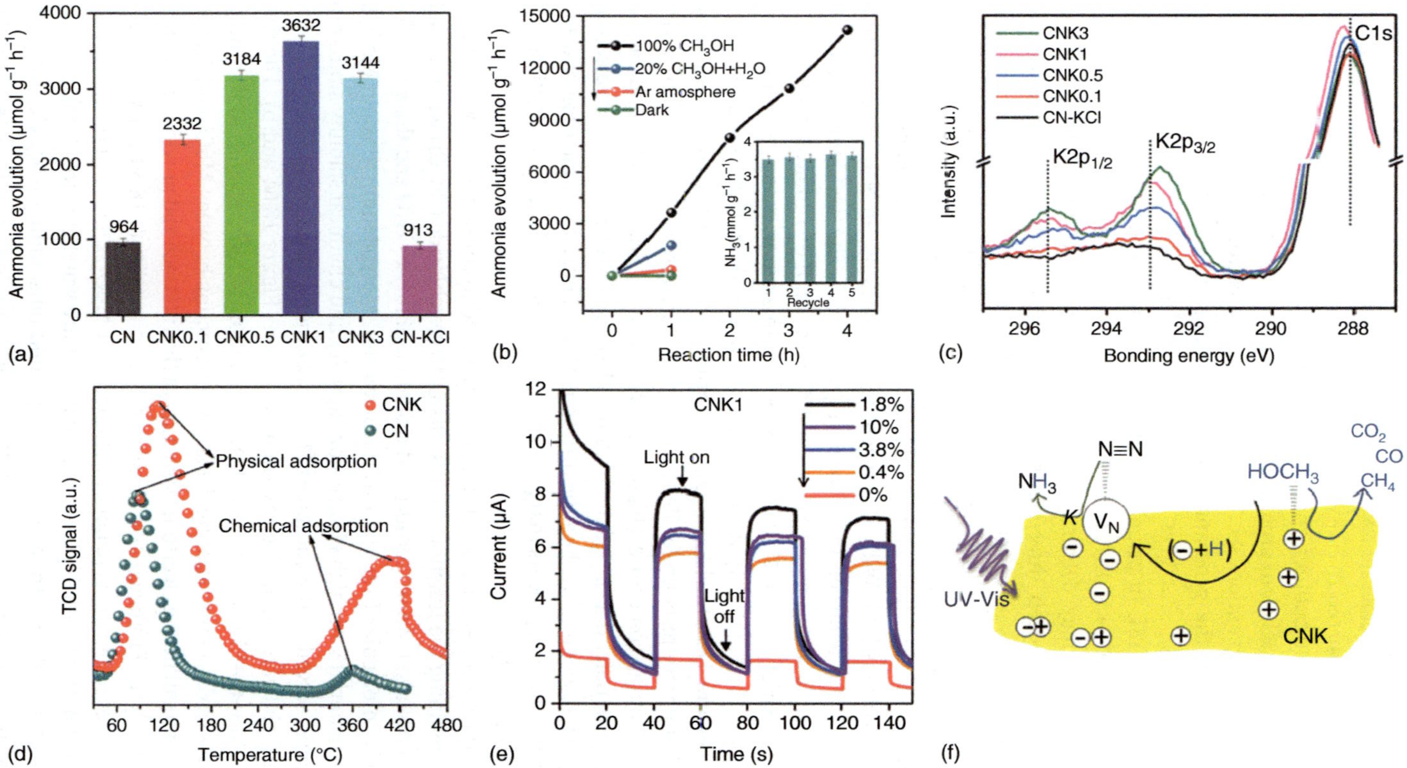

Figure 18.4 (a) Nitrogen fixation rate of the prepared samples under AM 1.5 G illumination; (b) Ammonia evolution with time of CNK1, inset shows the cycling experimental performance of CNK1; (c) XPS spectra of CN and CNKx; (d) N_2-TPD of CN and CNK1; (e) Comparison of different CH_3OH addition amounts in the Na_2SO_4 water electrolyte for CNK1 sample; (f) The proposed mechanism of the photocatalytic N_2 fixation model in which CH_3OH serves as both the solvent and proton source. Source: Reproduced with permission from Li *et al.* [14]; © The Royal Society of Chemistry 2018.

to increase the concentration of reactive electrons. On the other hand, the much lower surface energy of methanol (71.38 mJ m^{-2}) than that of C_3N_4 (115 mJ m^{-2}) made it easier to agglomerate on C_3N_4 and participate in N_2 reduction as a proton donor and part of the electron donor. Therefore, the considerable NH_3 production rate was attributed to both K^+ grafting for N_2 activation as well as NH_3 desorption and methanol as an efficient proton/electron supplier (Figure 18.4f).

During N_2 hydrogenation, the limited number of electron transfers in each step would bring about the generation of a high-energy intermediate (for example: $N_2 + H^+ + e^- \rightarrow N_2H$, −3.2 V *vs.* NHE; $N_2 + 2\ H_2O + 4\ H^+ + 2\ e^- \rightarrow 2\ NH_3OH^+$, −1.83 V *vs.* NHE) [15]. In principle, applying a multi-electron reduction process via $N_2 + 5\ H^+ + 4\ e^- \rightarrow N_2H_5{}^+$ (−0.23 V *vs.* NHE) or $N_2 + 8\ H^+ + 6\ e^- \rightarrow 2\ NH_4{}^+$ (0.274 V *vs.* NHE) may decrease the thermodynamic barrier for NH_3 production. Since the one-step three-electron N_2 reduction over the aforementioned BiO quantum dots was unique, we tried to design the multi-electron process from another perspective, *i.e.* charged excitons. Ultrathin molybdenum disulfide (MoS_2) was selected as a model photocatalyst (Figures 18.5a and b), in which the photoinduced electron-hole pairs could form tightly bound excitons to capture additional electrons [16]. With a sufficient feed of protons, an NH_3 concentration of 0.83 mg L^{-1} was achieved in 10 hours (Figure 18.5c). The well-known A and B absorption bands at 665 nm (1.86 eV) and 616 nm (2.01 eV) indicated the direct excitonic transitions at the K point of the Brillouin zone in mono- and few-layer MoS_2 (Figure 18.5d) [17, 18]. A photoluminescence peak around 633 nm was also observed (Figure 18.5e), which originated from the relaxation of direct excitons. Cyclic voltammetry tests under light irradiation were further conducted to verify the multi-electron conjecture (Figure 18.5f). By analyzing the variation of the reduction peak potential (E_p) along with the change in scan rate (ν), a six-electron transfer process was qualified on the ultrathin MoS_2. Thus, it was the trion-assisted six-electron reduction process that gave rise to an appreciable NH_3 production rate over the photocatalyst.

Bearing the essentials above in mind, we attempted to construct composite materials by combining the functional units. Bismuth oxychloride (BiOCl) was selected as the photocatalyst, prized for its good oxidation ability to H_2O and the π backdonation to N_2 molecules from the surface Bi species. Molybdenum dioxide (MoO_2), a metallic material with high electric conductivity and chemical stability similar to noble metals, served as the cocatalyst [19]. MoO_2 nanosheets were contacted with BiOCl nanoplates through electrostatic interaction (Figure 18.6a and b). The formation of Mo—O—Bi bonds rendered the generation of Mo^{6+} species at the interface (Figure 18.6c), demonstrating the strong interaction between MoO_2 and BiOCl, which could regulate the electronic structure of the interface and provide active sites for photocatalysis. Promoted charge carrier dynamics (Figure 18.6d), as well as N_2 activation (Figure 18.6e), were hence obtained over MoO_2/BiOCl composite, consequently the photocatalytic N_2 fixation activity in pure water (Figure 18.6f).

KOH-treated carbon nitride (K-C_3N_4) exhibited outstanding reactivity for photocatalytic NH_3 synthesis. However, its performance was highly dependent on methanol as a proton source, since K-C_3N_4 could barely provide the required

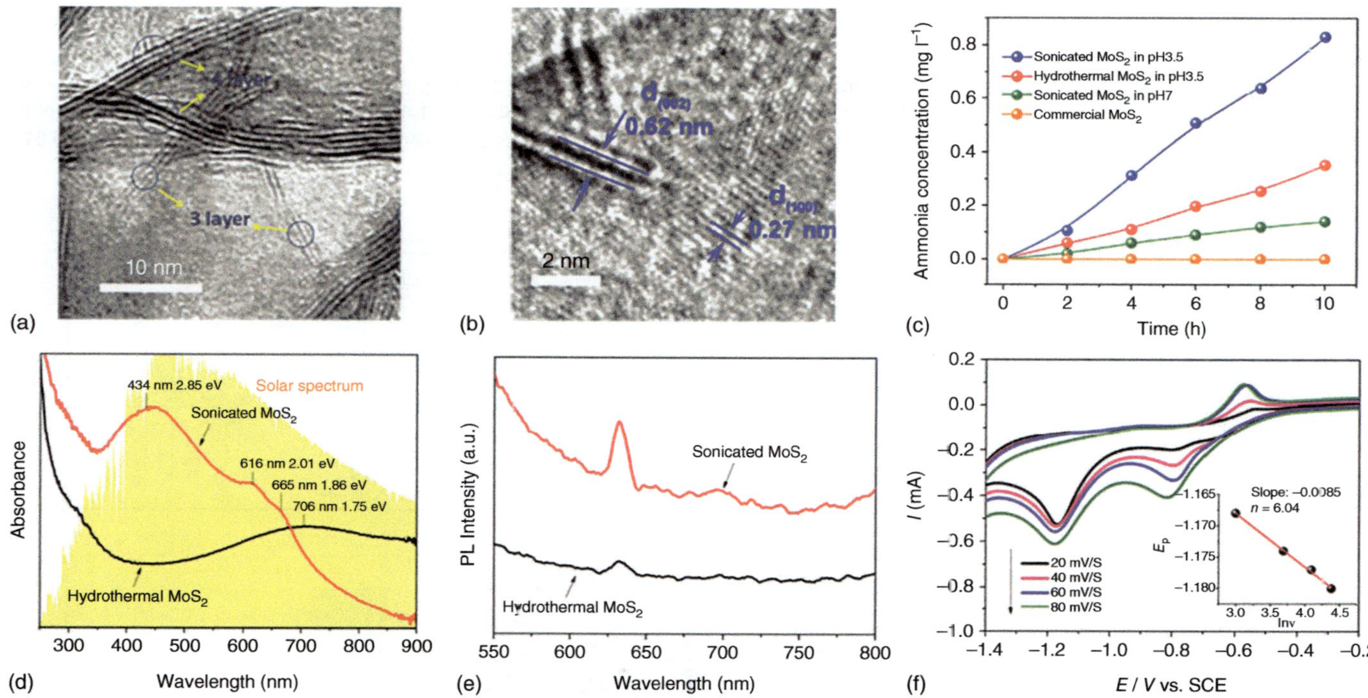

Figure 18.5 (a) TEM and (b) HRTEM image of sonicated ultrathin MoS_2; (c) Ammonia production efficiency of the as-prepared MoS_2 samples under different conditions; (d) UV–Vis absorption spectra of the as-prepared MoS_2 samples dispersed in pure water; (e) Photoluminescent spectra of the as-prepared MoS_2 samples excited at 420 nm at room temperature; (f) Cyclic voltammograms of sonicated ultrathin MoS_2 electrode at different scan rates in N_2 saturated 0.5 M Na_2SO_4 (pH = 3.5) under room temperature and simulated sun light irradiation. Source: Sun *et al.* [16] Reproduced with permission from Elsevier.

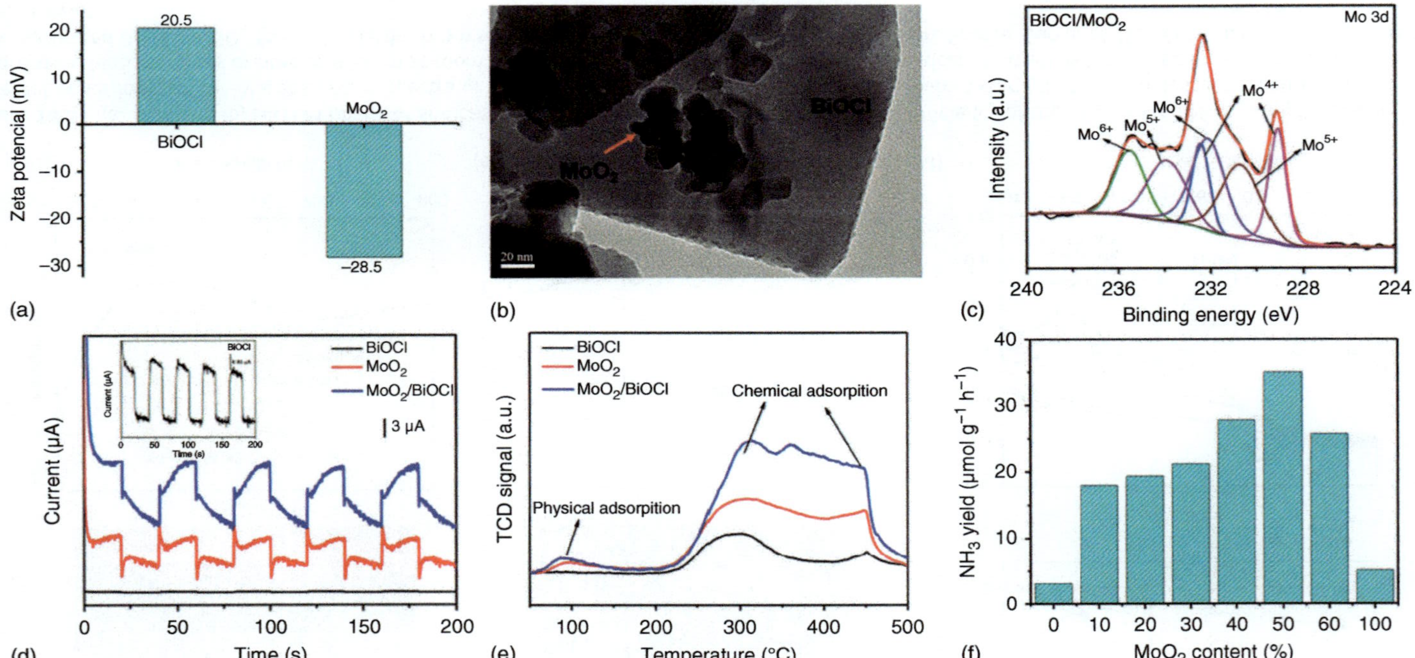

Figure 18.6 (a) Zeta potential of BiOCl and MoO_2; (b) TEM image of MoO_2/BiOCl; (c) Mo 3d XPS spectra of MoO_2/BiOCl; (d) Photocurrent curves of BiOCl, MoO_2, and MoO_2/BiOCl; (e) N_2-TPD profiles of the as-prepared BiOCl, MoO_2, and MoO_2/BiOCl; (f) Photocatalytic N_2 fixation activity of x-MoO_2/BiOCl samples with x equal to 0, 10, 20, 30, 40, 50, 60, 100 under Xe lamp irradiation. Source: Reproduced with permission from Xiao *et al.* [19]/Reproduced with permission from John Wiley & Sons.

protons from H_2O splitting. To make up for the deficiency, a polyoxometalate unit (more specifically, silicotungstic acid, denoted as SiW_{12}) with appropriate redox potential and activity was decorated as an electron/proton regulator, while phosphate anions were chosen as efficient adhesives and an "electron bridge" [20–24] (Figure 18.7a). With the successful bridging of SiW_{12} over K-C_3N_4 via phosphate, the onset potential in the linear sweep voltammogram (Figure 18.7b) shifted negatively and the current density was greatly enhanced, suggesting the improvement of H_2O oxidation. Furthermore, the transient current response of SiW_{12}/K-C_3N_4 (Figure 18.7c) not only confirmed the significance of the phosphate bridge in charge transfer but also revealed the electronic storage characteristics of SiW_{12} in the dark. Cyclic voltammetry measurements under the Ar atmosphere (Figure 18.7d) subsequently validated that the transformation of SiW_{12} made it a suitable container for the electrons and protons. Under N_2 exposure, SiW_{12} in the heteropoly blue form would inject the stored electrons and protons to the active sites on K-C_3N_4 for NH_3 production (Figure 18.7e). As a result, an NH_3 evolution rate of 353.2 $\mu mol\ g^{-1}\ h^{-1}$ was realized over SiW_{12}/K-C_3N_4 in pure water, getting rid of dependence on reactive proton source and sacrificial agent.

18.4 Challenges and Opportunities

NH_3 has for sure become one of the core components of renewable energy technology, sitting alongside H_2 and carbon-derived fuels. Its photocatalytic evolution under ambient conditions arises as an attractive alternative to sustain fertilizer and fuel production. Although the efficiency of photocatalytic N_2 fixation is still not satisfactory, the blueprint for solar farms based on this process has been validated and begun to take shape. Herein, beyond light absorption, separation of the charge carriers as well as surface reactions, we outline the critical roles of N_2 activation and electron/proton transfer. Our explorations and discussions are not meant to be authoritative; rather, we hope that the findings presented will encourage more insightful research.

N_2 activation is always the top priority in this process. Constructing surface metal sites with proper d-orbital configuration via crystal, defect, and surface/interface engineering is favorable to the interaction with N_2 molecules. The unique nitrogen exchange feature of nitride materials (carbon nitride for example) also needs to be highlighted. Specifically, the surface nitrogen species would participate in NH_3 formation directly as an "activated" nitrogen source and leave nitrogen vacancies to be compensated by N_2. For electron/proton transfer, the introduction of "electron promoter," "electride" or trions to enrich the electron density is an effective strategy that might induce a one-step multi-electron pathway. Besides, the proton-coupled electron transfer (PCET) process has great potential for facilitating reaction kinetics. Much experimental and theoretical effort should be made to establish an efficient PCET approach in the aqueous phase.

For photocatalyst design, we have noticed that most of the reported photocatalysts for N_2 fixation are multi-taskers, which would complicate the performance regulation and mechanism investigation. Modularization of the photocatalyst (like the traditional Haber–Bosch catalyst) is more recommended. The building blocks

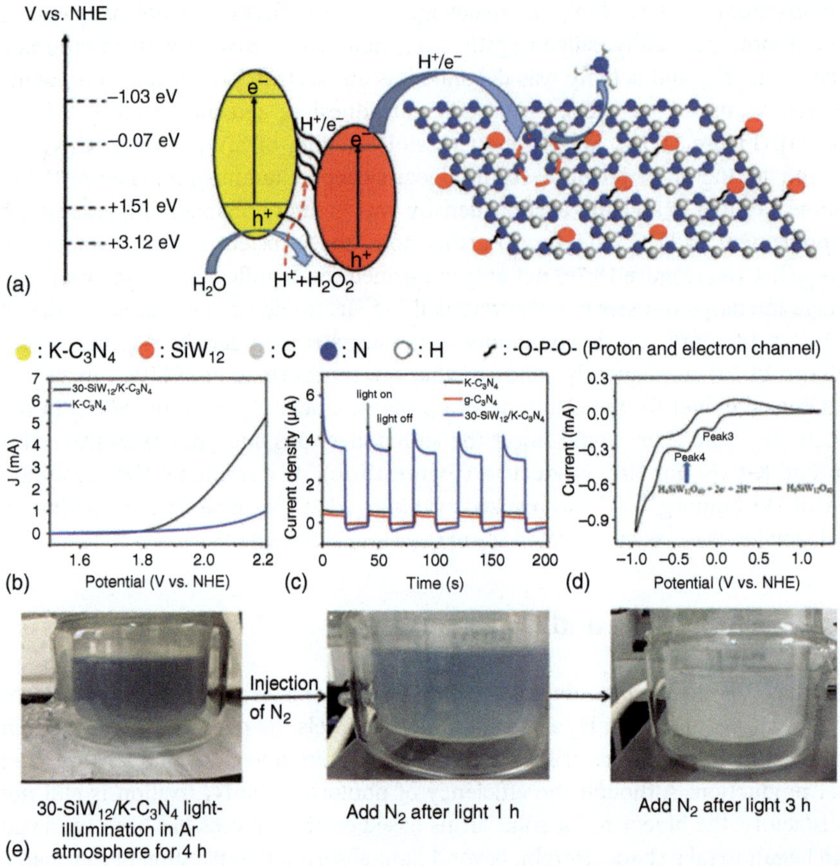

Figure 18.7 (a) Schematic illustration of the transfer and separation of photogenerated charges in the -O-P-O- bridged 30-SiW_{12}/K-C_3N_4 nanocomposite and mechanism of photocatalytic nitrogen fixation; (b) Linear sweep voltammogram curves of K-C_3N_4 and SiW_{12}/K-C_3N_4 in the dark; (c) Curves of photocurrent response for g-C_3N_4, K-C_3N_4, 30-SiW_{12}/K-C_3N_4 samples in 0.25 M Na_2SO_4 aqueous solution; (d) Cyclic voltammetry curve of 30-SiW_{12}/K-C_3N_4 (0.25 M Na_2SO_4) under Ar and at room temperature; (e) Contrast diagram of 30-SiW_{12}/K-C_3N_4 in N_2 and Ar atmosphere after illumination. Source: Xiao *et al.* [24]/Reproduced with permission from Elsevier.

include support, photosensitizer, sites of N_2 activation, H_2O oxidation, hydrogenation, *etc.*, and are coupled at the space-time scale by state-of-the-art synthesis and characterization techniques. This strategy has been applied to photocatalytic overall water splitting, resulting in a significant efficiency improvement [25]. Moreover, photocatalysis is intimately intertwined with thermocatalysis. Energy input from the photothermal effect and the localized surface plasmon resonance is critical to overcome the activation barrier of the N_2 molecule. A light-focusing device would also be an important part of this system in the future. In general, despite the challenges ahead, it is anticipated that photocatalytic N_2 fixation will blossom and lead to more convenient technology.

References

1 MacFarlane, D.R., Cherepanov, P.V., Choi, J. et al. (2020). A roadmap to the ammonia economy. *Joule* 4: 1186–1205.

2 Cheng, M., Xiao, C., and Xie, Y. (2019). Photocatalytic nitrogen fixation: the role of defects in photocatalysts. *J. Mater. Chem. A* 7: 19616–19633.

3 Li, R. (2018). Photocatalytic nitrogen fixation: an attractive approach for artificial photocatalysis. *Chin. J. Catal.* 39: 1180–1188.

4 Dhar, N., Seshacharyulu, E., and Biswas, N. (1941). New aspects of nitrogen fixation and loss in soils. *Proc. Natl. Acad. Sci., India* 7: 115–131.

5 Schrauzer, G.N. and Guth, T.D. (1977). Photolysis of water and photoreduction of nitrogen on titanium dioxide. *J. Am. Chem. Soc.* 99: 7189–7193.

6 Cheng, M., Xiao, C., and Xie, Y. (2021). Shedding light on the role of chemical bond in catalysis of nitrogen fixation. *Adv. Mater.* 33: e2007891.

7 Xu, Y.-J. and Li, J.-Q. (2005). The interaction of N_2 with active sites of graphite: a theoretical study. *Chem. Phys. Lett.* 406: 249–253.

8 Ganji, M.D. (2008). Behavior of a single nitrogen molecule on the pentagon at a carbon nanotube tip: a first-principles study. *Nanotechnology* 19: 025709.

9 Li, X., Wang, W., Jiang, D. et al. (2016). Efficient solar-driven nitrogen fixation over carbon-tungstic-acid hybrids. *Chem. Eur. J.* 22: 13819–13822.

10 Ertl, G. (2008). Reactions at surfaces: from atoms to complexity. *Angew. Chem. Int. Ed.* 47: 3524–3535.

11 Sun, S., An, Q., Wang, W. et al. (2017). Efficient photocatalytic reduction of dinitrogen to ammonia on bismuth monoxide quantum dots. *J. Mater. Chem. A* 5: 201–209.

12 Engvall, K., Holmlid, L., Kotarba, A. et al. (1996). Potassium promoter in industrial ammonia synthesis catalyst: studies by surface ionization. *Appl. Catal., A* 134: 239–246.

13 Iyngaran, P., Madden, D.C., King, D.A., and Jenkins, S.J. (2014). Infrared spectroscopy of ammonia on iron: thermal stability and the influence of potassium. *J. Phys. Chem. C* 118: 12184–12194.

14 Li, X., Sun, X., Zhang, L. et al. (2018). Efficient photocatalytic fixation of N_2 by KOH-treated g-C_3N_4. *J. Mater. Chem. A* 6: 3005–3011.

15 Bazhenova, T.A. and Shilov, A.E. (1995). Nitrogen fixation in solution. *Coord. Chem. Rev.* 144: 69–145.

16 Sun, S., Li, X., Wang, W. et al. (2017). Photocatalytic robust solar energy reduction of dinitrogen to ammonia on ultrathin MoS_2. *Appl. Catal., B* 200: 323–329.

17 Splendiani, A., Sun, L., Zhang, Y. et al. (2010). Emerging photoluminescence in monolayer MoS_2. *Nano Lett.* 10: 1271–1275.

18 Mak, K.F., Lee, C., Hone, J. et al. (2010). Atomically thin MoS_2: a new direct-gap semiconductor. *Phys. Rev. Lett.* 105: 136805.

19 Xiao, C., Wang, H., Zhang, L. et al. (2019). Enhanced photocatalytic nitrogen fixation on MoO_2/BiOCl composite. *ChemCatChem* 11: 6467–6472.

20 Hurst, J.K. (2010). In pursuit of water oxidation catalysts for solar fuel production. *Science* 328: 315–316.

21 Symes, M.D. and Cronin, L. (2013). Decoupling hydrogen and oxygen evolution during electrolytic water splitting using an electron-coupled-proton buffer. *Nat. Chem.* 5: 403–409.

22 Liu, C., Jing, L., He, L. et al. (2014). Phosphate-modified graphitic C_3N_4 as efficient photocatalyst for degrading colorless pollutants by promoting O_2 adsorption. *Chem. Commun.* 50: 1999–2001.

23 Long, M., Brame, J., Qin, F. et al. (2017). Phosphate changes effect of humic acids on TiO_2 photocatalysis: from inhibition to mitigation of electron-hole recombination. *Environ. Sci. Technol.* 51: 514–521.

24 Xiao, C., Zhang, L., Wang, K. et al. (2018). A new approach to enhance photocatalytic nitrogen fixation performance via phosphate-bridge: a case study of SiW_{12}/K-C_3N_4. *Appl. Catal., B* 239: 260–267.

25 Takata, T., Jiang, J., Sakata, Y. et al. (2020). Photocatalytic water splitting with a quantum efficiency of almost unity. *Nature* 581: 411–414.

Part IV

Applications of Photocatalysis

19

Photocatalysis for Pollution Remediation

Ren Su[1,2]

[1] *Soochow University, Soochow Institute for Energy and Materials InnovationS (SIEMIS), Key Laboratory of Advanced Carbon Materials and Wearable Energy Technologies of Jiangsu Province, Suzhou 215006, China*
[2] *Synfuels China Technology Co. Ltd., Leyuan South Street II, No.1, Yanqi Economic Development Zone C# Huairou District, Beijing 101407, China*

19.1 Basic Concept

Human beings mastered the use of the sun to disinfect water and clothes back in ancient times. The Greek God, Apollo, is associated with the sun and further inspires the Romans as a God of healing. This is the effect of photolysis, where toxic matters (i.e. molecules, viruses, and bacteria) absorb the light, excite to higher energy levels, and transform into human-safe matters. Photolysis is still extensively used in hospitals, where UV light is used to achieve a more efficient disinfection process. By introducing a catalyst that absorbs light, the photolysis process is "upgraded" into a photo-catalytic process, in which a catalytic step involving adsorption, surface reaction, and desorption takes place after the excitation of the photocatalyst. The presence of a proper photocatalyst is expected to accelerate the decomposition of toxic chemicals at lower excitation energy via an optimum reaction path; however, the thermodynamics of certain chemical reactions are still valid.

19.1.1 Consideration of Photocatalysts for Pollutant Remediation

The positions of the conduction band minimum (CBM) and valence band maximum (VBM) need to meet the redox potentials of the half reactions of aimed reactions. In organic decomposition, the energy level of the VBM must be greater than that of the organic molecules, whereas the CB electrons should have sufficient energy to reduce O_2. For the reduction of cations, the VB of the photocatalyst should satisfy the oxidation potential of water or the hole scavengers. Therefore, large bandgap photocatalysts (i.e. TiO_2 and $SrTiO_3$) are more suitable for the application of sterilization due to their stability and strong redox potentials under UV irradiations. Photocatalysts with smaller bandgaps (i.e. Ta_3N_5, CdS, and g-C_3N_4) are not good candidates for the decomposition of organic pollutants due to the mismatch between

UV-Visible Photocatalysis for Clean Energy Production and Pollution Remediation: Materials, Reaction Mechanisms, and Applications, First Edition. Edited by Xinchen Wang, Masakazu Anpo, and Xianzhi Fu.
© 2023 WILEY-VCH GmbH. Published 2023 by WILEY-VCH GmbH.

their VBM and the oxidation potential of the pollutant molecules. Additionally, they are subjected to photocorrosion under UV irradiation [1].

The photocatalysts also need to catalyze the reactions. This is a key step toward an efficient photocatalyst but has been often neglected. In the case of oxidative decomposition of organic pollutants, the amount of decomposed molecules should be stochiometric with the amount of reduced O_2. Most photocatalysts are incapable of achieving fast reaction kinetics of both oxidation and reduction half reactions, thus requiring surface engineering with a cocatalyst. Noble metal (i.e. Au and Pd) nanoparticles are good candidates for accelerating the oxygen reduction half reaction, whereas Fe species can be employed to facilitate the formation of hydroxyl radicals.

UV response semiconductor photocatalysts: TiO_2 is the most studied semiconductor photocatalyst in the remediation of pollutants owing to several advantages. It is a nontoxic inorganic compound to the organism and the environment and is stable under irradiation in either acidic or basic conditions. Another advantage is its abundance, making it an affordable photocatalyst. TiO_2-based photocatalysts also show a relatively high performance in the decomposition of organic molecules under UV irradiation owing to a reasonable band position and charge transfer kinetics. Additionally, TiO_2 has rich physiochemical properties, allowing easy modification to further enhance its photocatalytic performance. This includes polymorph and facet engineering, deposition of metal cocatalysts, doping, and vacancy engineering. Nevertheless, TiO_2 is an ideal model catalyst for fundamental studies, allowing researchers to acquire photo-induced surface physics and chemistry, and site-specific adsorption of reactants and intermediates at atomic scales [2].

In contrast, the investigations of ZnO-based photocatalysts are significantly less due to the stability issue, as ZnO dissolves in acidic and basic conditions. The perovskite-based photocatalysts (i.e. $SrTiO_3$) have also been reported for photocatalytic remediation of pollutants. The performance of engineered perovskite is on par or even better than that of TiO_2-based photocatalysts in the destruction of NO_x, formaldehyde, and isopropanol under UV irradiation [3].

Visible-light response photocatalysts: Great efforts have been made to extend the absorption wavelength to realize solar-driven degradation of pollutants. This includes narrowing the bandgap of TiO_2-based photocatalyst and finding alternative materials that have better light absorption in the visible region. Cation and anion dopants have been employed to introduce defects to TiO_2 (i.e. V, Fe, Cu, C, N, and S), which will form mid-gap states to absorb visible light [4]. Unfortunately, the enhancement in light absorption is more successful than the photocatalytic performance in most cases. There are two issues that need to be considered for choosing the identity and concentration of the dopants. The introduced mid-gap states should be aligned properly, that is, still energetically feasible for the targeted redox reactions. Second, the defect sites should have a minimum negative impact on the lifetime of the charge carriers.

There are also attempts to use relative-narrow bandgap materials (i.e. CdS, V_2O_5, and Bi_2O_3) for photocatalytic degradation of organic pollutants. However, these materials normally show a reduced oxidation power due to the up-shifted VBM, which is not ideal for the complete oxidation of organic molecules. Additionally, some of these semiconductors have a VBM that is more negative than the oxidation potential of water, thus resulting in severe photo corrosion under aerobic conditions.

Nonsemiconductor systems : A series of coinage metal (i.e. Au, Ag, and Cu) nanostructures display the capability in the decomposition of pollutants under visible light irradiation. This is the application of the localized surface plasmonic resonance (LSPR) effect of the metal. When the oscillation frequency of the valance electrons of the metal matches the frequency of an electromagnetic wave, absorption will occur. By size and shape engineering of the metal cluster, the absorption can take place in the visible-light region [5]. This high-frequency motion of the electrons around the atom results in transient high temperatures. However, it must be stressed that the oscillation of electrons is localized and thus cannot be transferred or injected into an electron acceptor. The LSPR-induced reaction is a photoinduced thermal process rather than a photocatalytic process.

19.1.2 Consideration of Reaction Conditions

Light sources: The gas discharge lamps (i.e. mercury-vapor lamp), arc lamps (i.e. the Xe arc lamps), and light-emitting diode (LED) lamps are the major light sources. The low-, medium- and high-pressure mercury-vapor lamps can produce characteristic emission lines in UV-C (184 and 253 nm), UV-A (400 nm), and blue–green regions (440–550 nm), respectively. The emission spectrum of Xe is very close to that of solar radiation, making the Xe lamp the most frequently used light source in the lab. However, the heating effect caused by infrared radiation should be considered. LED-based light sources are gradually employed owing to their high efficiency, low cost, long lifetime (>3000 hours), and convincing. The quantum efficiencies (QE) of LED can reach 10% and 80% for generating UV-A and blue light (420–500 nm), which significantly reduces the heating effect. A series of "monochromatic" light (i.e. 365, 380, 410, and 450 nm) can be easily achieved as the emission bandwidth is 10–20 nm (0.1–0.2 eV). This is convenient for research purposes to precisely calculate the QE of the photocatalytic remediation process. It is recommended to measure the light intensity of the light source for serious experiments.

Loading of the photocatalyst and pollutant: An optimum loading of the photocatalyst should allow full absorption of the incidence light. This should be verified experimentally as different materials exhibit distinct absorption coefficients (α) at specific wavelengths. For a photocatalyst aqueous suspension, the minimum depth for complete light absorption (Transmittance < 1%) can be estimated from the α and density of the photocatalyst. For example, the α of bulk TiO_2 powders ($\rho = 4.26\,g\,cm^{-3}$) at 365 nm is $4.89 \times 10^4\,cm^{-1}$, therefore the effective α of a $50\,mg\,l^{-1}$ suspension is estimated to be $0.58\,cm^{-1}$ by considering water has no absorption at this wavelength. By assuming that the number of scattered photons is equal to the absorbed photons (attenuation coefficient = 2α), complete absorption of the 365 nm light for this suspension requires a minimum depth of 3.97 cm. This estimation matches the experimental observation well and thus can be used for other systems for practical applications, i.e. to estimate the minimum thickness of a photocatalyst film.

Photocatalytic removal of organic compounds is ideally operated at a relatively low concentration in general (i.e. ppm or sub mM level) due to the slow kinetics of the oxygen reduction reaction. The other consideration is to avoid complicated chemical reactions between pollutant molecules that generate more toxic chemicals. Therefore, photocatalysis should be coupled with other techniques and used as the last procedure for fine polishing. For the removal of toxic heavy cations, the deposition of the metal on the photocatalyst is inevitable and will gradually reduce the performance of the photocatalyst. The maximum loading of the metal species should be tested to set up procedures for the regeneration of photocatalysts.

Addition of chemicals: For the oxidation of organic pollutants, an equal amount of oxidants should be reduced. This ideal should be O_2 in the atmosphere. However, additional oxidizers are sometimes used to enhance the performance (e.g. H_2O_2) due to the slow reduction kinetics of O_2. The presence of H_2O_2 will promote the formation of hydroxyl radicals ($^{\cdot}OH$), which is beneficial for the oxidation of organic molecules. A higher concentration of O_2 also has a positive effect on the overall catalytic performance. This can be realized by employing vigorous mixing or using concentrated O_2 gas. For the reduction of toxic anions in an aqueous solution, hole scavengers (i.e. critic acid and aniline) are often used to consume the oxidative radical species (h^+, $^{\cdot}OH$, $^{\cdot}OOH$). This is, in fact, a key addition to avoiding the challenging water oxidation reaction, but has been underestimated in most cases.

Temperature, pH, and humidity: The reaction rates of photocatalytic remediation processes increase upon increasing the reaction temperature. A significant reduction in photocatalytic performance is expected for the outdoor removal of VOCs in winter in high latitude regions, which should be considered for practical applications. The pH will influence the surface charging properties of the photocatalyst powders, causing the change of the zeta potential (ζ) of the particle, thus resulting in agglomeration or well dispersion of the colloidal. The pH also influences the adsorption of the reactants, thus alternating the reaction kinetics and paths in pollutant remediation. Relative humidity (RH) sometimes influences the gas-phase photodecomposition of VOCs. More $^{\cdot}OH$ tends to form at a high RH under irradiation due to the high surface coverage of water, whereas superoxide anion radicals ($^{\cdot}O_2$) are more likely to be the major species at low RH. For catalyst screening, identical reaction conditions should be used.

19.2 Reactants, Products, and Intermediates Analysis and Reaction Mechanisms

Direct monitoring of the evolution of CO_2, the final oxidation product in the photocatalytic decomposition of most organic compounds, is the most straightforward and strict method to evaluate the remediation efficiency and the performance of the photocatalyst. The infrared-based CO_2 sensing method is a commercially available and mature protocol; however, the relatively slow kinetics and a limited amount of produced CO_2 require scientists to use multiple techniques to study the reaction and the catalyst. It is also of great interest to understand the decomposition paths of complicated organic molecules to design better photocatalysts and to avoid the evolution of toxic intermediates.

19.2.1 Direct Analysis of Decomposed Products

Mass spectrometry (MS) is a good choice for the quantitative analysis of evolved CO_2. It is also possible to use labeled molecules to probe the reaction mechanisms. Since MS consumes gas molecules, the evolution rates of CO_2 and other gaseous chemicals during photocatalysis should be significantly faster than the consumption rate by the spectrometer. This includes the inlet consumption and the ionization of gas molecules in the vacuum chamber of the MS, in which the prior one counts for >99% of the usage of the gas molecules. A typical commercialized capillary inlet requires a flow rate of 1–2 standard cubic centimeters per minute, which is not suitable for the continuous detection of gas-phase products in a lab-scale photocatalytic reactor. This is solved by employing a leak valve to directly connect a leak-tight reactor under ambient pressure with the MS under high vacuum (Figure 19.1a), in which the "leak rate" of gas from the reactor into the vacuum chamber of the MS is in the range of 10^{-12} to 10^{-8} mbar l s^{-1}. Such a low flow rate of gases allows a pseudo-steady-state measurement of the evolution of gas-phase components, including CO_2 and H_2O formation and O_2 consumption spontaneously, which is important for understanding the dissociation kinetics and mechanisms of organic molecules to design high-performance photocatalyst.

A case study of the reusability of sol-immobilized Au and Pd on TiO_2 (Au/TiO_2 and Pd/TiO_2) in the photocatalytic decomposition of phenol in water by MS is discussed in Figure 19.1b,c [6]. Here, a 50 ml phenol solution (200 μM) was used for the analysis, resulting in a theoretical amount of 60 mmol CO_2. The freshly prepared photocatalysts (50 mg) are directly used without any treatment for the first cycle, and nine cycles with eight-hour irradiation are performed for both Au/TiO_2 and Pd/TiO_2. Clear decays of both CO_2 generation rate and quantity are observed for Au/TiO_2 after the first and fourth cycles, implying Au/TiO_2 is suffering a deactivation. This is caused by the formation of benzoquinone and the polymerization of phenolic compounds in the liquid phase. Second, the total amount of evolved CO_2 exceeds the theoretical number for the first cycle for both Au/TiO_2 and Pd/TiO_2, which is associated with the dissociation of the residual polyvinyl alcohol (PVA) that is used as a ligand during the synthesis of Au and Pd colloidal solutions. Finally, the Pd/TiO_2 shows a full conversion of phenol into CO_2 with a stable evolution rate within nine cycles, revealing high stability and efficiency for applications. Note that the dissolved CO_2 in the liquid phase should be calculated and counted using Henry's law.

Another case study is the photocatalytic decomposition of VOC mixtures in the gas phase (Figure 19.1d–f) [7]. Pristine TiO_2 and Pd/TiO_2 show a similar photocatalytic removal rate of a formaldehyde-toluene mixture according to an electrochemical sensor-based method, however, Pd/TiO_2 presents a much faster CO_2 evolution

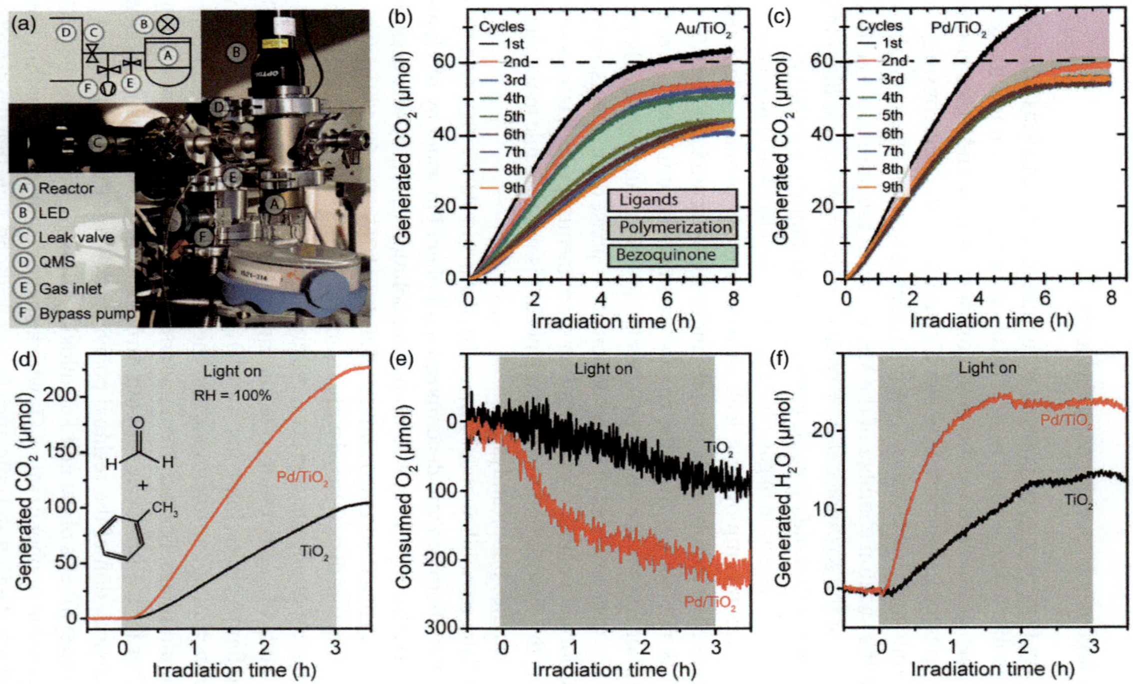

Figure 19.1 (a) Sketch and image of the *in situ* MS system. (b) and (c) Time-resolved CO_2 evolution in liquid-phase phenol decomposition using Au/TiO_2 and Pd/TiO_2 under UV irradiation. Source: Su et al. [6]/Reproduced with permission from American Chemical Society. (d)–(f) Time-resolved CO_2, O_2, and H_2O evolution in photocatalytic decomposition of gas-phase formaldehyde-toluene mixture using TiO_2 and Pd/TiO_2. Source: Reproduced with permission from Wu et al. [7] © 2021 Elsevier.

rate than the pristine TiO_2 determined by *in situ* MS (Figure 19.1d). The *in situ* MS also shows that Pd/TiO_2 exhibits more rapid O_2 consumption and a faster water formation than pristine TiO_2 (Figure 19.1e,f), which agrees well with the fast CO_2 evolution. The gas chromatograph-mass spectrometry (GC-MS) analysis reveals that toxic hydroxylated toluene species are generated when pristine TiO_2 is used as the photocatalyst, whereas benzaldehyde is the only intermediate species for Pd/TiO_2. The hydroxylated toluene species are more stable than benzaldehyde, thus resulting in a poor CO_2 evolution rate of TiO_2.

For industrial applications, total organic carbon (TOC) and chemical oxygen demand (COD) analysis are often used methodologies [8]. There are standard operating procedures (SOP) and equipment, which will not be discussed here.

19.2.2 Indirect: Consumption of Dye Molecules

A boom in photocatalytic environmental treatment with the development of enor mous photocatalysts can be dated back to the 1990s. Dye tests have been developed and have been frequently used due to the simplicity of employing a UV–Vis spectrometer to measure the bleaching kinetics of dye molecules with distinct light absorption at certain wavelengths (i.e. methylene blue and congo red). This is a user-friendly protocol that allows internal and external comparisons of the photocatalysts and greatly promotes the research at a moment.

However, the photobleaching of dye does not require a complete decomposition of the molecule, thus the decay of light absorption observed in UV–Vis absorption spectra is only relevant to the change of chromophore and auxochrome groups. A famous work by A. Mills shows that methylene blue (MB, $C_{16}H_{18}ClN_3S$, peak absorption at 660 nm) can be bleached even under deaerated conditions due to the formation of colorless leuco-form methylene blue (LMB), as described in Eqs. (19.1) and (19.2) [9]:

$$MB + e^-_{CB} \xrightarrow{TiO_2,\ h\nu \geq 3.2\ eV} MB^{\cdot-} \tag{19.1}$$

$$2MB^{\cdot-} \rightarrow MB + LMB \tag{19.2}$$

$$2LMB + O_2 \rightarrow 2MB + 2H_2O \tag{19.3}$$

The LMB turns back to bluish MB under aerated conditions in the dark (Eq. (19.3)). In fact, a few studies show that the dye molecules are quite stable under UV irradiation even using the most powerful TiO_2-based photocatalysts. Nevertheless, the adsorption of dyes on the photocatalysts has been often ignored. This can be significant and miscounted as decomposition, especially for porous-structured photocatalysts.

19.2.3 Radicals Species

The formation of radical species under irradiation via the interaction of surface adsorbed molecules with photo-generated charge carriers (e^- and h^+) is the initial

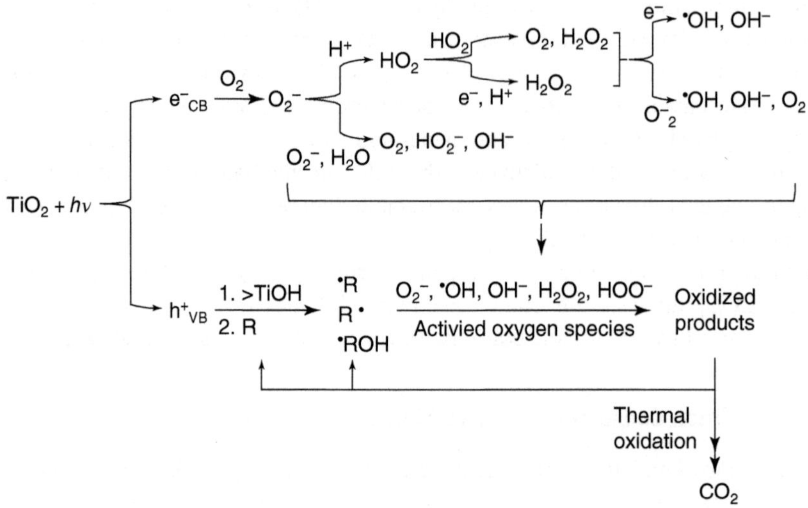

Figure 19.2 A roadmap of possible radical species involved in photocatalytic decomposition of organics using TiO_2 photocatalyst. Source: Hoffmann et al. [10]/Reproduced with permission from American Chemical Society.

chemical step in a photocatalytic process. Figure 19.2 shows the possible reactive oxygen species (ROS) generated using a TiO_2 photocatalyst under radiation [10]. Two criteria need to be considered for the design of a high-performance photocatalyst. First, this important step occurs in the time regime of 10–100 ns, which is in competition with the recombination of the charge carriers but significantly faster than the surface redox chemistry initiated by the radicals (μs to ms). Second, the identity of photogenerated radicals is dependent on the surface properties of the catalysts (i.e. under-coordinated sites, vacancies, and metals) and the surface adsorbed chemicals (i.e. O_2 and H_2O). Therefore, a fast generation of preferred radical species and the fast consumption rate of the preferred acceptors are critical to achieving optimum performance.

The electron paramagnetic resonance (EPR) is a widely used tool to analyze radical species. The classic case study is the origin of the high reactivity of the Degussa P25 TiO_2 photocatalyst by D. Hurum et al. [11], in which the identification of surface hole sites and lattice electron trapping sites in anatase and rutile is successfully achieved at 10 K under deaerated conditions. For the case of aerobic oxidation of organic molecules, the oxidative radicals and ROS can be analyzed at 77 K, as shown in Figure 19.3 of a case study of Pd supported on anatase TiO_2 with different crystallite sizes and crystallinity [12]. Ti^{3+}, CB electrons ($e^-{}_{CB}$), and $O^{\cdot-}$ are formed on pristine TiO_2 under irradiation at a high vacuum (10^{-5} mbar), due to charge separation (Figure 19.3a):

$$e^- + Ti^{4+} \rightarrow Ti^{3+} \tag{19.4}$$

$$O^{2-} + h^+ \rightarrow O^{\cdot-} \tag{19.5}$$

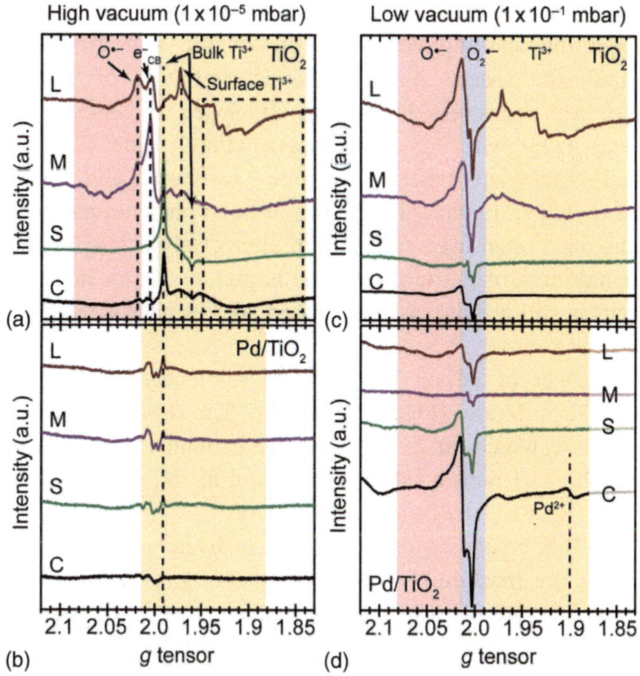

Figure 19.3 EPR spectra of TiO_2 and Pd/TiO_2 under UV irradiation for 30 minutes at 77 K under high vacuum (a and b) and low vacuum (c and d), respectively. Source: Su et al. [12] Reproduced with permission from American Chemical Society.

For poorly crystalline and small size anatase TiO_2 (C: 13% and S: 9 nm), bulk Ti^{3+} ($g^{\perp} = 1.992$, $g_{//} = 1.960$) are the major radical species, whereas the e^{-}_{CB} are the dominant species on medium-sized anatase (M: 16.4 nm). For crystalline TiO_2 with a larger crystallite size (L: 25 nm), near-surface Ti^{3+} ($g = 1.93$), e^{-}_{CB}, and $O\cdot^{-}$ are all observed. The radical signal almost vanished when Pd NPs are immobilized on the TiO_2 (Figure 19.3b), suggesting an efficient electron transfer from TiO_2 to Pd sites.

Under low vacuum conditions (10^{-1} mbar), the presence of O_2 results in a change in the radical species upon irradiation (Figure 19.3c). $O_2\cdot^{-}$ is the only radical species for poorly crystalline and small anatase (C and S) with poor intensity. In comparison, surface Ti^{3+}, $O_2\cdot^{-}$, and $O\cdot^{-}$ were observed for larger anatase (M and L) with stronger intensity. The weakening of $O_2\cdot^{-}$ and vanishing of surface Ti^{3+} signals are observed for Pd on larger anatase, confirming a fast charge transfer from TiO_2 to Pd even in the presence of O_2 (Figure 19.3d). In contrast, the intensity of $O_2\cdot^{-}$ radicals remains high for the Pd on TiO_2 with low crystallinity (C) and small crystallite size (S). Additionally, weak fingerprints of Pd^{2+} signals are observed for sample C. This suggests a poor charge separation between TiO_2 and the Pd, thus resulting in poor catalytic performance.

The identity and quantity of photogenerated radicals in the liquid phase can be analyzed under ambient conditions by using spin traps (a type of diamagnetic molecule) to yield a relatively stable paramagnetic spin adduct with a longer lifetime at room temperature (RT). By deconvolution of the spectra to extrapolate the hyperfine splitting constants, line width, and symmetry, it is possible to identify the species and relative concentration of radicals [13]. A series of spin trap molecules have been developed to analyze the C-centered radicals and ROS, with 5,5-dimethylpyrrolidine-1-oxide (DMPO), 3,4-dihydro-2-methyl-1,1-dimethylethyl ester-2*H*-pyrrole-2-carboxylic acid-1-oxide (BMPO), 2,2,6,6-tetramethylpiperidinooxy (TEMPO), and alpha-phenyl-*n*-*tert*-butyl nitrone (PBN) being the most often used spin trap in photocatalysis [14]. A spin trap presents a longer half-time of the target radical is preferred [15]; however, immediate measurement is still a must, as the trapped species have a relatively short half-time (minutes).

A case study is the effect of Pd on TiO_2 in the photodegradation of a formaldehyde-toluene VOC mixture (Figure 19.4) [7]. The EPR analysis is performed in the liquid phase, which can be considered an extreme of high humidity reaction conditions. The $^{\cdot}OH$ radical ($a[^1H] = 6.41$ and $a[^{14}N] = 38.83$ MHz) that originates from the hydroxylated surface (–OH) of TiO_2 is observed in the dark (Figure 19.4a). This results in the formation of hydroxylated toluene (methyl-phenolics) during the reaction. For Pd/TiO_2, the very weak signals of the $^{\cdot}OH$ radical indicate that the surface –OH on TiO_2 is effectively removed (Figure 19.4b). After irradiation, the evolution of $^{\cdot}CH_3$ radical ($a[^1H] = 8.67$ and

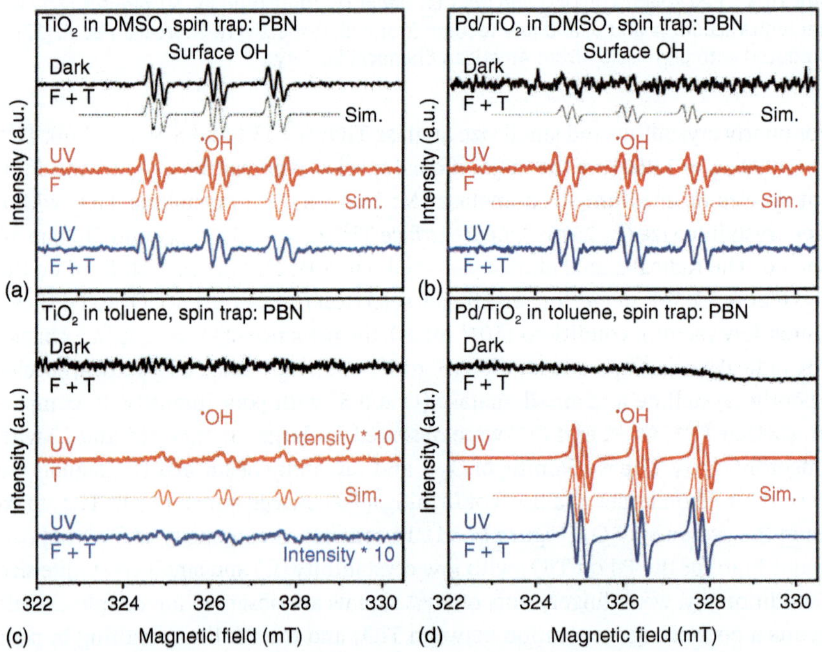

Figure 19.4 EPR spectra of irradiated photocatalyst suspensions with formaldehyde (F, 5 mM), toluene (T, 1.5 mM) and formaldehyde-toluene mixture (F + T) under aerated conditions at RT. (a) and (b) TiO_2 and Pd/TiO_2 in DMSO; (c) and (d) TiO_2 and Pd/TiO_2 in toluene. Source: Reproduced with permission from Wu et al. [7] © 2021 Elsevier.

$a[^{14}N] = 42.60$ MHz) is observed for both TiO_2 and Pd/TiO_2 (Figure 19.4a,b) due to the conversion of $^{\cdot}OH$ into $^{\cdot}CH_3$ via reaction with DMSO:

$$(CH_3)_2SO + {}^{\cdot}OH \rightarrow CH_3SO_2H + {}^{\cdot}CH_3 \tag{19.6}$$

A significant change in the EPR spectra is observed when toluene is employed as the solvent (Figure 19.4c,d). No detectable EPR signal is observed for both photocatalysts in the dark. Upon irradiation, very intense ·OH radicals ($a[^1H] = 6.23$ and $a[^{14}N] = 39.05$ MHz) are observed for Pd/TiO_2 with/without formaldehyde (Figure 19.4d), whereas in comparison, the pristine TiO_2 only generates very few $^{\cdot}OH$ radicals (Figure 19.4c). This suggests that the Pd cocatalyst is crucial to activating O_2 under high pollutant concentrations.

19.2.4 Reaction Intermediates

GC, GC-MS, and HPLC-MS are straightforward approaches to analyze relatively stable intermediates and by-products that evolve during the photocatalytic decomposition of organic compounds, e.g. aromatic and aliphatic carbonyls. This requires: (i) all intermediates desorb from the surface of the photocatalyst, (ii) good knowledge of all possible intermediates and by-products, and (iii) fast and prompt analysis of the aliquots extracted from the reaction systems.

Infrared spectroscopy is a frequently used technique to analyze surface adsorbed intermediate species. The removal and formation of certain functional groups, the adsorption configuration of reactants, and the adsorption sites can be determined to probe the reaction paths and kinetics. Besides, it is also possible to combine IR spectrometry with mass spectrometry to control the gas phase and monitor gaseous reactants (e.g. O_2 consumption) and products (i.e. CO_2, CO, and H_2O). Figure 19.5 shows a case study of photodissociation of ethylene glycol (EG) on TiO_2 under ambient conditions at monolayer surface coverage (3 ML) [15]. The decomposition of EG can be analyzed using the hydroxyl and methyl vibrational peaks (labeled as species i, $\nu[OH] = 3333\,cm^{-1}$, $\nu_{as}[CH_3] = 2940\,cm^{-1}$, and $\nu_s[CH_3] + \nu_s[CH_2] = 2877\,cm^{-1}$), whereas the formation of formaldehyde can be followed using the carbonyl vibrational peak (species ii, $\nu[CO] = 1717\,cm^{-1}$). This indicates that the EG undergoes a C—C bond cleavage under irradiation (Eqs. (19.7)–(19.9)):

$$HOH_2C - CH_2OH \xrightarrow{h\nu,\ TiO_2} 2HCHO + 2H^+_{ads} + 2e^-_{CB} \tag{19.7}$$

$$e^-_{CB} + O_2 \rightarrow {}^{\cdot}O_2^- \tag{19.8}$$

$$4H^+_{ads} + e^-_{CB} + O_2\left({}^{\cdot}O_2^-\right) \rightarrow 2H_2O \tag{19.9}$$

The concentration of formaldehyde decreases while four additional peaks evolve at a longer irradiation time (3410 and $1640\,cm^{-1}$, red dashed lines; 1186 and $1120\,cm^{-1}$, purple dashed lines). This matches well with the IR spectra of Formalin solution (40% of formaldehyde) dosed on TiO_2, indicating that both decomposition of formaldehyde and the polymerization of formaldehyde into paraformaldehyde [$H(CH_2O)_nH$, species iii] with the photogenerated water (species iv) take place

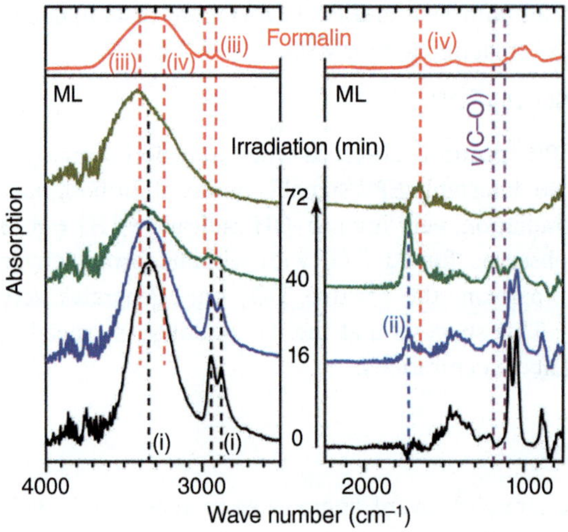

Figure 19.5 Photodegradation of ethylene glycol using pristine TiO_2 under aerated conditions studied by *in situ* FT-IR. A ~ML surface coverage of ethylene glycol is dozed on the TiO_2 film before the reaction. The marked species are (i): EG; (ii): formaldehyde; (iii): paraformaldehyde; and (iv): water. The ν(C–O) vibrations of paraformaldehyde according to a reference are also labeled. Source: Reproduced with permission from Li et al. [15] © 2018 Elsevier.

(Eqs. (19.10) and (19.11)). The paraformaldehyde is gradually decomposed/desorbed at a longer reaction time with only water on the TiO_2 surface (72 minutes).

$$HCHO + O_2\left(\cdot O_2^-\right) \xrightarrow{h\nu} CO_2 + H_2O \tag{19.10}$$

$$n\text{HCHO} + H_2O \rightarrow OH(CH_2O)_nH \tag{19.11}$$

UV–Vis can also realize the quantitative analysis of reaction intermediates in a complicated photocatalytic reaction with delicate control of the reaction and a measurement of the absorption coefficient of all intermediates. Take the photocatalytic decomposition of phenol as an example, which undergoes a step-wise ring opening process via the formation of phenolic intermediates (i.e. hydroquinone and benzoquinone) during irradiation with the presence of a photocatalyst (Figure 19.6a) [6]. The generation of benzoquinone during photocatalytic decomposition of phenol should be avoided, as O_2 and photo-generated electrons may be wasted in the dead loop between the redox couple (Steps III and IV). This requires a designed photocatalyst that can decompose the produced phenolic intermediates rapidly. The UV–Vis spectra recorded during the photocatalytic decomposition of phenol using an Au/TiO_2 photocatalyst are shown in Figure 19.6b. Though the absorption spectra of phenol, benzoquinone, hydroquinone, and catechol are slightly overlapped, it is possible to extrapolate their concentrations using the Beer–Lambert law by measuring their absorption coefficients at four different wavelengths, respectively. The wavelength-dependent Beer–Lambert law is described as follows:

$$A_{WL} = \varepsilon_P^{WL} \times C_P + \varepsilon_B^{WL} \times C_B + \varepsilon_H^{WL} \times C_H + \varepsilon_C^{WL} \times C_C \tag{19.12}$$

where A is the apparent absorbance at specific wavelengths (WL), which corresponds to the sum of the absorbance contributed by each compound; ε is the molar absorption coefficient of phenol (P), benzoquinone (B), hydroquinone (H), and catechol (C) at specific wavelengths; C is the concentration. An accumulation of hydroquinone is observed for tTiO$_2$ (Figure 19.6c), which reduces the overall

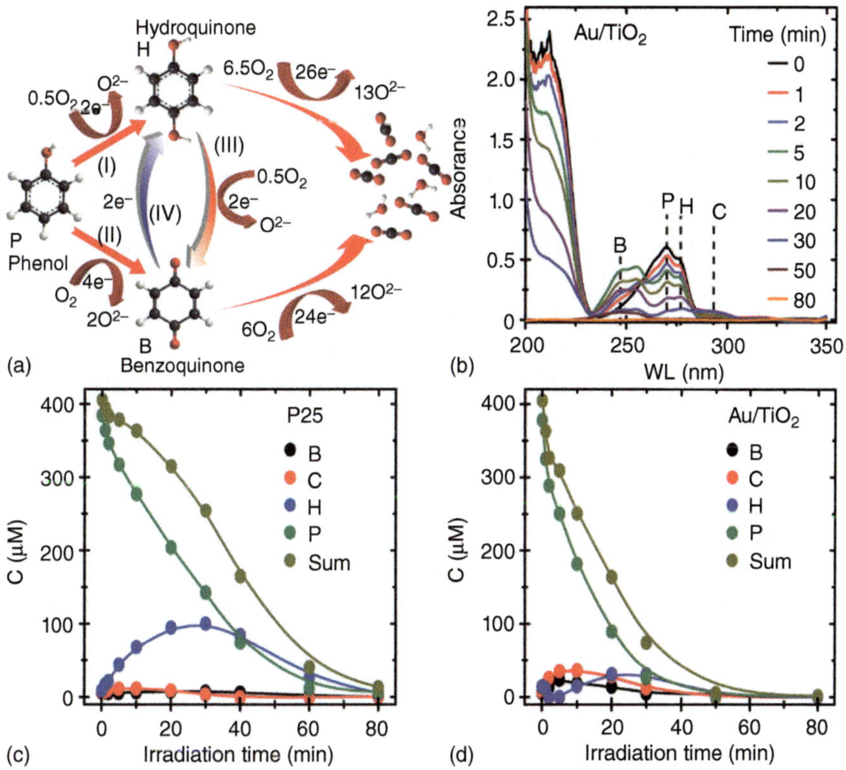

Figure 19.6 (a) Photodegradation pathways of phenol. (b) UV–Vis spectroscopy of the photocatalytic decomposition of phenol using Au/TiO_2. (c) and (d) Evolution of phenolic intermediates determined by UV–Vis spectrometry during photocatalytic decomposition of phenol using P25 and Au/TiO_2. Source: Reproduced with permission from Su et al. [6]; © 2012 American Chemical Society.

photodecomposition rate of phenol. The Au NPs effectively suppress the formation or accelerate the removal of hydroquinone, thus resulting in enhanced catalytic performance (Figure 19.6d). This allows fast analysis of the photo-decomposition paths and kinetics of a wide range of aromatic compounds in the liquid phase without the need for product separation. However, it should be aware that any unknown species with a UV–Vis response will cause an overestimation of the concentration.

19.3 Concluding Remarks and Perspectives

We have discussed the selection criteria of a suitable photocatalyst for pollutant remediation applications and the effect of reaction conditions. Methods for precise analysis of reactants, products, and intermediates are provided with case studies. Their advantages and limitations are discussed as well.

The following challenges must be done to boost the applications of photocatalytic remediation. First, the performance, intermediates, and safety evaluation of the photocatalytic remediation process should be analyzed in real-life environments. Second, case investigations should be performed to design specific photocatalysts, as the types of pollutants vary from in-house to industrial environments (i.e. oil refineries and pharmaceutical synthesis). Third, the surface chemistry of the photocatalyst during the remediation of organic molecules should also be investigated, as the deposition of complicated residuals and polymers may take place and eventually block the active sites. It is important to understand the deactivation mechanism of the photocatalyst and provide solutions for prolonging its stability. Nevertheless, a complete reaction system, including the reactor, the types of radiation, and the loading of the catalyst, should be carefully designed. Flow chemistry-based techniques may be a promising solution for applications.

Acknowledgment

The author would like to thank the financial support from the National Natural Science Foundation of China (NSFC, project number: 21972100).

References

1 Chen, P., Dong, X., Huang, M. et al. (2022). Rapid self-decomposition of g-C_3N_4 during gas–solid photocatalytic CO_2 reduction and its effects on performance assessment. *ACS Catal.* 12: 4560–4570.

2 (a) Diebold, U. (2003). The surface science of titanium dioxide. *Surf. Sci. Rep.* 48: 53–229. (b) Bechstein, R., Kristoffersen, H.H., Vilhelmsen, L.B. et al. (2012). Packing defects into ordered structures: strands on TiO_2. *Phys. Rev. Lett.* 108: 236103.

3 (a) Wang, J., Yin, S., Zhang, Q. et al. (2004). Influences of the factors on photocatalysis of fluorine-doped $SrTiO_3$ made by mechanochemical method. *Solid State Ionics* 172: 191–195. (b) Li, X. and Zang, J. (2009). Facile hydrothermal synthesis of sodium tantalate ($NaTaO_3$) nanocubes and high photocatalytic properties. *J. Phys. Chem. C* 113: 19411–19418.

4 (a) Dong, F., Guo, S., Wang, H. et al. (2011). Enhancement of the visible light photocatalytic activity of C-doped TiO_2 nanomaterials prepared by a green synthetic approach. *J. Phys. Chem. C* 115: 13285–13292. (b) Su, R., Bechstein, R., Kibsgaard, J. et al. (2012). High-quality Fe-doped TiO_2 films with superior visible-light performance. *J. Mater. Chem.* 22: 23755–23758.

5 Langhammer, C., Schwind, M., Kasemo, B., and Zorić, I. (2008). Localized surface plasmon resonances in aluminum nanodisks. *Nano Lett.* 8: 1461–1471.

6 Su, R., Tiruvalam, R., He, Q. et al. (2012). Promotion of phenol photodecomposition over TiO_2 using Au, Pd, and Au-Pd nanoparticles. *ACS Nano* 6: 6284–6292.

7 Wu, Q., Ye, J., Qiao, W. et al. (2021). Inhibit the formation of toxic methylphenolic by-products in photo-decomposition of formaldehyde–toluene/xylene mixtures by Pd cocatalyst on TiO_2. *Appl. Catal., B* 291: 120118.

8 (a) Zhao, W. and Xie, H.Y. (2013). COD removal of textile printing and dyeing wastewater by photocatalysis. *Adv. Mater. Res.* 704: 51–54. (b) Waters, A. (1993). Photocatalysis of TOC measurements. *Filtr. Sep.* 30: 533–535.

9 Mills, A. and Wang, J. (1999). Photobleaching of methylene blue sensitized by TiO_2: an ambiguous system? *J. Photochem. Photobiol., A* 127: 123–134.

10 Hoffmann, M.R., Martin, S.T., Choi, W., and Bahnemann, D.W. (1995). Environmental applications of semiconductor photocatalysis. *Chem. Rev.* 95: 69–96.

11 Hurum, D.C., Agrios, A.G., Gray, K.A. et al. (2003). Explaining the enhanced photocatalytic activity of degussa P25 mixed-phase TiO_2 using EPR. *J. Phys. Chem. B* 107: 4545–4549.

12 Su, R., Dimitratos, N., Liu, J. et al. (2016). Mechanistic insight into the interaction between a titanium dioxide photocatalyst and Pd Co-catalyst for improved photocatalytic performance. *ACS Catal.* 7: 4239–4247.

13 Dąbrowski, J.M. (2017). *Advances in Inorganic Chemistry*, vol. 70, 343–394. Academic Press.

14 Zhao, H., Joseph, J., Zhang, H. et al. (2001). Synthesis and biochemical applications of a solid cyclic nitrone spin trap: a relatively superior trap for detecting superoxide anions and glutathiyl radicals. *Free Radical Biol. Med.* 31: 599–606.

15 Li, C., Wang, X., Cheruvathur, A. et al. (2018). *In-situ* probing photocatalytic C—C bond cleavage in ethylene glycol under ambient conditions and the effect of the metal cocatalyst. *J. Catal.* 365: 313–319.

20

Biomimetic Photocatalytic Wastewater Treatment: From Lab-scale to Commercial Operation

Jiaqiang Wang, Xiaoqian Ma, Liang Jiang, Jiao He, Daomei Chen, and Yongjuan Chen

Yunnan University, Yunnan Province Engineering Research Center of Photocatalytic Treatment of Industrial Wastewater, School of Materials and Energy, School of Chemical Sciences & Technology, School of Engineering, 2 Cuihu North Road, Kunming, 650091, P.R. China

20.1 Introduction

Photocatalytic processes involving TiO_2 semiconductor particles have been shown to be potentially advantageous in wastewater treatment [1]. Particularly, since water resources become more strained, wastewater regulations become more stringent, and the benefits of photocatalytic processes become more appealing. However, in comparison to the intensive studies about the synthesis of a variety of photocatalysts and their different potential applications, pilot and demonstration plants are still countable [2]. In spite of substantial research carried out in recent decades, the application of photocatalysis in practical water treatment systems has been limited compared with other advanced oxidation processes [1]. Given these trends, how should the research community look at the technological horizon of photocatalytic water treatment?

The difficulties that hamper the commercial success of photocatalytic water treatment have resulted in only a few systems currently being used in practice. This contradicts the massive research literature on the application of photocatalysis technology in the treatment of surface pollutants and common groundwater [3]. As is common, cost-effectiveness, photocatalyst stability, ecological safety, easy recovery and reuse, and scalable production have to be considered in the synthesis of photocatalysts. Although significant advances in photocatalyst synthesis and applications have been achieved, the design and manufacture of photoreactors are urgently needed to fill the gap between scientific efforts and industrial applications [4].

In this chapter, we reviewed our progress in the synthesis of photocatalysts with different morphologies via the biotemplating method. A vision for the building-scale implementation of photocatalytic reactors will be proposed, with the goal of clarifying the opportunities and challenges in this context.

UV-Visible Photocatalysis for Clean Energy Production and Pollution Remediation: Materials, Reaction Mechanisms, and Applications, First Edition. Edited by Xinchen Wang, Masakazu Anpo, and Xianzhi Fu.
© 2023 WILEY-VCH GmbH. Published 2023 by WILEY-VCH GmbH.

20.2 Biotemplated Photocatalysts

Natural biological templates could be harvested in large amounts at low costs and have little requirement for equipment [5], thus they have been extensively used as building blocks to form various structures and their corresponding unique functions [6]. Up to now, a number of TiO_2 photocatalysts with novel morphologies have been synthesized using biotemplates, including bacterial cellulose membranes [7], plant skins [8], leaves [6], natural rubber latex [9], hypocrellins [10], diatoms [11], and cyanobacteria [12].

The visible-light photocatalytic activity of TiO_2 was effectively enhanced through self-doping elements from biotemplate materials containing nitrogen [13], carbon [14], and other elements. For example, we used preserved frustules of diatom (Cocconeis placentula) cells as both hard templates and silicon sources to synthesize TiO_2-coated SiO_2 with biogenic C self-doped in it (C-TiO_2/SiO_2, Figure 20.1) [11]. The silica exoskeleton of a diatom, as a biophotonic crystal, has spatially ordered and periodic nanostructures and may control the propagation of light by only allowing certain wavelengths to pass through the crystal. The unique morphology and structure of C-TiO_2/SiO_2 in conjunction with C self-doping produced efficient photocatalytic efficiency for the photodegradation of rhodamine B under visible light. This promoted a new avenue for using smart biological structures with superior solar light-harvesting abilities to synthesize photocatalysts.

However, we found that the majority of research on the hard-biotemplated synthesis of TiO_2 has concentrated solely on reproducing the macrostructural morphology of templates, with little attention paid to the impact of the microstructural building blocks on macrostructure fabrication [13]. Generally, only tightly packed TiO_2 nanoparticles were obtained during the duplication of macrostructural morphology of hard biotemplates via the sol-gel method [15]. It is expected that the photocatalytic

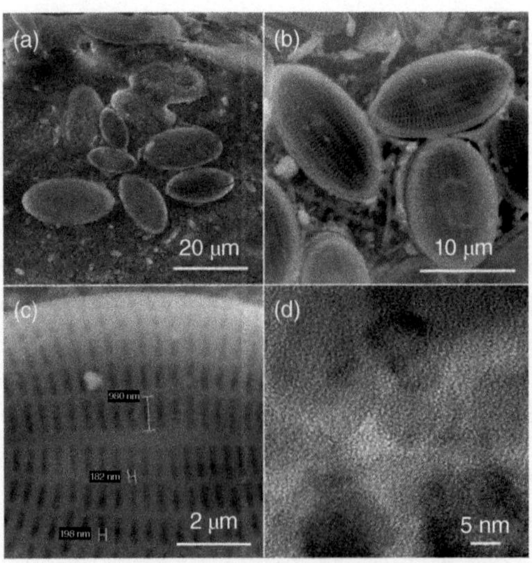

Figure 20.1 SEM images of (a) diatom cells, (b, c) diatom-templated-C-TiO_2/SiO_2, (d) HR-TEM image of diatom-templated-C-TiO_2/SiO_2. Source: He et al. [11]/Reproduced with permission from Springer Nature.

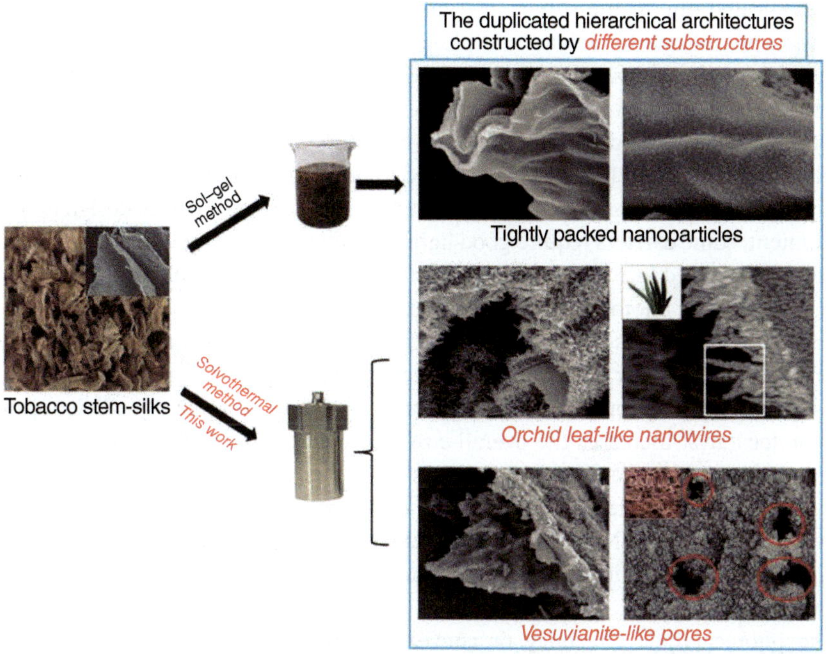

Figure 20.2 The biotemplated TiO_2 synthetic processes of conventional solvothermal and newly developed solvothermal processes. Source: Jiang et al. [18]/Reproduced with permission from Elsevier.

activities of hard-biotemplated TiO_2 could be improved by building the macrostructure with designed microstructural building units [16]. However, studies on biotemplated TiO_2 macrostructural morphology constructed by designed microstructural building units are largely missing [17]. More recently, we developed a new strategy to achieve the leap of biotemplating method from macrostructural morphology replication to microstructural building unit design and assemble the macrostructural morphology of hard biotemplates with designed microstructural building units that could exhibit enhanced visible-light photocatalytic performance for degrading tetracycline [18]. Interestingly, the biotemplated TiO_2 with concave porous microstructural building units is 22.0, 5.5, and 4.4 times higher than those of pure TiO_2, and the two biotemplated TiO_2 samples were prepared via the conventional sol-gel method and solvothermal method (Figure 20.2).

These findings would open new strategies and perspectives for further processing of biotemplates in a multitude of potential applications, including the design of a variety of materials with different morphologies and advanced functionalities.

20.3 Photocatalytic Reactors

It is not a trivial process to design a TiO_2 photocatalytic reactor; it involves wastewater, solid photocatalysts, and light sources. Photocatalytic reactor engineering is complicated due to the sufficient irradiation that needs to be provided within the reaction volume to activate the photocatalyst, uniform reactor irradiation to avoid treatment dead zones, efficient mass transfer, and easy catalyst separation after treatment [19]. As the key component of water treatment equipment, the photocatalyst-supported carrier is directly related to the treatment efficiency of the equipment. Generally, the carrier is required to have stable chemical properties and a large specific surface area [20]. Considering the specific performance design of the equipment, some carriers require good light transmission properties. Specifically, it is normally divided into stationary phase photocatalytic water treatment equipment and mobile phase photocatalytic water treatment equipment [21].

Too many continuous flow reactors employing suspensions of photocatalysts have been reported, but their practical feasibility is still far from the target [22]. Actually, the expensive separation of the treated solution and low recyclability efficiency of the photocatalyst increases the overall cost of the process. And TiO_2 or any other photocatalyst supported on some materials is often used [23]. For this reason, thin films of TiO_2 are simpler to prepare and are suitable for the continuous operation of photocatalysis [24]. Nonetheless, the stability of photocatalyst film and its subsequent reuse, along with the increased time and operational costs of treatment in these types of reactors.

The photocatalytic efficiency or conversion efficiency of the system will be enhanced if the continuous flow reactor is designed in such a manner that it approaches plug-flow conditions. Degradation efficiency as a function of system throughput is a powerful indicator for comparing the performance of photocatalytic reactors of different types and geometries at different development scales as well [25].

Moreover, since the conditional wastewater treatment device has a large construction area and high- project investment costs it is, unsuitable for remote areas and suitable for small-scale enterprises. Additionally, some customers lacking enough space or finances to construct the wastewater treatment equipment would give priority to financial costs, material consumption, and manpower. If wastewater has a very high chemical oxygen demand (COD) ($\geq$1000 mg L^{-1}), the situation will be more serious as the large-scale photocatalytic treatment system for the treatment of this wastewater with a high COD is still rare [1, 20, 21].

Interestingly, a new type of photocatalytic water treatment equipment in which the photocatalyst and adsorbent were coated in a cylindrical reactor rotating around the axis realized a complete water treatment process from adsorption to photocatalysis, and improved the efficiency of the photocatalytic reaction [26]. The ultraviolet lamp tube was arranged in the center of the reactor, and an aeration plate was arranged at the bottom. The lower part of the separation zone was inclined, and in terms of the characteristics of the photocatalyst, magnets could be set or not set. These photocatalytic reactors can run continuously and have a high utilization efficiency of ultraviolet light [27].

It is known that skid-mounted devices are extensively applied to various industries due to the following characteristics of facilitating installation and transfer: The production and assembly of a skid-mounted device are performed in the factory with

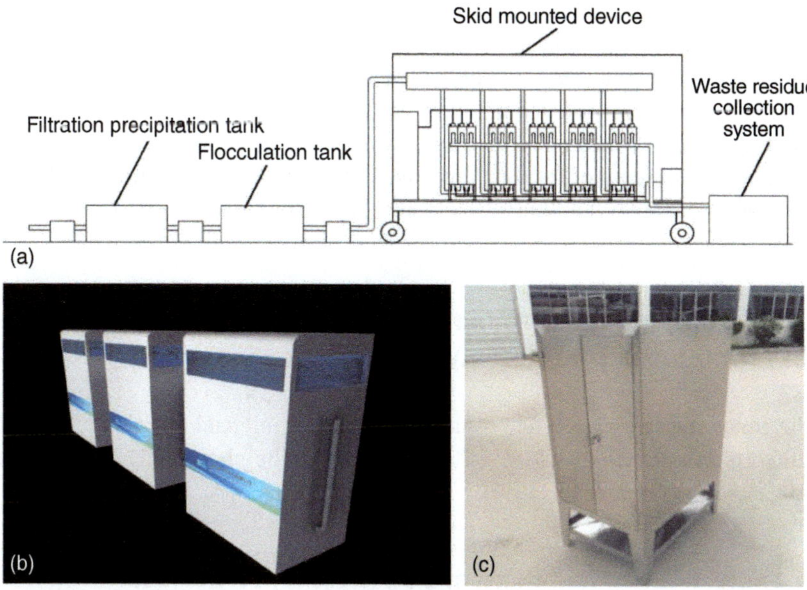

Figure 20.3 (a) Skid-mounted device for the upper-spreading internal diffusion horizontal plug flow photocatalytic wastewater treatment, (b) rendering of skid-mounted equipment, (c) the pictures of skid-mounted equipment. Source: Wang et al. [29]/Reproduced with permission from U.S Patent.

very little on-site installation work, such that the skid-mounted device can work after connecting to pipes and external electrical equipment; the functional components are integrated into an entire base so that the skid-mounted device can be easily transferred as a whole; and the skid-mounted device is compact in structure and occupies less space than a traditional installation [28]. We developed a skid-mounted device for upper-spreading internal diffusion vertical plug flow photocatalytic wastewater treatment, including internal diffusion vertical plug flow photocatalytic reaction tank groups and a skid base (Figure 20.3). The internal diffusion vertical plug flow photocatalytic reaction tank groups are arranged on the skid base. The internal diffusion vertical plug flow photocatalytic reaction tank groups are connected to each other in series, in parallel, or in series parallel. Each internal diffusion vertical plug flow photocatalytic reaction tank group consists of two or more photocatalytic reaction tanks connected in series, in parallel, or in series–parallel. The wastewater pipes connected to the photocatalytic reaction tanks in each internal diffusion vertical plug flow photocatalytic reaction tank group are mounted above the photocatalytic reaction tanks [29].

The speed and efficiency of wastewater treatment are improved by utilizing the series–parallel connection between the skid-mounted internal diffusion horizontal plug flow photocatalytic reaction tanks. The series–parallel multistage reaction mode between the internal diffusion vertical plug flow photocatalytic reaction tanks improved the efficacy of wastewater treatment. Concurrently, due to the synergistic effect of the main catalyst and the cocatalyst, the efficiency of the photocatalytic reaction was further improved. Moreover, the construction investment of the wastewater treatment system has been reduced by half, and the occupied volume of the device is only one-third that of the fixed treatment system, which effectively solves the problem of constructing the wastewater treatment device in small enterprises and in remote areas [30]. In our group, a few skid-mounted photocatalytic reactors with a capacity of 100 tons/day have been successfully constructed. And it will be easy to increase the capacity by increasing the number of modules.

20.4 Examples for Commercial Operations of Skid-mounted Photocatalytic Reactors

In the following, we give a few examples of the commercial operations of skid mounted photocatalytic reactors. Different shapes of the photocatalysts shown in Figure 20.4 were produced.

20.4.1 Treatment of Wastewater at the Expressway Service Area

The operation of the expressway service area produced a considerable amount of sewage, which is filled with organic matter arising out of the activities related to washing water from engineering vehicular (automobile), restaurants, and gas stations. Therefore, proper treatment of this sewage is required to meet stringent effluent quality standards, even though the amount of wastewater is not large. Comparing quite a few studies on biological methods to treat this sewage, there are only a few reports on the lab-scale photocatalytic treatment of wastewater in highway service areas, let alone a scalable photocatalytic reactor before we started manufacturing. We have fabricated one with a capacity of 120 tons/day, and the other three with a capacity of 60 tons/day on skid-mounted photocatalytic

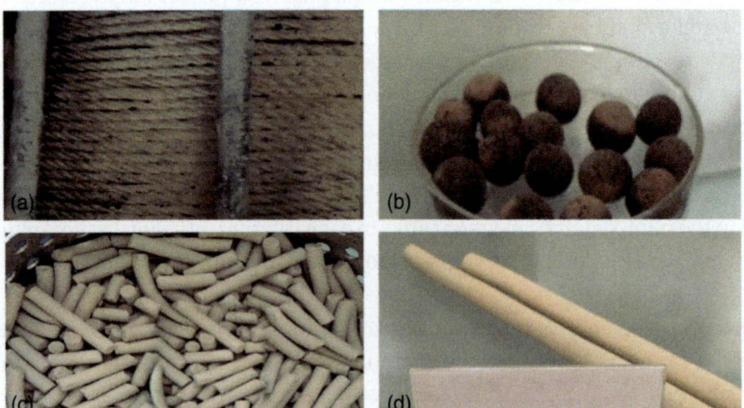

Figure 20.4 The photos of photocatalysts in different shapes: coated (a), spherical (b), bar (c), and tubular (d).

Figure 20.5 Photos of typical photocatalytic reactors used in the expressway service areas. The outdoor scene of a service area (a), the outside (b), and inside (c) of a container with a skid-mounted photocatalytic reactor, and the electric controller (d).

Table 20.1 Summarization of wastewater treatment by using photocatalytic reactors used in the expressway service areas.

Sample	COD ($mg\,L^{-1}$)		Removal rate (%)	NH_3-N ($mg\,L^{-1}$)		Removal rate (%)
	Before treatment	After treatment		Before treatment	After treatment	
1	174.2	13.4	92.3	9.25	1.64	82.3
2	159.1	29.9	81.2	7.17	3.61	49.7
3	129.1	25.4	80.3	5.99	2.63	56.1
4	174.2	45.0	74.2	7.95	3.63	54.3

reactors, respectively, in four service areas of the expressway in Yunnan Province (Figure 20.5). The results are summarized in Table 20.1. It is seen that after photocatalytic treatment, the effluent quality was superior to the COD and NH_3-N in effluent water and met the First Class A Standard of the "Discharge Standard of Pollutants for Municipal Wastewater Treatment Plant" in China (GB18918-2002). The treated water was recycled for landscape pool water or directly discharged.

Figure 20.6 Photos of typical skid-mounted photocatalytic reactors at a hydropower station. The outside (a) and inside (b) of a container with a skid mounted photocatalytic reactor.

20.4.2 Treatment of Wastewater at the Hydropower Stations

The main water quality of the wastewater in the hydroelectric power stations is similar to general domestic sewage, although it is surely small in view of the quantity of water to be treated. The source water protection region includes hydroelectric units as well, so the standards for wastewater discharge are quite strict. In addition, it is far from the city, and it is difficult to build a sewer system to collect and transport waste to a fixed wastewater treatment station for treatment. Particularly, it is very difficult to install the photocatalytic reactors due to the narrow and steep locations. We designed and manufactured three skid-mounted photocatalytic reactors with capacities of 5, 10, and 10 tons/day, respectively (Figure 20.6).

20.4.3 Advanced Treatment of Wastewater from Lignite Gasification After Biological Processes

Although a large number of attempts have been made to improve the water quality of lignite gasification wastewater after biological processes, the high organic loads (COD $\geq$ 1000 mg L^{-1}) often cause serious pollution. We developed one skid-mounted photocatalytic reactor with a capacity of 3 tons/hour for the advanced treatment of wastewater from lignite gasification after biological processes (Figure 20.7). The results are summarized in Table 20.2.

20.5 Challenges and Opportunities

In this chapter, a few large-scale photocatalytic reactors have been used in wastewater treatment, including expressway service areas, hydropower stations, lignite gasification after biological processes, and the skid-mounted photocatalytic reactors, have shown great potential as an alternative wastewater treatment

Figure 20.7 Photos of typical skid-mounted photocatalytic reactors for advanced treatment of wastewater from lignite gasification after biological processes. The inset in (a) is the touchscreen control system of photocatalytic reactors; The inset in (b) is a comparison photo of the wastewater before and after treatment.

Table 20.2 The characteristics of before and after treated wastewater samples.

Composition	Before treatment	After treatment
pH	7.01	7.05
COD ($mg\,L^{-1}$)	1571	93
NH_3-N ($mg\,L^{-1}$)	122.4	32

technique to traditional methods. Overall, the skid-mounted photocatalytic reactors promise great returns in the form of extraordinarily fast, low-cost, easily operated, and environmentally friendly wastewater treatment. The continued growth and advancement of new skid-mounted photocatalytic reactors with better performance are to be expected. However, there is still a long way to go before the current outcomes can be translated into more real-world results. This is in part because of the limited options for feasible photocatalysts and their immobilization methods, light sources, and reactor types. Significant improvements to the treatment capacity still need to be made before they are scaled up for larger-scale industrial productions such as centralized water treatment plants, considering the higher economic feasibility and maneuverability. In reality, however, the existence of natural organic matter, ions, and suspended particulates may greatly lower the photocatalytic efficiency. The applicability of the skid-mounted photocatalytic reactors to a variety of pollutants and complex practical wastewater environments needs to be further investigated. Investigations of the reaction process in different water matrixes will surely promote the development of real-world applications of skid-mounted photocatalytic reactors. Nevertheless, a systematic understanding of the effects of water quality parameters on the reactor durability and the mechanisms behind them is still required. Long-term monitoring of the water quality and photocatalysts during the commercialization is also needed to keep going.

Because the use of solar energy as a primary source of energy will exponentially increase in the foreseeable future, operating expensive water desalination facilities using fossil fuels to treat wastewater is simply not sustainable and could lead to severe problems if other options are not implemented to take over. Furthermore, for wastewater with low COD and less pollution, it is more meaningful to directly pass the photocatalytic reaction without a biochemical reaction. Thus, we propose a combined strategy based on solar-powered skid-mounted photocatalytic reactors and integrated wastewater treatment plants as the pillars of a sustainable solution, along with sound water management strategies and strict regulations.

On the other hand, we need to have an intimate understanding of customer behavior during the commercialization of photocatalytic reactors and really tailor our products around customer needs. This requires much more than simply interviewing customers. We still very much believe in those long-term applications, but are strong advocates for "small miracles" along the way. Business partners who can bridge the gap to develop commercial products are highly needed.

Acknowledgments

The authors would like to thank the National Natural Science Foundation of China (22062026), the Yunling Scholar (YNWR-YLXZ-2019-002), New Research & Development Organization of Yunnan Province, the Program for Innovation Team of Yunnan Province, Industrialization Cultivation Project of Department of Yunnan Education (2016CYH04), and the Key Laboratory of Advanced Materials for Wastewater Treatment of Kunming for financial support.

References

1 Loeb, S., Alvarez, P., Brame, J. et al. (2018). The technology horizon for photocatalytic water treatment: sunrise or sunset? *Environ. Sci. Technol.* 53: 2937–2947.
2 Mecha, A.C. and Chollom, M.N. (2020). Photocatalytic ozonation of wastewater: a review. *Environ. Chem. Lett.* 18: 1491–1507.
3 Van Lente, H., Spitters, C., and Peine, A. (2013). Comparing technological hype cycles: towards a theory. *Technol. Forecasting Social Change*. 80: 1615–1628.
4 Wang, D., Mueses, M.A., Márquez, J.A.C. et al. (2021). Engineering and modeling perspectives on photocatalytic reactors for water treatment. *Water Res.* 202: 117421–117443.
5 Pan, D., Luo, S., Feng, Y. et al. (2021). Highly thermally conductive 3D BN/MWCNTs/C spatial network composites with improved electrically insulating

and flame retardancy prepared by biological template assisted method. *Composites Part B.* 222: 109039–109050.
6 Xie, C., Xiong, Q., Jiang, L. et al. (2022). Carbon supported copper catalyst prepared in situ by one-pot pyrolysis of Bougainvillea glabra: an efficient and stable catalyst for selective oxidation of cyclohexane. *Appl. Surf. Sci.* 576: 151833–151847.
7 Zhang, D. and Qi, L. (2005). Synthesis of mesoporous titania networks consisting of anatase nanowires by templating of bacterial cellulose membranes. *Chem. Commun.* 21: 2735–2737.
8 Miao, Y., Zhai, Z., He, J. et al. (2010). Synthesis, characterizations and photocatalytic studies of mesoporous titania prepared by using four plant skins as templates. *Mater. Sci. Eng., C.* 30: 839–846.
9 Li, J., Chen, Y., Wang, Y. et al. (2015). Synthesis and photocatalysis of mesoporous titania templated by natural rubber latex. *RSC Adv.* 5: 21480–21486.
10 Wang, J., Ou, E., Li, J. et al. (2015). Synthesis of mesoporous titania-graphite composite templated by hypocrellins for visible-light photocatalytic degradation of acetaldehyde. *Mater. Sci. Semicond. Process.* 31: 397–404.
11 He, J., Chen, D., Li, Y. et al. (2013). Diatom-templated TiO_2 with enhanced photocatalytic activity: biomimetics of photonic crystals. *Appl. Phys. A.* 113: 327–332.
12 He, J., Zi, G., Yan, Z. et al. (2014). Biogenic C-doped titania templated by cyanobacteria for visible-light photocatalytic degradation of rhodamine B. *J. Environ. Sci.* 26: 1195–1202.
13 Gao, L., Gao, W., Qiu, Z. et al. (2018). Biomorphic carbon-doped TiO_2 for photocatalytic gas sensing with continuous detection of persistent volatile organic compounds. *ACS Applied. Nano Material.* 1: 1766–1775.
14 Zhang, D., Qi, L., Cheng, H., and Ma, J. (2002). Preparation of ZnS nanorods by a liquid crystal template. *J. Colloid Interface Sci.* 246: 413–416.
15 Fattakhova-Rohlfing, D., Zaleska, A., and BeinT. (2014). Three-dimensional titanium dioxide nanomaterials. *Chem. Rev.* 114: 9487–9558.
16 Song, R., Chi, H., Ma, Q. et al. (2021). Highly efficient degradation of persistent pollutants with 3D nanocone TiO_2-based photoelectrocatalysis. *J. Am. Chem. Soc.* 143: 13664–13674.
17 Wang, C., Li, J., Paineau, E. et al. (2020). A sol–gel biotemplating route with cellulose nanocrystals to design a photocatalyst for improving hydrogen generation. *J. Mater. Chem. A.* 8: 10779–10786.
18 Jiang, L., He, J., Yang, Y. et al. (2022). Enhancing visible-light photocatalytic activity of hard-biotemplated TiO_2: From macrostructural morphology replication to microstructural building units design. *J. Alloys Compd.* 898: 162886–162898.
19 Manassero, A., Satuf, M., and Alfano, O. (2017). Photocatalytic reactors with suspended and immobilized TiO_2: comparative efficiency evaluation. *Chem. Eng. J.* 326: 29–36.
20 Visan, A., Van Ommen, J., Kreutzer, M., and Lammertink, R. (2019). Photocatalytic reactor design: guidelines for kinetic investigation. *Ind. Eng. Chem. Res.* 58: 5349–5357.

21 Natali Sora, I. and Bekbolet, M. (2021). New perspectives in photocatalytic water treatment. *Water*. 13: 3083–3086.

22 Chang, S., Yang, X., Sang, Y., and Liu, H. (2016). Highly efficient photocatalysts and continuous-flow photocatalytic reactors for degradation of organic pollutants in wastewater. *Chem. Asian J.* 11: 2352–2371.

23 Mukherjee, P. and Ray, A. (1999). Major challenges in the design of a large-scale photocatalytic reactor for water treatment. *Chem. Eng. Technol.* 22: 253–260.

24 Das, S. and Mahalingam, H. (2019). Reusable floating polymer nanocomposite photocatalyst for the efficient treatment of dye wastewaters under scaled-up conditions in batch and recirculation modes. *J. Chem. Technol. Biotechnol.* 94: 2597–2608.

25 Żukowski, S. (2020). Performance of TiO_2/Fly ash cenospheres as a catalytic film in a novel type of periodic air-sparged photocatalytic reactor. *Materials*. 13: 1691–1703.

26 Zhou, Z., Wang, H., Shi, H., Zhang, S., and Wang, X. (2015) New type of photocatalytic water treatment device prepare. C.N. patent201520253454.X[P].

27 Kozlowski, M. and Yoon, T. (2016). Editorial for the special issue on photocatalysis. *J. Org. Chem.* 81: 6895–6897.

28 Chen, N., Lin, B., and Chen, Y. (2013). Study on control system for skid-mounted wastewater treatment device. *China Water Wastewater*. 29: 74–76.

29 Wang, J., Wang, W., Li, Yi., and Jiang, L. (2021) Skid mounted device for upper-spreading internal diffusion vertical plug flow photocatalytic wastewater treatment. U.S.11,155,475B2[P].

30 Malik, A., Younan, N., Rao, B., and Mousa, K. (1989). Skid mounted mobile brackish water reverse osmosis plants at different sites in Kuwait. *Desalination*. 75: 341–361.

21

Preparation of Highly Functional TiO_2-Based Thin-Film Photocatalysts by Ion Engineering Techniques, Photocatalysis, and Photo-Induced Superhydrophilicity

Masato Takeuchi[1] *and Masakazu Anpo*[1,2]

[1] *Osaka Metropolitan University, Graduate School of Engineering, Department of Applied Chemistry, 1-1, Gakuen-cho, Naka-ku, Sakai, Osaka 599-8531, Japan*
[2] *Fuzhou University, State Key Laboratory of Photocatalysis on Energy and Environment, Fuzhou, Fujian 350116, P. R. China*

21.1 Introduction

The discovery of the photoinduced superhydrophilicity of TiO_2 thin films [1] has opened up various possibilities for TiO_2 photocatalysts to be applied in self-cleaning, antifogging, and antibacterial agents [2]. Much attention has been directed toward the applications of photocatalysts to obtain clean environments [3–8]. To apply TiO_2 photocatalysts for the purification of toxic compounds, the fixation of TiO_2 powders onto various substrates with strong adhesion is desired. The TiO_2 thin films coated on various substrates have been investigated not only for photocatalytic reactivity but also for their unique surface wettability [1, 2]. TiO_2 coatings are generally carried out by a wet process using precursor solutions containing titanium alkoxide or TiO_2 sol [9]. The wet process necessitates a post-calcination at high temperatures after coating the precursor solutions to obtain a strong adhesion of thin films. In contrast, we applied a dry coating process such as ionized cluster beam (ICB) [10–13] or radio-frequency-magnetron sputtering (RF-MS) [14–17] deposition methods in preparing highly functional TiO_2 thin-film photocatalysts. The preparation of thin films in a vacuum chamber has some potential advantages: (i) Contamination of the thin films with impurities is prevented; (ii) A dry preparation method does not require the use of any organic solvents, so it is an "environmentally friendly process"; (iii) Thin films with high crystallinity and strong adhesion are easily prepared without post-calcination at high temperatures; (iv) Easy control of the various physical and chemical properties.

Furthermore, we have reported that Ti/Si binary oxides show unique photocatalytic properties, which are caused by tetrahedral Ti-oxide species within the SiO_2 [18, 19] or zeolite matrices [20, 21]. Such tetrahedral TiO_4 units are known to show photocatalytic reactivity for partial oxidation or epoxidation reactions of olefins. Based on these considerations, we applied the ion engineering techniques to prepare TiO_2, TiO_2–SiO_2, and TiO_2–B_2O_3 thin-film photocatalysts [22, 23].

UV-Visible Photocatalysis for Clean Energy Production and Pollution Remediation: Materials, Reaction Mechanisms, and Applications, First Edition. Edited by Xinchen Wang, Masakazu Anpo, and Xianzhi Fu.
© 2023 WILEY-VCH GmbH. Published 2023 by WILEY-VCH GmbH.

Then, the effects of the local structure of Ti-oxide species on the photocatalytic reactivity and the surface wettability of binary oxide thin films were discussed.

For widespread applications of photocatalysts, the immobilization of TiO_2 films on thermally unstable substrates such as plastics, polymers, or textiles is desired. Coating solutions containing TiO_2 sols and inorganic/organic binders were commercialized. However, since the major part of the TiO_2 powder was buried in the binder layers, sufficiently high photocatalytic performance was challenging. We reported a coating of highly transparent TiO_2 thin films on polycarbonate substrates by an RF-MS deposition method [24].

21.2 Ion Engineering Techniques to Prepare Thin-Film Photocatalysts

We applied an ICB deposition method to prepare transparent TiO_2-based thin films [10–13]. A schematic diagram of the ICB method with multi-ion sources is illustrated in Figure 21.1a. Titanium vapor obtained by heating Ti grains at approximately 2000 °C is introduced into a vacuum chamber, producing titanium clusters. The titanium clusters react with O_2 (pressure: 2.7×10^{-5} kPa) in the vacuum chamber, forming TiO_2 clusters. The TiO_2 clusters were ionized by electron beam irradiation, accelerated by a high voltage of 0.5 kV. The ionized clusters were then bombarded onto the substrates, resulting in the formation of TiO_2 thin films. TiO_2-based binary oxide thin films, TiO_2–SiO_2, and TiO_2–B_2O_3 were prepared via SiO_2 and B_2O_3 grains and were used simultaneously as the ion source, respectively [22, 23].

Furthermore, we applied a RF-MS deposition method to prepare TiO_2 thin-film photocatalysts [14–17]. A schematic diagram of the RF-MS deposition method is illustrated in Figure 21.1b. In the present chapter, we used a TiO_2 plate as a sputtering target and Ar gas as a sputtering gas. When a magnetic field is vertically applied to an electric field in the presence of sputtering gas, ring-state Ar plasma is induced on the target material. Ar^+ in the gas plasma sputters the TiO_2 target surface to produce Ti^{4+} and O^{2-} ions. These ions are uniformly rearranged on the substrates to form TiO_2 thin films. The photocatalytic properties of the TiO_2 thin films are strongly affected by the preparation conditions such as the induced RF power, substrate temperatures, distances between the target and substrates (D_{TS}), and sputtering gas flow rates. Thus, we fixed the induced RF power at 300 W and the D_{TS} at 80 mm. The substrate temperatures were changed from 100 to 700 °C, and post-calcination treatments were not carried out.

21.2.1 Transparent TiO_2 Thin-Film Photocatalysts Prepared by Ionized Cluster Beam (ICB) Deposition Method

In this section, we deal with the preparation of highly transparent TiO_2 thin-film photocatalysts on quartz substrates by applying the ICB deposition method [10–13]. From XRD measurements, the TiO_2 thin films were confirmed to be a mixture of anatase and rutile structures. Since the TiO_2 thin films showed a similar

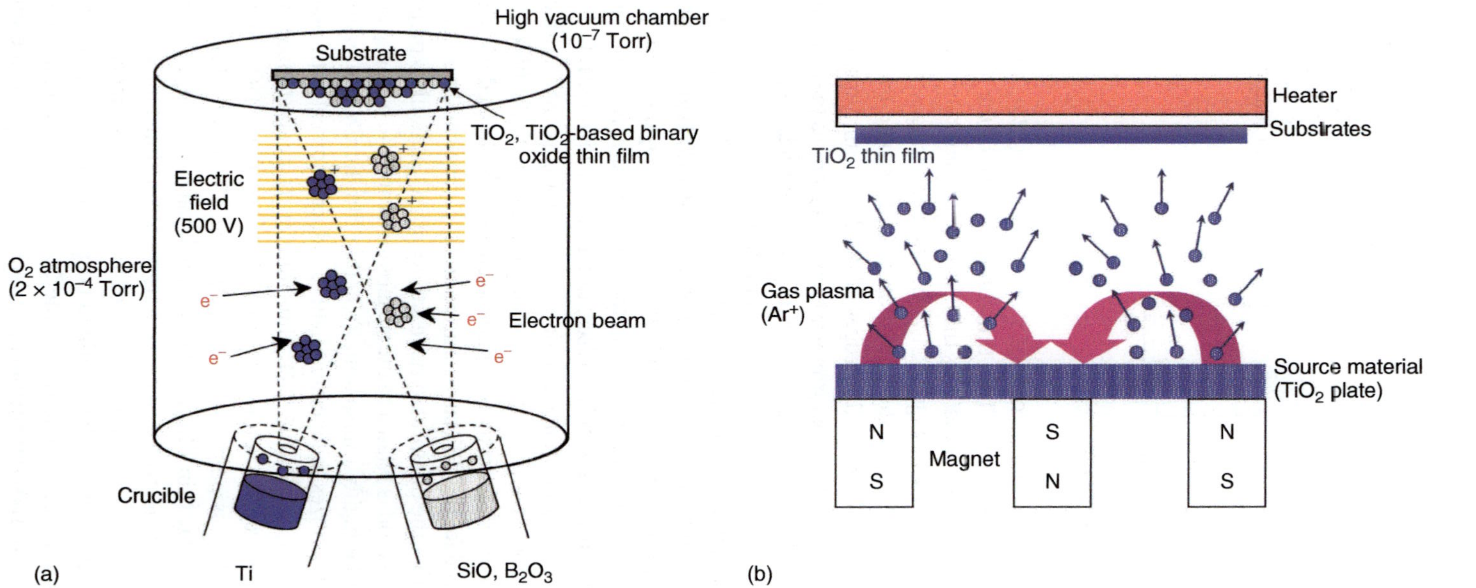

Figure 21.1 Schematic diagram of the ionized cluster beam (ICB) deposition method using multi-ion sources (a) and the RF-magnetron sputtering (RF-MS) deposition method (b). Source: Reproduced with permission from Takeuchi et al. [22]. Copyright 2003 American Chemical Society. Reprinted with permission from Dohshi et al. [23]/Elsevier. Reproduced with permission from Takeuchi et al. [17]/Springer Nature.

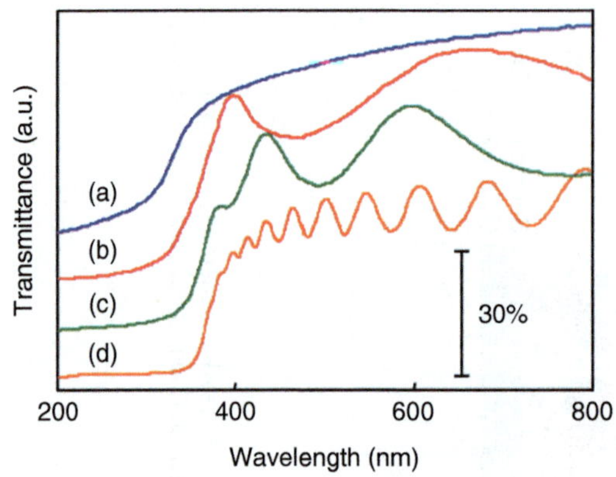

Figure 21.2 UV–Vis absorption (transmittance) spectra of the TiO_2 thin films prepared by the ICB deposition method. Film thickness: (a) 20, (b) 100, (c) 300, (d) 1000 nm. Source: Reproduced with permission from Takeuchi et al. [10]/Springer Nature.

ratio of anatase to rutile to commercial TiO_2 photocatalysts (P25), high photocatalytic reactivity was expected for the TiO_2 thin films prepared by the ICB deposition method. The optical properties of the TiO_2 thin films were characterized by UV–Vis absorption spectra, as shown in Figure 21.2. Interference fringes were observed in visible-light regions, indicating the formation of uniform and transparent thin films on the substrates. The absorption edges of the TiO_2 thin films, which were observed at 350–380 nm, shifted to shorter wavelength regions as the film thickness decreased. This phenomenon can be explained by a quantum size effect caused by nanosized TiO_2 particles composed of transparent thin films [18–21].

The TiO_2 thin films showed a photocatalytic reactivity for the decomposition of NO into N_2 and N_2O under UV light irradiation ($\lambda > 270$ nm) [10]. The photocatalytic reactivity of the thin films was even higher than the TiO_2 thin films prepared by the sol–gel method. Furthermore, the photocatalytic reactivity of the TiO_2 thin films was dependent on the film thicknesses, as shown in Figure 21.3. The thinner TiO_2 films showed higher photocatalytic reactivity, a higher BET surface area, as well as an absorption edge in shorter wavelength regions. As the film thickness increased, the photocatalytic reactivity slightly decreased and leveled off. The same tendency was observed for the BET surface areas and the wavelengths of the absorption edge. It means that the photocatalytic reactivity of TiO_2 thin films prepared by the ICB deposition method is related to the surface areas and the bandgap energies. Furthermore, since the highly transparent TiO_2 thin film hardly scatters incident UV light, it effectively utilizes the incident light to induce photocatalytic reactivity.

The surface morphologies of the TiO_2 thin films were characterized by SEM observations [13]. Figure 21.4 shows the SEM images of the TiO_2 thin films: (a) deposited at 350 °C (as-prepared); (b) calcined at 550 °C; and (c) at 700 °C. Because as-prepared TiO_2 thin film consists of small particles with a diameter of approximately 20–50 nm. However, when the as-prepared thin film was calcined at 550 °C, the crystal-grain growth was observed but the surface morphology was almost similar to the as-prepared films (without post-calcination). In contrast, when the as-prepared film was calcined at 700 °C, the TiO_2 particles were aggregated to grain sizes in the range of 100–150 nm and its surface morphology became smoother as compared to the as-prepared film. The crystal structures of the TiO_2 thin films

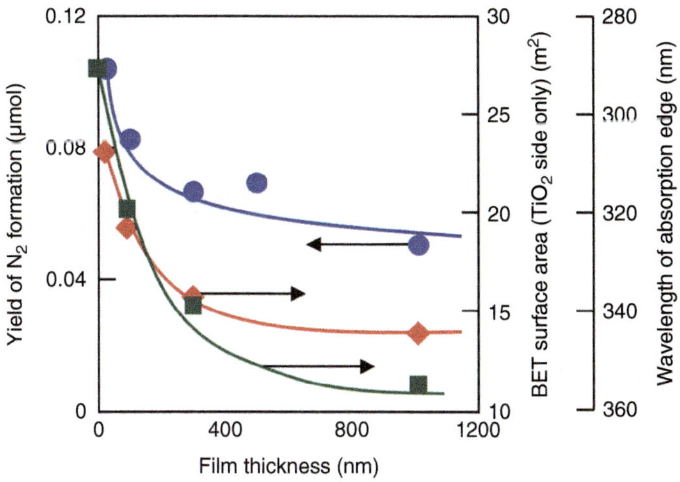

Figure 21.3 Relationship between the photocatalytic reactivities for NO decomposition (circle plots), BET surface areas (diamond plots), and wavelengths of the absorption edges (square plots) of the TiO_2 thin films prepared by the ICB deposition method. Source: Reproduced with permission from Takeuchi et al. [10]/Springer Nature.

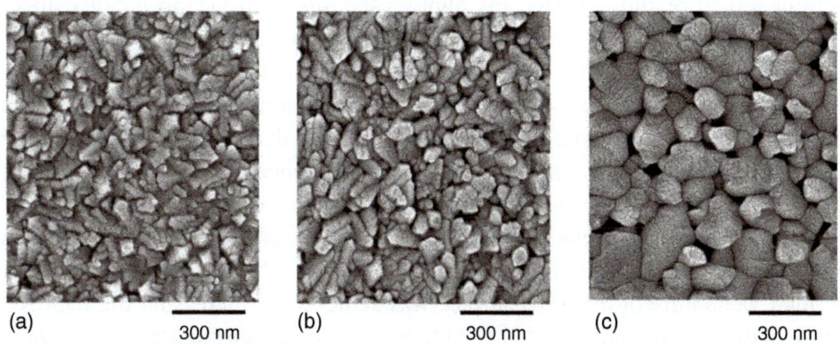

Figure 21.4 SEM images of the top surfaces of the TiO_2 thin films prepared by the ICB deposition method: (a) as-prepared; (b) calcined at 550 °C; and (c) 700 °C. Source: Reproduced with permission from Takeuchi et al. [13]/Springer Nature.

were then characterized by XRD measurements. Without any post-calcinations, the thin film that was formed at 350 °C displayed diffraction patterns that were attributed to the blending of rutile and anatase structures. As mentioned above, the ratio of anatase to rutile was approximately 70 %, similar to the commercial TiO_2 photocatalysts (P25). This result suggests that the ICB deposition method has the potential to prepare highly crystalline TiO_2 thin films even at low temperatures of 350 °C. When the as-prepared film was calcined at 550 °C, the ratio of the anatase to rutile phases decreased to approximately 40 %. Furthermore, the thin films calcined at 700 °C were completely converted to the rutile structure.

The photoinduced surface wettability of the TiO_2 thin films was evaluated by the measurement of the water contact angle (WCA). UV light irradiation was

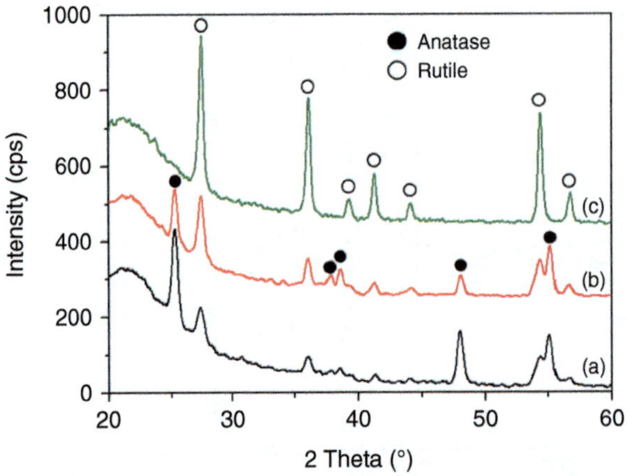

Figure 21.5 XRD patterns of the TiO_2 thin films prepared by the ICB deposition method: (a) as-prepared; (b) calcined at 550 °C; and (c) 700 °C. Source: Reproduced with permission from Takeuchi et al. [13]/Springer Nature.

carried out using a commercial fluorescent bulb under controlled light intensity (10–20 μW cm^{-2}). Figure 21.5 shows the time courses of the contact angles of water droplets on the TiO_2 thin films under UV light irradiation. The as-prepared TiO_2 film and the film calcined at 550 °C showed high wettability (WCA < 5°) under UV light irradiation. However, the TiO_2 thin film calcined at 700 °C, which consisted of its rutile structure, did not have any photo-responses in surface wettability under UV light irradiation. Furthermore, the photocatalytic reactivity for oxidation of acetaldehyde under UV light irradiation was evaluated (data not shown here). Among the TiO_2 thin films having different crystal structures, the as-prepared film showed higher photocatalytic reactivity as compared to the films calcined at 550 and 700 °C. The tendency was closely related to the differences in surface morphology, surface area, crystal structure, and photoinduced surface wettability properties.

The mechanism of the change in the photoinduced surface wettability was then discussed using near-infrared spectroscopy [25, 26]. Until then, it was claimed that the high wettability of the TiO_2 surface was achieved by increasing the hydroxyl groups of the surface under UV light irradiation. However, we reported that the high wettable state of the TiO_2 surface readily disappeared by outgassing the thin film at room temperature. It means that the phenomenon cannot be explained by the newly formed hydroxyl groups of the TiO_2 surface by UV light irradiation. Furthermore, TPD studies of TiO_2 single crystals under ultrahigh vacuum settings by White et al. revealed that wetting processes are not dependent on the breakdown of H_2O into OH groups [27]. Moreover, we revealed the ratio of H-bond-free water to H-bonded water on the TiO_2 surface increased from 6.3 % to 11.3 % by UV light irradiation. This result indicates that the surface tension of H_2O clusters on the TiO_2 surface lessens during light irradiation. The changes in the surface tension of the H_2O clusters are associated with a driving force behind the H_2O clusters spreading out on the TiO_2 surfaces to show a high wettability. Interestingly, the photo-functionality of the TiO_2 surface is applied as biomaterials for fracture treatments or restorative treatments such as dental implants. Ogawa et al. reported that UV-C light irradiation (λ < 254 nm) on titanium dental implants dramatically enhanced the osseointegration (attachment between the dental implants and bone tissues) at least eightfold as compared to commercial samples [28–30] . This phenomenon was reasonably explained by the photo-functionality of the TiO_2 layer formed on the titanium implants. That is, since the TiO_2 layers show high wettability against blood by UV light irradiation, the osteoblast cells are efficiently cultivated on the implant surface cells (a cell-philic property), resulting in the formation of thick bone tissues on the implant surface.

21.2.2 Functional TiO_2–SiO_2 and TiO_2–B_2O_3 Binary Oxide Thin-Film Photocatalysts Prepared by Multi-Ion Source Ionized Cluster Beam (ICB) Deposition Method

TiO_2–SiO_2 and TiO_2–B_2O_3 binary oxide thin-film photocatalysts were prepared using the multi-ion source ICB deposition method [22, 23]. The optical properties of thin-film photocatalysts, such as absorption edges and transparency in visible-light regions, are some of the most important factors in determining photocatalytic reactivity. Figure 21.6 shows the UV–Vis absorption (transmittance) spectra of the TiO_2–SiO_2 (left) and TiO_2–B_2O_3 (right) thin films (film thickness: 700 nm) with different Ti-oxide compositions. Since pristine SiO_2 and B_2O_3 thin films did not show any significant absorption in measurable ranges (200–800 nm), the absorption observed for the binary oxide thin films could be, thus, attributed to the Ti-oxide species dispersed in the SiO_2 or B_2O_3 matrices. That is, binary oxide thin films, which contained TiO_2 of more than 50 %, showed typical interference fringes in visible-light regions, indicating the transparent and uniform TiO_2 thin layers on a quartz substrate. On the other hand, when the TiO_2 content became smaller than 10 %, the interference fringes were barely observed. Furthermore, the blue shift of absorption edges toward shorter wavelength regions was also observed. Especially, the TiO_2–SiO_2 thin film having a TiO_2 content of 6.6 % showed an absorption edge at approximately 250 nm. These results indicate that the Ti-oxide species are dispersed within the transparent SiO_2 or B_2O_3 matrices. On the other hand, the TiO_2–B_2O_3

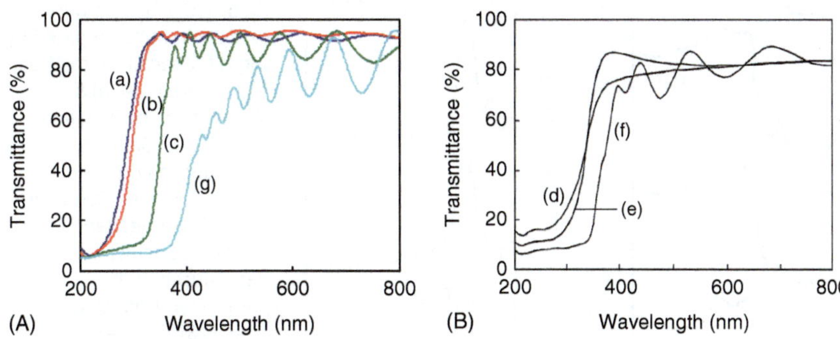

Figure 21.6 UV–Vis absorption (transmittance) spectra of: (a–c) TiO_2–SiO_2, (d–f) TiO_2–B_2O_3 binary oxide thin films, and (g) TiO_2 thin films prepared by the ICB deposition method. TiO_2 content: (a) 6.6 %, (b) 9.5 %, (c) 50.1 %, (d) 5.0 %, (e) 10.0 %, (f) 50.0 %. Source: Reproduced with permission from Takeuchi et al. [22]/American Chemical Society. Reproduced with permission from Dohshi et al. [23]/Elsevier.

thin films having low TiO_2 content exhibited an absorption edge at 350–370 nm, indicating that the Ti-oxide species are aggregated within the B_2O_3 matrices.

The local structures of Ti-oxide species were further discussed by XAFS measurements of the binary oxide thin films. Figure 21.7 shows the Ti K-edge XANES (left) spectra of the TiO_2–SiO_2 and TiO_2–B_2O_3 thin films with different TiO_2 contents. The TiO_2–SiO_2 films having TiO_2 contents of lower than 10 % showed a sharp pre-edge peak at 4.97 keV attributed to a tetrahedrally coordinated TiO_4 unit (Figure 21.7A). As the TiO_2 content increased, the intensity of the sharp pre-edge peak decreased, indicating the formation of an octahedrally coordinated TiO_6 unit (Figure 21.7B,C). On the other hand, the TiO_2–B_2O_3 thin film having a TiO_2 content of 5 % showed the typical pre-edge peaks attributed to an anatase structure, indicating that small Ti-oxide clusters having anatase structure are formed within the B_2O_3 matrices. Furthermore, the Fourier transforms of the EXAFS oscillation of the TiO_2–SiO_2 and TiO_2–B_2O_3 thin films are also shown in Figure 21.7 (right). The TiO_2–SiO_2 thin films showed a single peak attributed to the Ti–O bond at 1.6–1.7 Å, indicating the formation of tetrahedrally coordinated TiO_4 structure within the SiO_2 matrices. On the other hand, the TiO_2–B_2O_3 films showed a peak due to the Ti–O bond and a peak due to the Ti–O–Ti bond at 2.6–2.9 Å, indicating the formation of aggregated TiO_2 particles within the B_2O_3 matrices. The coordination numbers and the Ti–O bond distance obtained from the curve fitting analysis are summarized in Table 21.1. The Ti–O bond distance of the TiO_4 unit in the TiO_2–SiO_2 films was estimated to be 1.81–1.82 Å. This value is slightly longer than the isolated tetrahedral Ti-oxide species incorporated into zeolite frameworks (1.78 Å) [20, 21]. On the other hand, the TiO_2–B_2O_3 films showed a Ti–O bond distance of approximately 1.90 Å, which is close to the anatase TiO_2 structure. The detailed measurements of the binary oxide

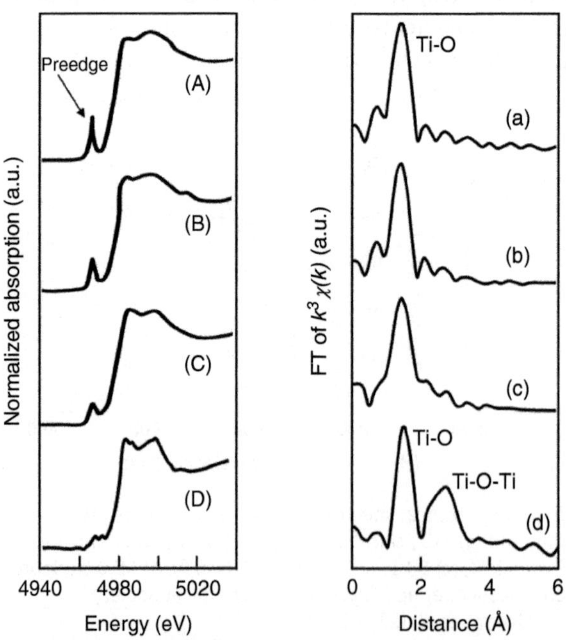

Figure 21.7 Ti K-edge XANES spectra (left) and Fourier transforms of the EXAFS oscillation (right) of: (A–C, a–c) TiO_2–SiO_2 and (D, d) TiO_2–B_2O_3 binary oxide thin films prepared by the ICB deposition method. TiO_2 content: (A, a) 6.6, (B, b) 9.5, (C, c) 50.1, and (D, d) 5.0 %. Source: Reproduced with permission from Takeuchi et al. [22]/American Chemical Society. Reproduced with permission from Dohshi et al. [23]/Elsevier.

Table 21.1 Coordination number and bond length of the Ti-oxide species in the TiO_2-SiO_2 and $TiO_2-B_2O_3$ binary oxide thin films as determined by the curve fitting for the Fourier transform of the EXAFS oscillation.

Catalysts	Shell	Bond distance (Å)	Coordination number
TiO_2–SiO_2 (6.6/93.4)	Ti–O	1.81	4.3
TiO_2–SiO_2 (9.5/90.5)	Ti–O	1.82	4.4
TiO_2–SiO_2 (50.1/49.9)	Ti–O	1.85	4.9
TiO_2–B_2O_3 (5/95)	Ti–O	1.91	5.98
TiO_2–B_2O_3 (10/90)	Ti–O	1.90	5.97
TiO_2–B_2O_3 (50/50)	Ti–O	1.90	5.97

Source: Reproduced with permission from Takeuchi et al. [22]/American Chemical Society. Reproduced with permission from Dohshi et al. [23]/Elsevier.

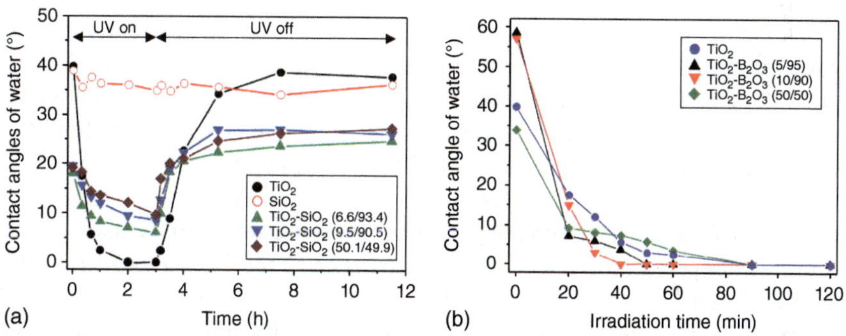

Figure 21.8 Time courses for the changes in contact angles of water droplets on (a) the TiO_2-SiO_2 and (b) $TiO_2-B_2O_3$ binary oxide thin films prepared by the ICB deposition method under UV light irradiation. Source: Reproduced with permission from Takeuchi et al. [22]/American Chemical Society. Reproduced with permission from Dohshi et al. [23]/Elsevier.

thin films having different TiO_2 contents clarified that the optical properties and local structures are dependent on the type and nature of the combined oxide materials. In turn, the effects of the local structure of Ti-oxide species on the photocatalytic reactivity and the surface wettability of binary oxide thin films were then discussed.

The surface wettability of binary oxide thin films containing tetrahedral TiO_4 units or TiO_2 nanoclusters was investigated, as shown in Figure 21.8. For comparison, the results of the pristine TiO_2 and SiO_2 thin films prepared by the ICB deposition method were illustrated. The SiO_2 thin film did not show any photoresponses for the surface wettability. The pristine TiO_2 thin film readily showed the highly wettable property, in which the WCA of 0° was achieved after two hours irradiation of UV light. However, in the case of the TiO_2–SiO_2 binary oxide thin films containing the tetrahedral TiO_4 species; the WCA leveled off at approximately 5° (did not reach 0°). These results suggest the tetrahedral TiO_4 species are not effective for exhibiting photoinduced wettability conversion under UV light irradiation. This interpretation closely corresponds to the fact that the tetrahedral TiO_4 species incorporated within zeolite frameworks are effective for the partial oxidation or epoxidation of olefins but not for the complete oxidation of organic compounds. Furthermore, the TiO_2–SiO_2 thin films maintained a high wettability under dark conditions as compared to the TiO_2 thin films. On the other hand, the TiO_2–B_2O_3 thin films showed higher efficiency in the photo-conversion of surface wettability as compared to the TiO_2 film. This result is reasonably explained by the highly hydrophilic property of the B_2O_3 domains of the binary oxide thin films.

21.2.3 Preparation of Crystalline TiO_2 Thin-Film Photocatalysts on the Polycarbonate Substrate by an RF-Magnetron Sputtering (RF-MS) Method

For the widespread application of TiO_2 photocatalysts, a coating of TiO_2 thin films on thermally unstable substrates, such as plastics, polymers, or textiles, is greatly desired. Based on these considerations, we applied the RF-MS deposition method to prepare crystalline TiO_2 thin films at temperatures lower than 100 °C [24]. Some of the TiO_2 thin films prepared at temperatures higher than 100 °C readily peeled off by mechanical scratching. However, as shown in Figure 21.9, the TiO_2 thin films prepared at temperatures less than 100 °C exhibited high transparency and mechanical stability against scratching and bending. Furthermore, as shown in Figure 21.10, the TiO_2 thin films deposited on the polycarbonate substrates at 100 °C showed typical diffraction peaks at 25°, 48°, and 55° attributed to the (101), (200), and (211) phases of the anatase structure, respectively. The polycarbonate substrates showed high transmittance of 90 % in visible-light regions. The TiO_2 thin films coated on the polycarbonate substrates showed an absorption edge attributed to the bandgap excitation of the TiO_2 semiconductor at 370 nm. Furthermore, the transmittance was maintained to be approximately 80 %, indicating that the highly uniform TiO_2 thin films could be obtained without any loss of the high transparency of polycarbonates.

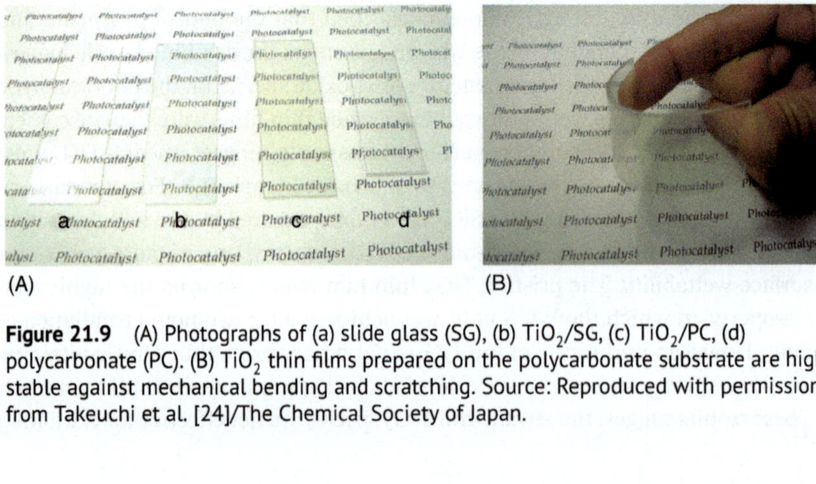

(A) (B)

Figure 21.9 (A) Photographs of (a) slide glass (SG), (b) TiO_2/SG, (c) TiO_2/PC, (d) polycarbonate (PC). (B) TiO_2 thin films prepared on the polycarbonate substrate are highly stable against mechanical bending and scratching. Source: Reproduced with permission from Takeuchi et al. [24]/The Chemical Society of Japan.

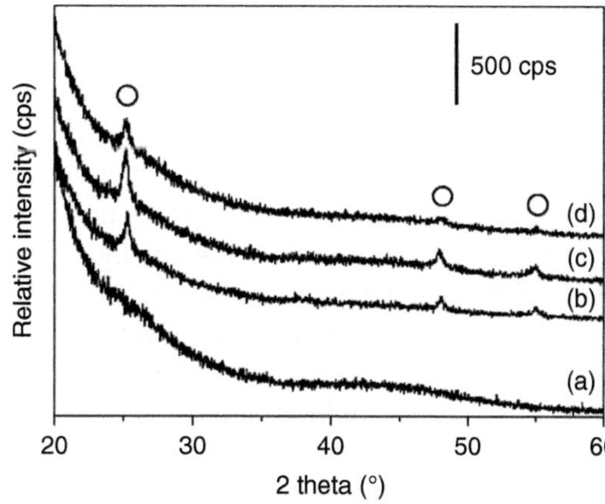

Figure 21.10 XRD patterns of: (a) polycarbonate substrates and (b)–(d) TiO_2 thin films prepared by the RF-MS deposition method. Sputtering gas pressures: (b) 1.0 Pa, (c) 2.0 Pa, and (d) 3.0 Pa. Source: Reproduced with permission from Takeuchi et al. [24]/The Chemical Society of Japan.

The surface wettability was evaluated by the contact angles of water droplets under UV light irradiation. However, since the TiO_2 thin films were prepared in vacuum chambers, the as-prepared thin films showed high wettability (low contact angles of water droplets). The thin films were, thus, stored in a clean room without light irradiation for at least two weeks. These TiO_2 thin films showed contact angles of approximately 100° before UV light irradiation. A BLB (black light blue) lamp was then irradiated on the TiO_2 thin films with different intensities of 0.05–1 mW cm^{-2}. The water contact angles gradually decreased by continuous UV light irradiation and then reached less than 5°. Although prolonged UV light irradiation of at least one day is necessary, it was noted that the TiO_2 thin films coated on plastic substrates showed high wettability by weak UV light intensity of 0.05 mW cm^{-2}. These results indicate that the crystalline TiO_2 thin-film photocatalysts are successfully prepared on thermally unstable polycarbonate substrates at low temperatures by the RF-MS deposition method.

21.3 Conclusions

In the present chapter, we have reviewed the preparation of the TiO_2 and TiO_2-based binary oxide thin-film photocatalysts by applying ion engineering techniques. Highly transparent TiO_2 thin films were successfully prepared using the ICB deposition method. The TiO_2 thin films showed photocatalytic reactivity for the decomposition of NO under UV light irradiation. Furthermore, the TiO_2 thin films showed high wettability under UV light irradiation, depending on the surface morphologies, crystalline structures, and surface areas. As a next step, we prepared the TiO_2–SiO_2 and TiO_2–B_2O_3 binary oxide thin films using the multi-ion source ICB deposition method. Interestingly, in the case of the TiO_2–SiO_2 thin films, the tetrahedrally coordinated TiO_4 species were highly dispersed within the SiO_2 matrices. The tetrahedral TiO_4 species did not show highly efficient surface wettability under UV light irradiation as compared with the pristine TiO_2 thin films, because the tetrahedral TiO_4 species were not effective for the complete oxidation of organic compounds. In contrast, the TiO_2–B_2O_3 thin films contained the TiO_2

nanoclusters having octahedral coordination within the host B_2O_3 matrices and showed an efficient photoresponse in the surface wettability as compared with the pristine TiO_2 thin films.

Then, we introduced the preparation of crystalline TiO_2 thin films on thermally unstable plastic substrates at temperatures lower than 100 °C by the RF-MS deposition method. The TiO_2 thin films were highly transparent and mechanically stable with no peeling. Furthermore, the thin films showed a photoresponse in the surface wettability under UV light irradiation from a BLB lamp (0.05–1 mW cm^{-2}).

We obtained many innovative insights for the preparation of highly transparent TiO_2 and TiO_2-based binary oxide thin-film photocatalysts using ion engineering techniques. These findings have been passed down to the present application of photocatalysts to purify our environments. Thus, the environmentally friendly photocatalytic system will provide new approaches for the "Sustainable Development Goals (SDGs)" in the future.

References

1 Wang, R., Hashimoto, K., Fujishima, A. et al. (1997). Light-induced amphiphilic surfaces. *Nature* 388: 431–432.

2 Fujishima, A., Rao, T.N., and Tryk, D.A. (2000). Titanium dioxide photocatalysis. *J. Photochem. Photobiol., C* 1: 1–21.

3 Pelizzetti, E. and Schiavello, M. (1991). *Photochemical Conversion and Storage of Solar Energy*. Amsterdam: Kluwer Academic Publishers.

4 Anpo, M. and Takeuchi, M. (2003). The design and development of highly reactive titanium oxide photocatalysts operating under visible light irradiation. *J. Catal.* 216: 505–516.

5 Anpo, M. (2004). Preparation, characterization, and reactivities of highly functional titanium oxide-based photocatalysts able to operate under UV–visible light irradiation: approaches in realizing high efficiency in the use of visible light. *Bull. Chem. Soc. Jpn.* 77: 1427–1442.

6 Anpo, M. and Kamat, P.V. (2010). *Environmentally Benign Photocatalysts: Applications of Titanium Oxide-based Materials*. New York: Springer.

7 Schneider, J., Matsuoka, M., Takeuchi, M. et al. (2014). Understanding TiO_2 photocatalysis: mechanisms and materials. *Chem. Rev.* 114: 9919–9986.

8 García-López, E.I. and Palmisano, L. (2021). *Materials Science in Photocatalysis*. Amsterdam: Elsevier.

9 Negishi, N., Takeuchi, K., and Ibusuki, T. (1998). Surface structure of the TiO_2 thin film photocatalyst. *J. Mater. Sci.* 33: 5789–5794.

10 Takeuchi, M., Yamashita, H., Matsuoka, M. et al. (2000). Photocatalytic decomposition of NO on titanium oxide thin film photocatalysts prepared by an ionized cluster beam technique. *Catal. Lett.* 66: 185–187.

11 Takeuchi, M., Yamashita, H., Matsuoka, M. et al. (2000). Photocatalytic decomposition of NO under visible light irradiation on the Cr-ion-implanted TiO_2 thin film photocatalyst. *Catal. Lett.* 67: 135–137.

12 Zhou, J., Takeuchi, M., Zhao, X.S. et al. (2006). Photocatalytic decomposition of formic acid under visible light irradiation over V-ion-implanted TiO_2 thin film photocatalysts prepared on quartz substrate by ionized cluster beam (ICB) deposition method. *Catal. Lett.* 106: 1–2.

13 Takeuchi, M., Sakamoto, K., Tsujimaru, K., and Anpo, M. (2009). Photo-induced superhydrophilicity on TiO_2 thin films prepared by an ionized cluster beam deposition method. *Catal. Lett.* 131: 189–193.

14 Kitano, M., Takeuchi, M., Matsuoka, M. et al. (2005). Preparation of visible light-responsive TiO_2 thin film photocatalysts by an RF magnetron sputtering deposition method and their photocatalytic reactivity. *Chem. Lett.* 34 (4): 616–617.

15 Kitano, M., Tsujimaru, K., and Anpo, M. (2006). Decomposition of water in the separate evolution of hydrogen and oxygen using visible light-responsive TiO_2 thin film photocatalysts: effect of the work function of the substrates on the yield of the reaction. *Appl. Catal., A* 314: 179–183.

16 Kitano, M., Takeuchi, M., Matsuoka, M. et al. (2007). Photocatalytic water splitting using Pt-loaded visible light-responsive TiO_2 thin film photocatalysts. *Catal. Today* 120: 133–138.

17 Takeuchi, M., Sakai, S., Matsuoka, M., and Anpo, M. (2009). Preparation of the visible light responsive TiO_2 thin film photocatalysts by the RF magnetron sputtering deposition method. *Res. Chem. Intermed.* 35: 973–983.

18 Anpo, M., Nakaya, H., Kodama, S. et al. (1986). Photocatalysis over binary metal oxides. Enhancement of the photocatalytic activity of TiO_2 in titanium-silicon oxides. *J. Phys. Chem.* 90: 1633–1636.

19 Yamashita, H., Kawasaki, S., Ichihashi, Y. et al. (1998). Characterization of titanium-silicon binary oxide catalysts prepared by the sol-gel method and their photocatalytic reactivity for the liquid-phase oxidation of 1-octanol. *J. Phys. Chem. B* 102: 5870–5875.

20 Yamashita, H., Ichihashi, Y., Zhang, S.G. et al. (1997). Photocatalytic decomposition of NO at 275 K on titanium oxide catalysts anchored within zeolite cavities and framework. *Appl. Surf. Sci.* 121, 122: 305–309.

21 Ikeue, K., Yamashita, H., Anpo, M., and Takewaki, T. (2001). Photocatalytic reduction of CO_2 with H_2O on Ti-β zeolite photocatalysts: effect of the hydrophobic and hydrophilic properties. *J. Phys. Chem. B* 105: 8350–8355.

22 Takeuchi, M., Dohshi, S., Eura, T., and Anpo, M. (2003). Preparation of titanium-silicon binary oxide thin film photocatalysts by an ionized cluster beam deposition method. Their photocatalytic activity and photoinduced super-hydrophilicity. *J. Phys. Chem. B* 107: 14278–14282.

23 Dohshi, S., Takeuchi, M., and Anpo, M. (2003). Effect of the local structure of Ti-oxide species on the photocatalytic reactivity and photo-induced super-hydrophilic properties of Ti/Si and Ti/B binary oxide thin films. *Catal. Today* 85: 199–206.

24 Takeuchi, M., Yamasaki, T., Tsujimaru, K., and Anpo, M. (2006). Preparation of crystalline TiO_2 thin film photocatalysts on polycarbonate substrates by a RF-magnetron sputtering deposition method. *Chem. Lett.* 35: 904–905.

25 Takeuchi, M., Sakamoto, K., Martra, G. et al. (2005). Mechanism of photoinduced superhydrophilicity on the TiO_2 photocatalyst surface. *J. Phys. Chem. B* 109: 15422–15428.

26 Takeuchi, M., Martra, G., Coluccia, S., and Anpo, M. (2007). Verification of the photoadsorption of H_2O molecules on TiO_2 semiconductor surfaces by vibrational absorption spectroscopy. *J. Phys. Chem. C* 111: 9811–9817.

27 White, J.M., Szanyi, J., and Henderson, M. (2003). The photon-driven hydrophilicity of titania: a model study using TiO_2(110) and adsorbed trimethyl acetate. *J. Phys. Chem. B* 107: 9029–9033.

28 Aita, H., Hori, N., Takeuchi, M. et al. (2009). The effect of ultraviolet functionalization of titanium on integration with bone. *Biomaterials* 30: 1015–1025.

29 Att, W., Hori, N., Takeuchi, M. et al. (2009). Time-dependent degradation of titanium osteoconductivity: an implication of biological aging of implant materials. *Biomaterials* 30: 5352–5363.

30 Hirota, M., Sugita, Y., Ishijima, M. et al. (2021). UV photocatalytic activity of titanium dioxide (TiO_2) surface contaminated with bacterial biofilm: implications for photorestoration of osteoconductivity. *Mater. Today Adv.* 12: 100182.

22

The Surface-related Photocatalysis and Superwettability

Jing Pan and Fan Xia

State Key Laboratory of Biogeology and Environmental Geology, Faculty of Materials Science and Chemistry, China University of Geosciences, Wuhan, Hubei, China

22.1 Introduction

In view of the pressing environmental and energy problems, the evolution of green energy conversion technology and pollution treatment methods is essential to the sustainable development of society. Photocatalytic reactions, which could sustainably convert solar energy into renewable energy and degrade organic pollutants, attract more and more attention in the areas of organic synthesis, liquid separation, air treatment, water splitting, and self-cleaning [1]. Considering that the photocatalytic reaction happens on the surface of the photocatalyst, the catalytic activity of the photocatalyst is mainly determined by the surface [2]. Thus, various surface-controlling methods have been developed to fully utilize solar energy. These methods are usually in terms of photogenerated carriers' separation efficiency improving, recombination mitigating, and transfer ability enhancing, as well as surface area-increasing. However, the other properties of the photocatalysts' surface could also affect their catalytic performance, for example, the wettability [3].

It has been reported that surfaces with superwettability could exhibit the enhanced photocatalytic activity of photocatalysts [4]. For instance, a superhydrophobic surface could protect the photocatalysis reactions from the disturbance of inorganic contamination or pollutants such as dust, dirt, and so on. That is because the inorganic contamination or pollutants can be rinsed off by the water droplets rolling from the superhydrophobic surface. In addition, the superhydrophilic surfaces usually exhibit good degradation performance for water-soluble contaminants and pollutants on account of the sufficient accessibility of the catalytic sites. Thus, building a supperwettable surface could be one efficient way to improve the performance of some photocatalysts [5]. Moreover, the surface with superwettability is also applied to other areas, such as biochemical sensors, microfluidic devices, oil spill capture, and separation [6].

On the other hand, the photocatalytic activity of the surface could also have an influence on its wettability. One reason lies in the free radicals produced during the

UV-Visible Photocatalysis for Clean Energy Production and Pollution Remediation: Materials, Reaction Mechanisms, and Applications, First Edition. Edited by Xinchen Wang, Masakazu Anpo, and Xianzhi Fu.
© 2023 WILEY-VCH GmbH. Published 2023 by WILEY-VCH GmbH.

photocatalytic reaction, which would increase the surface energy and benefit the formation of a hydrophilic surface. Thus, it is difficult to build one surface with both superhydrophobic and photocatalytic activity. For example, the superhydrophobic surface of some TiO_2 (the most commonly used photocatalysts) could turn into superhydrophilic under light illumination [7]. Therefore, some researchers even thought this kind of surface could not exist. Thanks to the development of nanotechnology and modification methods, there are more and more publications that report this kind of surface until recently.

Herein, the recent progress and our efforts in developing surfaces with superwettability and photocatalytic activity have been introduced. Based on the formation and applications, surfaces with only photocatalytic activity or superwettability, as well as surfaces with both photocatalytic activity and superwettability, are successively discussed. Finally, perspectives and prospects for the current and future developments of surfaces with both photocatalytic activity and superwettability are demonstrated. We hope this introduction will bring more interest and attention to the superwettability of photocatalysts and provide valuable insights for researchers to improve the performance of photocatalysts.

22.2 Surfaces with Photocatalytic Activity

A photocatalytic reaction usually follows four steps (Figure 22.1) [8]. The first step is generating charge carriers (electron-hole pairs) after light harvesting as well as optical absorption on the irradiated photocatalysts. Then, the photogenerated electrons and holes would separate. Thereafter, the separated electrons and holes would migrate to the surface of the photocatalysts. During this period, the recombination of some photogenerated charge carriers may happen. Finally, the photogenerated electrons and holes would initiate or accelerate the reduction or oxidation on the surface of the photocatalysts. As shown by the four steps, the catalytic performance of a photocatalyst is closely related to its surface properties.

Therefore, researchers have developed various methods to improve the surfaces and expect their photocatalytic performances to be optimized (Figure 22.2). Although the specific pathways are different, most of the methods are focused on three aspects. The first one is to maximize the absorption of solar energy. The second one is improving the quality and number of active sites. The third one is mitigating the recombination of the charge carriers. For example, morphology control and structural engineering have been widely used to obtain photocatalysts with excellent performance. Employing proper morphology and structure, the specific area of the surface could be increased. The increased specific surface area could not only enhance the amount of active sites, but the simultaneously improved illumination area could also take better advantage of reflection as well as refraction [9]. Additionally, some specific morphology and structure could mitigate the recombination of the photogenerated charge carriers, such as ultrathin nanosheets or nanowires. The reason lies in the short migration distance between the interior and the catalysts' surface, Moreover, the surface with sufficient exposure

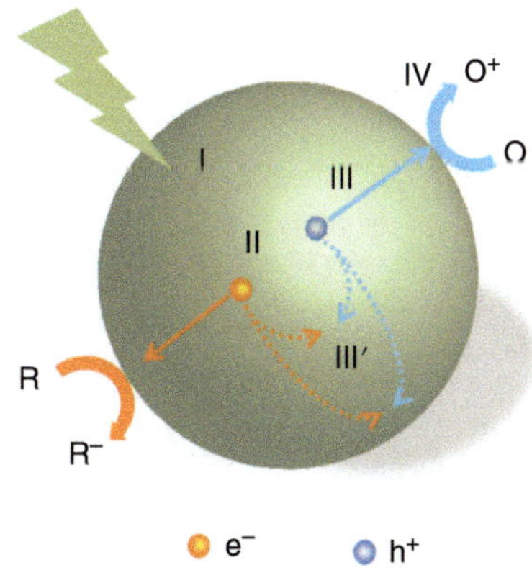

Figure 22.1 The illustration of the four steps of the photocatalytic reaction process: (I) charge carriers generation after the light illumination; (II) separation of the photogenerated charge carriers; (III) migration of the separated charge carriers to the surface of the photocatalysts; (III′) recombination of some photogenerated charge carriers; (IV) acceleration of the reduction or oxidation reactions. R refers to the chemicals in the reduction reactions. While O stands for the chemicals in oxidation reactions. Source: Reproduced with permission from Zhu et al. [8]/John Wiley & Sons.

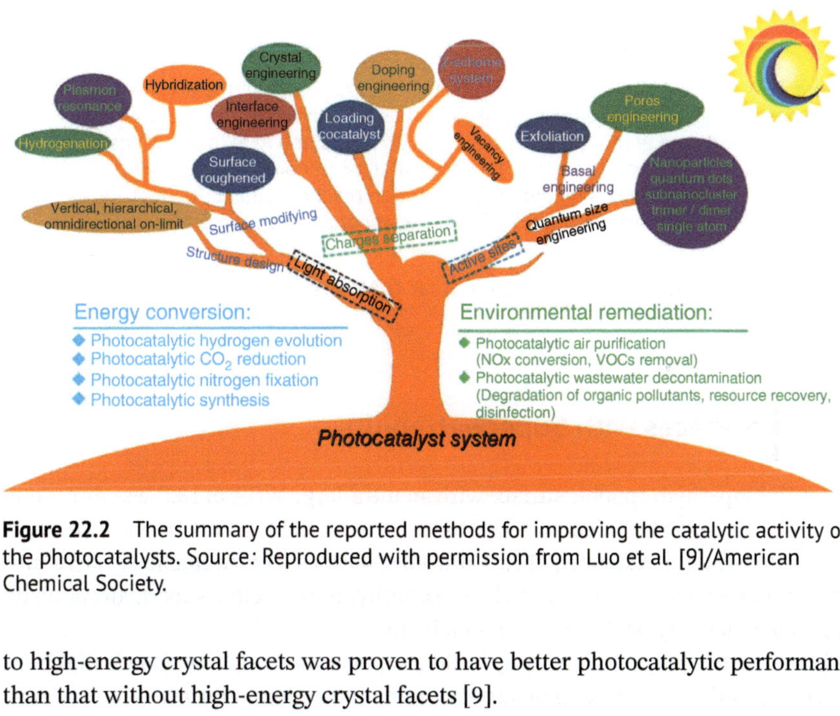

Figure 22.2 The summary of the reported methods for improving the catalytic activity of the photocatalysts. Source: Reproduced with permission from Luo et al. [9]/American Chemical Society.

to high-energy crystal facets was proven to have better photocatalytic performance than that without high-energy crystal facets [9].

Based on the above consideration, we designed and prepared ultrathin $CuInP_2S_6$ nanosheets with abundant sulfur vacancies as photocatalysts for hydrogen evolution reactions [10]. The abundant sulfur vacancies on the surface of the nanosheets could adjust the bandgap and therefore improve the light absorption of the catalysts (Figure 22.3a). Moreover, they can work as charge separation centers and further improve charge carrier separation. In addition, the few-layered thickness (4–7 nm, Figure 22.3b) could not only result in a large specific surface area but also mitigate the recombination of the photogenerated charge carriers based on the short migration distance between the interior and the catalysts' surface. In this way, the hydrogen production rate of the $CuInP_2S_6$ nanosheets could be as high as 804 $\mu mol\ g^{-1}\ h^{-1}$

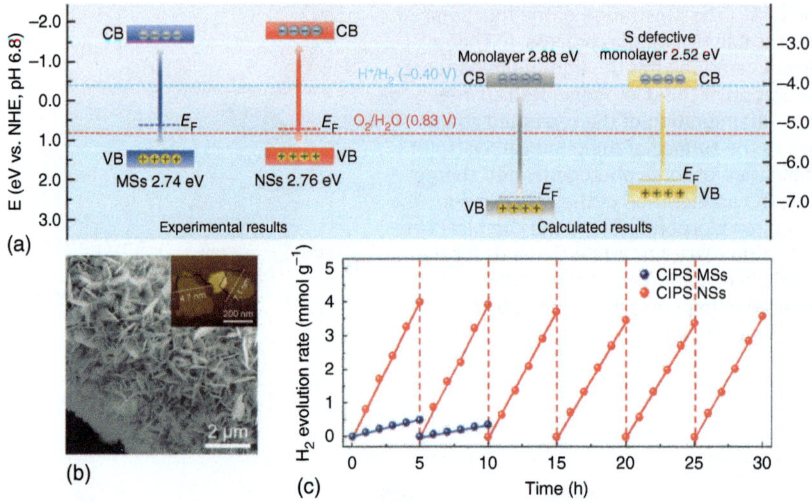

Figure 22.3 (a) The UV–Vis diffuse reflectance spectra results of the prepared $CuInP_2S_6$ nanosheets (NSs) and microsheets (MSs) as well as calculated results of $CuInP_2S_6$ monolayers with and without S vacancies; (b) SEM image and AFM image of the $CuInP_2S_6$ nanosheets on the carbon fibers; (c) The photocatalytic performance of $CuInP_2S_6$ nanosheets and microsheets for hydrogen evolution reaction. Source: Yu et al. [10]/Reproduced with permission from Royal society of chemistry.

(Figure 22.3c), which is much faster than that of the corresponding microsheets (100 $\mu mol\, g^{-1}\, h^{-1}$).

22.3 Surfaces with Superwettability

Since the superhydrophobic surface with an ultra-large WCA of 180° was discovered in 1907, the superwettable surfaces have aroused tremendous interest and have been intensively investigated [11]. Until now, various superwettable surfaces have been reported (Figure 22.4), such as superhydrophobicity, superhydrophilicity, superoleophobicity, and superoleophilicity in the air [12].

There are two ways to commonly build a superwettable surface. One is the chemical compositions controlling or surface functionalization to form a specific surface energy. For example, low surface energy is needed for a superhydrophobic surface. The other is to employ specific geometric morphologies and structures to make a rough surface. However, if only employing composition control or surface functionalization, the surface wettability improvement is limited [13]. Thus, the two methods are usually combined to obtain further enhanced superwettability. For instance, for man-made superhydrophobic surfaces, the low surface energy is usually made by Polytetrafluoroethylene (PTFE), fluoroalkyl silane, thiol, oil, and wax. Specifically, the WCA on smooth PTFE is only about 120° [4]. In a contrast, the cooperation of PTFE and rough surface could result in a much improved WCA of 165° [14]. The other example is the lotus leaf, whose surface superhydrophobicity is derived from

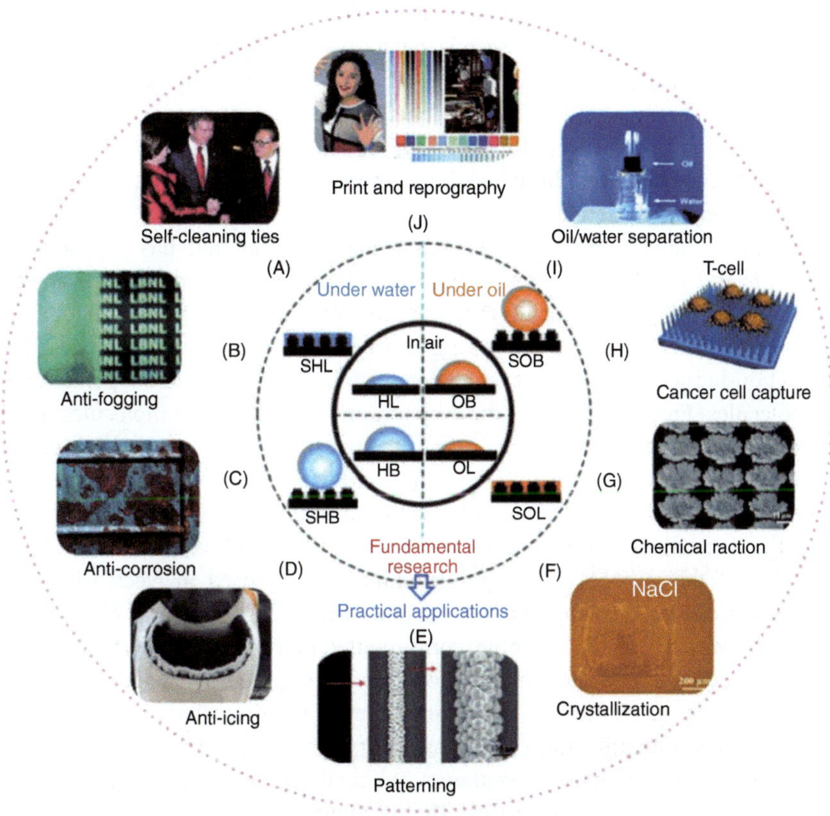

Figure 22.4 The examples of the superwettablity systems and the practical applications. The inner circle shows examples of the superwettablity systems. HL, HB, OB, and OL represent hydrophilicity, hydrophobicity, oleophilicity, and oleophobicity on flat substrates in air, respectively. SHL, SHB, SOB, and SOL refer to superhydrophilicity, superhydrophobicity, superoleophilicity, and superoleophobicity on rough substrates in air, respectively. (a–j) The practical applications of superwettablity systems. Source: Wen et al. [12]/Reproduced with permission from John Wiley & Sons.

the combination of micropapillary surface structures and the existence of hydrophobic epicuticular wax.

However, the extra use of organics could increase the cost, complicate the operation process, and raise the environmental risk [15]. Hence, developing a superwettable surface without the extra use of organics is meaningful. Recently, we found some biomaterials. For example, osmanthus flowers could be directly used as superhydrophobic materials for oil cleanup and other application fields [16]. The dry osmanthus flowers could display low-adhesion superhydrophobicity with a WCA of $158.3 \pm 2.7°$ and superoleophilicity with an oil contact angle (OCA) of 0°. The superhydrophobicity and superoleophilicity can remain even when the dry osmanthus flowers are crushed into powder. In addition, bioinspired nanostructured surfaces could also be employed to obtain superwettable without organics [17].

The surface with superwettability has wide and promising applications (Figure 22.4a–j). As examples, use our related works. By employing inorganic–organic superhydrophobic paint, the problem of roof leakage can be resolved [18]. Inspired by the surface of Namib desert beetles, the efficient water collection was realized by homemade hydrophilic/superhydrophobic patterned surfaces [19]. Our team has utilized superwettability outside of these conventional applications in the area of biochemical detection.

Aggregation-induced emission (AIE) is a phenomenon that has a promising prospect in the areas of biochemical detection, bioimaging, drug delivery as well as theranostics [20]. Considering that the AIE phenomenon has only happened when the related molecules are in the aggregate form or solid-state, we suppose the evaporation-induced enrichment on a superwettable surface could cooperate with AIE molecules (Figure 22.5). After the evaporation process, the AIE molecules could realize the fluorescence enhancement by the enrichment and aggregation within the superhydrophilic microwell. Based on this AIE-based superwettable microchip, the detection of microRNA-141 can be achieved with a very low detection limit of 1 pM and excellent reproducibility and selectivity [21]. Moreover, the antifouling ability of the surface with superwettability can be used in electrochemical aptamer-based sensors to improve the baseline stability of the sensors [22].

Visual biosensing, which is simple, convenient, and low-cost, is another method to realize biochemical detection. However, the development of visual biosensors still faces enormous challenges. For example, most existing visual biosensors are based on the color identification of the inspectors, which has limitations, especially for users with color blindness or weakness. Taking this into consideration, a surface with contact angle (CA) and critical sliding angle (CSA)-dependent wettability may be a promising solution. Besides, the interaction between oil molecules and short single-stranded DNA (ssDNA) or long ssDNA is different. Additionally, the long ssDNA molecules could be generated by the rolling-circle amplification (RCA) of the short ssDNA. Thus, we designed a target-triggered RCA for tuning the

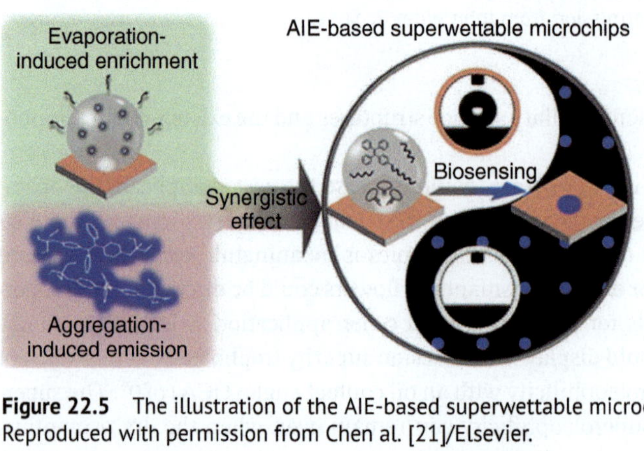

Figure 22.5 The illustration of the AIE-based superwettable microchips. Source: Reproduced with permission from Chen al. [21]/Elsevier.

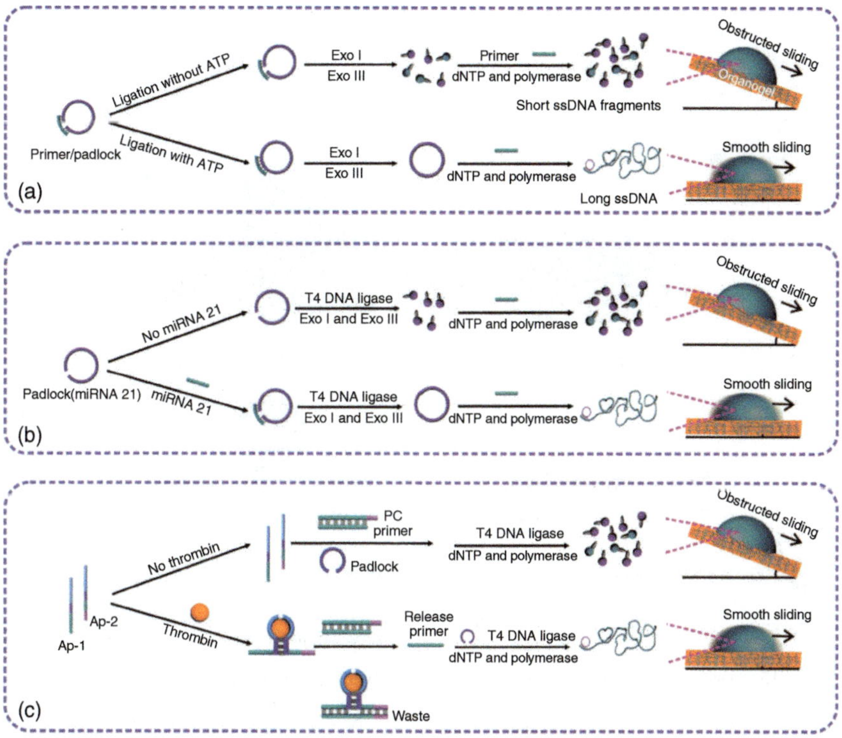

Figure 22.6 The visual biosensor based on RCA triggered different wettability on the oil-swollen organogel surface. (a–c) The detection mechanism for ATP, miRNA 21, and thrombin, respectively. Source: Reproduced with permission from Gao et al. [23]; Springer Nature.

hydrophobicity of DNA and realized the visual detection of adenosine triphosphate (ATP), microRNA, and thrombin on the surface of the oil-swollen organogel surface (Figure 22.6) [23].

In addition, some surfaces exhibit a stimuli-responsive wettability, which means the wettability can be adjusted by pH, temperature, or other stimuli [24]. The stimuli-responsive wettability could be directly employed in visual biosensors. For example, our group utilized a surface with pH-responsive superwettability to build a WCA-based naked-eye testing platform (Figure 22.7). When the pH of the droplet is 1, the CA of the droplet on this surface approached 0°. While, when the pH is increased to 13, the corresponding CA would reach $161.4° \pm 6.2°$. This platform could not only detect the pH of the droplet but also show the potential to monitor the concentration of some other targets, such as urea and glucose. Urea, which is an important biomarker for gout, can increase the pH of the droplet by the urease-catalyzed reaction. On the other hand, glucose can decrease the pH of the solution by the glucose oxidase-catalyzed reaction and the glucose levels in sweat can be successfully monitored by this platform [25].

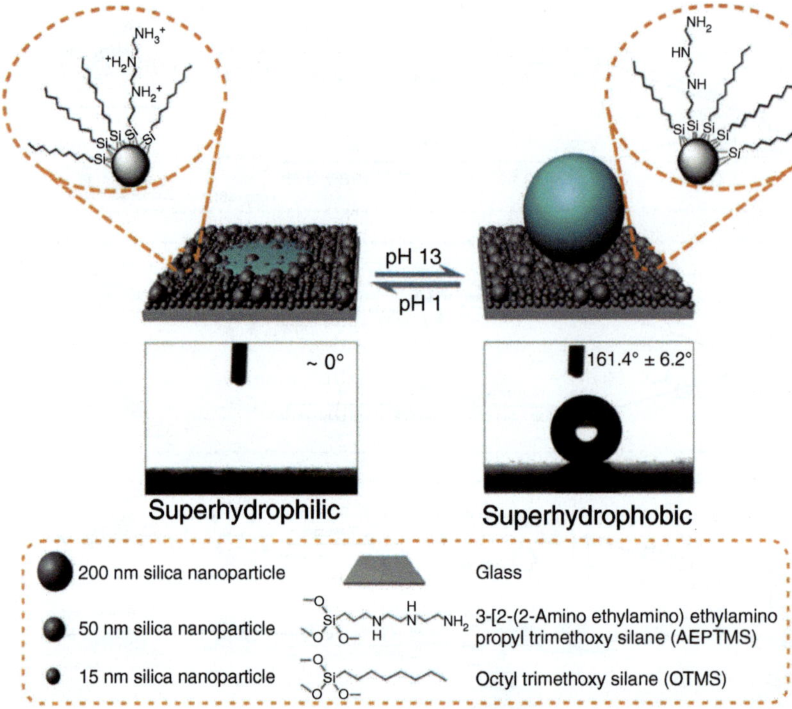

Figure 22.7 The illustration of the pH-responsive surface. Source: Reprinted with permission from Gao et al. [25]/Springer Nature/CC BY-4.0.

22.4 Surfaces with Both Photocatalytic Activity and Superwettability

It has been reported that surfaces with superwettability could exhibit the enhanced photocatalytic activity of photocatalysts [4]. On the other hand, the photocatalytic activity of the surfaces may also be beneficial to the superhydrophilcity surface. However, the influence of photocatalytic activity is much more complicated for the superhydrophobic surface. As mentioned above, the free radicals formed during the photocatalysis reaction would increase the surface energy and result in decreased superhydrophobicity. However, in some cases, the photocatalytic activity of the man-made surface may be beneficial to the stability of superhydrophobic stability [9]. For naturally occurring surfaces with superhydrophobicity, their restoration capacity can ensure the durability of superhydrophobic. For instance, the lotus leaves could continuously secrete epicuticular wax to maintain the water repellent capability of the surface. However, there is no maintenance performed for manmade surfaces. Their superhydrophobicity would gradually degrade on account of the outdoor exposure as well as the accumulation of contamination. This problem could be resolved if the surface is photocatalytically active. The organic stains could be degraded immediately by the photocatalytically active surface and prevent accumulation.

Surfaces with both photocatalytic activity and superhydrophobicity have wide applications. The most direct one is for self-cleaning [4]. For photocatalysis, one of the most important applications is degrading organic contaminations or pollutants by the free radicals formed on the surface of photocatalysts after light illumination. However, the common contaminants or pollutants are not only organic materials but also inorganic materials, which cannot be degraded, such as dust, dirt, and so on. The existence of these kinds of contamination or pollutants would block the surface of photocatalysts and hinder the contact between the photocatalysts and light illumination, which could result in much reduced photocatalytic performance. Thus, weakening the disturbance of these kinds of contamination and pollutants is of great importance to photocatalysts. One of the most effective ways is to form a superhydrophobic surface. With a superhydrophobic surface, the inorganic contamination or pollutants can be rinsed off by the water droplets rolling from the photocatalysts.

In addition, there are many other application potentials for this kind of surface. For example, they can be used for freshwater harvesting by the combination of water collection and purification. Although there are plenty of reports related to water harvesting, almost all of them are just focused on optimizing the harvesting efficiencies. The availability and safety of the obtained water are rare enough to be discussed. Taking this into consideration, we integrated superhydrophilic wedge-shaped patterns with P25 TiO_2 nanoparticles and a superhydrophobic candle soot@polydimethylsiloxane coating (Figure 22.8) [26]. This integrated surface could obtain a high water collection rate of $14.9 \pm 0.2\,mg\,min^{-1}\,cm^{-2}$, which is much larger than that of uniformed superhydrophilic or superhydrophobic surfaces. What's more important, the existence of P25 TiO_2 nanoparticles in superhydrophilic patterns endowed them with the additional photocatalytic activity for the degradation of nine kinds of pesticides under UV light illumination. In the collected and simultaneously cleaned water, ten zebrafish can live for more than 72 hours. While, in the untreated harvested water, the samples almost die after seven hours.

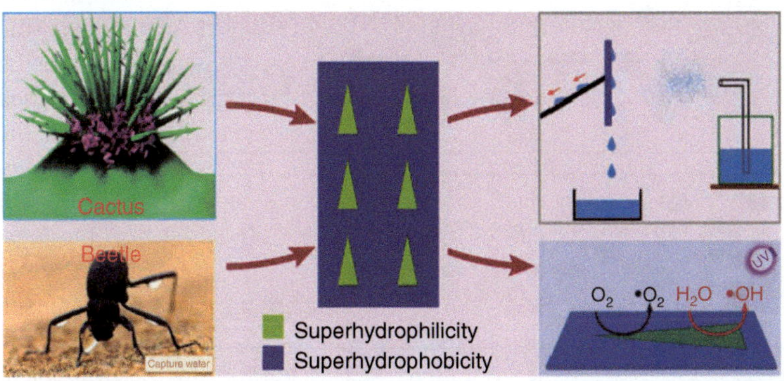

Figure 22.8 The illustration of the biomimetic surface with patterned wettability, which could simultaneously realize the water collection and purification. Source: Zhu et al. [26]/Reproduced with permission from Elsevier.

Although some semiconductors can be hydrophobic in the dark, their wettability would switch to superhydrophilic because of the increased surface energy caused by the formation of free radicals [27]. Thus, it usually needs low-free-energy organics to ensure the superhydrophobicity of the surface. For instance, we combined polydimethylsiloxane and P25 nanoparticles on the alkaline treated copper mesh to fabricate a surface with both photocatalytic activity and superhydrophobicity [28]. On the one hand, the use of P25 nanoparticles offers the surface good photocatalytic activity with degradation ability for organic dyes, such as methyl blue, Nile red, and methyl orange. On the other hand, the cooperation of polydimethylsiloxane and the formation of nanowire arrays provides the surface sufficient superhydrophobicity with a large WCA of 155.5° as well as a small rolling-off angle of 6.8°. In addition, this kind of surface could be changed to superhydrophilic (WCA = 0°) by O_2 plasma. After 12 hours, the superhydrophobicity can be recovered itself and results in a reversible wetting process.

Nevertheless, as mentioned above, the use of superhydrophobic polymers could lead to increased cost, complicated operation processes, and raised environmental risk. Therefore, it is of great significance to develop superhydrophobic and photocatalytically active surfaces without the use of organics. Recently, some semiconductor photocatalysts have been reported to have size-dependent wettability. These reports offer an opportunity. For example, the wettability of Ag_2O particles can be tuned from superhydrophobicity to superoleophilicity just by the particle size (Figure 22.9). The prepared superhydrophobic Ag_2O particles can float on the surface of the water and display outstanding photodegradation properties for various floating oils [27].

The photocatalytic activity of different supperwettable surfaces would be different in some cases. Set the degradation of water-soluble contaminants and pollutants as an example. The superhydrophobic surfaces always exhibit poor degradation performance because of the less accessibility of the catalytic sites. On the other hand, the degradation performance of the superhydrophilic surfaces is usually better on account of the sufficient accessibility of the catalytic sites. Besides, the

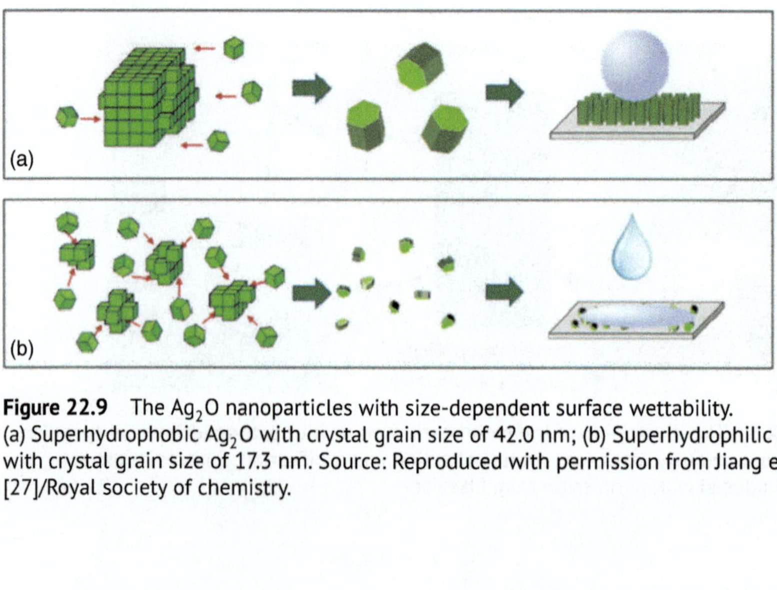

Figure 22.9 The Ag_2O nanoparticles with size-dependent surface wettability. (a) Superhydrophobic Ag_2O with crystal grain size of 42.0 nm; (b) Superhydrophilic Ag_2O with crystal grain size of 17.3 nm. Source: Reproduced with permission from Jiang et al. [27]/Royal society of chemistry.

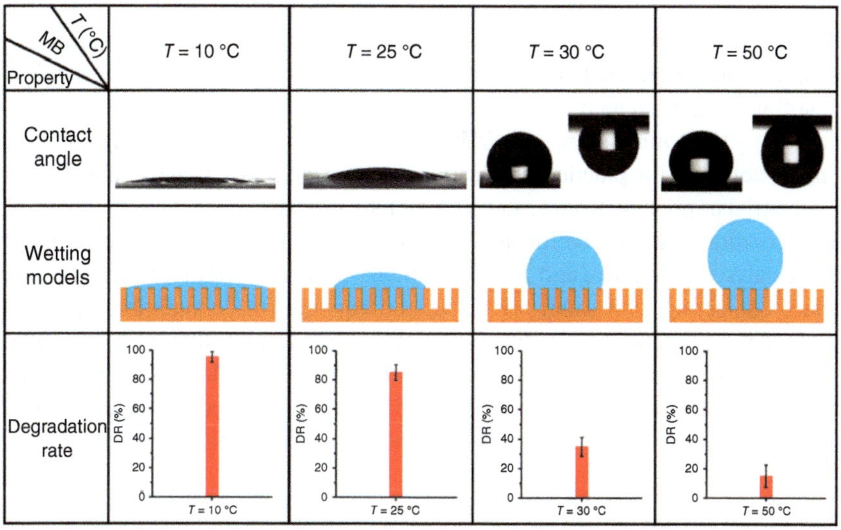

Figure 22.10 The temperature-dependent wettability and photocatalytic activity of the obtained surfaces. Source: Reproduced with permission from Zhu et al. [29]/Elsevier.

stimuli-responsive wettability could also result in tunable photocatalytic activity of the surface. For example, we designed and obtained a temperature-controlled wettability and photodegradation for water-based dyes (methyl blue and methyl orange) on one surface (Figure 22.10). At a relatively high temperature of 50 °C, the surface is superhydrophobic and shows a low degradation rate of 15.0% for methyl blue. When the temperature is decreased to 10°C, the surface would turn superhydrophilic and exhibit a high degradation rate of 95.5% [29].

The surface with both photocatalytic activity and superhydrophobicity could also be employed in biochemical detection. Employing perfluoro silane-coated TiO_2 nanoparticles, we built a biochemical detection platform that could realize naked-eye ATP detection [30]. On account of the photocatalytically active TiO_2 nanoparticles, the organic compound would be degraded by the UV irradiation, and the gradient UV irradiation could lead to a gradient wettability. In this way, a droplet sorting ability could be achieved for liquid droplets with different ATP concentrations based on ATP-triggered RCA.

22.5 Conclusion and Outlook

To sum up, we have shown the advances in our recent work on building surfaces with superwettability and photocatalytic activity. The formation methods and applications of surfaces with only photocatalytic activity or superwettability, as well as surfaces with both photocatalytic activity and superwettability, have been discussed. Although a lot of progress has been made, there are still lots of challenges that need to be addressed. For instance, to construct a superhydrophobic surface, the related polymers are usually used. However, they could lead to increased cost, complicated operation processes, and raised environmental risk. Besides, for some reactions, for example, hydrogen evolution reactions, the most efficient photocatalysts are still expensive noble metals. Thus, one of the emphases of superhydrophobic photocatalysts should be low-cost, robust, and environmentally friendly compositions. In addition, the practical applications require long-term usage of the surfaces in industrial or poor environmental conditions. Therefore, their stability under different conditions should be further investigated and improved. Furthermore, the photocatalysis mechanism and reaction pathway may be different for superwettable surface contacting with water for oil. In consequence, more attention should be paid to the understanding and observation at the atomic/molecular level by employing advanced characterization techniques or theoretical calculations. The intensive efforts on surfaces with superwettability and photocatalytic activity are expected to result in continuous innovations and increased industrial application potential.

Acknowledgement

This work was financially supported in part by the National Natural Science Foundation of China under Grants 22090050, 22004112, and 21874121, in part by GuangDong Basic and Applied Basic Research Foundation under Grant 2021A1515110036, and in part by the National Key R&D Program of China under Grants 2021YFA1200403 and 2018YFE0206900.

References

1 Hu, C., Tu, S.C., Tian, N. et al. (2021). Photocatalysis enhanced by external fields. *Angew. Chem. Int. Ed.* 60: 16309–16328.

2 Li, X., Yu, J.G., and Jaroniec, M. (2016). Hierarchical photocatalysts. *Chem. Soc. Rev.* 45: 2603–2636.

3 Xiong, X.Y., Wang, Z.P., Zhang, Y. et al. (2020). Wettability controlled photocatalytic reactive oxygen generation and Klebsiella pneumoniae inactivation over triphase systems. *Appl. Catal., B* 264: 118518.

4 Kamegawa, T., Shimizu, Y.K., and Yamashita, H. (2012). Superhydrophobic surfaces with photocatalytic self-cleaning properties by nanocomposite coating of TiO_2 and polytetrafluoroethylene. *Adv. Mater.* 24: 3697–3700.

5 Liu, J., Ye, L.J., Sun, Y.L. et al. (2020). Elastic superhydrophobic and photocatalytic active films used as blood repellent dressing. *Adv. Mater.* 32: 1908008.

6 Nagappan, S. and Ha, C.S. (2015). Emerging trends in superhydrophobic surface based magnetic materials: fabrications and their potential applications. *J. Mater. Chem. A* 3: 3224–3251.

7 Tadanaga, K., Morinaga, J., Matsuda, A., and Minami, T. (2000). Superhydrophobic-superhydrophilic micropatterning on flowerlike alumina coating film by the sol–gel method. *Chem. Mater.* 12: 590.

8 Zhu, S.S. and Wang, D.W. (2017). Photocatalysis: basic principles, diverse forms of implementations and emerging scientific opportunities. *Adv. Energy Mater.* 7: 1700841.

9 Luo, J., Zhang, S.Q., Sun, M. et al. (2019). A critical review on energy conversion and environmental remediation of photocatalysts with remodeling crystal lattice. *Surf. Interface ACS Nano* 13: 9811–9840.

10 Yu, P., Wang, F.M., Meng, J. et al. (2021). Few-layered $CuInP_2S_6$ nanosheet with sulfur vacancy boosting photocatalytic hydrogen evolution. *Crystengcomm* 23: 591–598.

11 Tian, Y., Su, B., and Jiang, L. (2014). Interfacial material system exhibiting superwettability. *Adv. Mater.* 26: 6872–6897.

12 Wen, L.P., Tian, Y., and Jiang, L. (2015). Bioinspired super-wettability from fundamental research to practical applications. *Angew. Chem. Int. Ed.* 54: 3387–3399.

13 Nishino, T., Meguro, M., Nakamae, K. et al. (1999). The Lowest surface free energy based on $-CF_3$ alignment. *Langmuir* 15: 4321–4323.

14 Jafari, R., Menini, R., and Farzaneh, M. (2010). Superhydrophobic and icephobic surfaces prepared by Rf-sputtered polytetrafluoroethylene coatings. *Appl. Surf. Sci.* 257: 1540–1543.

15 Afzal, S., Daoud, W.A., and Langford, S.J. (2014). Superhydrophobic and photocatalytic self-cleaning cotton. *J. Mater. Chem. A* 2: 18005–18011.

16 Zhu, H., Huang, Y., and Xia, F. (2020). Environmentally friendly superhydrophobic osmanthus flowers for oil spill cleanup. *Appl. Mater. Today* 19: 100607.

17 Zhu, H., Huang, Y., Lou, X., and Xia, F. (2019). Beetle-inspired wettable materials: from fabrications to applications. *Materials Today Nano* 6: 100034.

18 Zhu, H., Huang, Y., Zhang, S.W. et al. (2021). A universal, multifunctional, high-practicability superhydrophobic paint for waterproofing grass houses. *NPG Asia Mater.* 13: 47.

19 Zhu, H., Duan, R.L., Wang, X.D. et al. (2018). Prewetting dichloromethane induced aqueous solution adhered on cassie superhydrophobic substrates to fabricate efficient fog-harvesting materials inspired by Namib Desert beetles and mussels. *Nanoscale* 10: 13045–13054.

20 Li, J., Wang, J.X., Li, H.X. et al. (2020). Supramolecular materials based on AIE luminogens (AIEgens): construction and applications. *Chem. Soc. Rev.* 49: 1144–1172.

21 Chen, Y.X., Min, X.H., Zhang, X.Q. et al. (2018). Aie-based superwettable microchips for evaporation and aggregation induced fluorescence enhancement biosensing. *Biosens. Bioelectron.* 111: 124–130.

22 Li, S.G., Wang, Y.Y., Zhang, Z.S. et al. (2021). Exploring end-group effect of alkanethiol self-assembled monolayers on electrochemical aptamer-based sensors in biological fluids. *Anal. Chem.* 93: 5849–5855.

23 Gao, Z.F., Liu, R., Wang, J.H. et al. (2020). Manipulating the hydrophobicity of DNA as a universal strategy for visual biosensing. *Nat. Protoc.* 15: 316–337.

24 Lou, X.D., Huang, Y., Yang, X. et al. (2020). External stimuli responsive liquid-infused surfaces switching between slippery and nonslippery states: fabrications and applications. *Adv. Funct. Mater.* 30: 1901130.

25 Gao, Z.F., Sann, E.E., Lou, X.D. et al. (2018). Naked-eye point-of-care testing platform based on a Ph-responsive superwetting surface: toward the non-invasive detection of glucose. *NPG Asia Mater.* 10: 177–189.
26 Zhu, H., Cai, S., Zhou, J. et al. (2021). Integration of water collection and purification on cactus-and beetle-inspired eco-friendly superwettable materials. *Water Res.* 206: 117759.
27 Jiang, W., Fu, H.Y., Zhu, Y.M. et al. (2018). Floatable superhydrophobic Ag_2O photocatalyst without a modifier and its controllable wettability by particle size adjustment. *Nanoscale* 10: 13661–13672.
28 Zhang, Y.B., Chen, Y., Wang, C. et al. (2022). A multifunctional composite membrane with photocatalytic, self-cleaning, oil/water separation and antibacterial properties. *Nanotechnology* 33: 355703.
29 Zhu, H., Tu, Y.D., Luo, C.H. et al. (2021). Temperature-triggered switchable superwettability on a robust paint for controllable photocatalysis. *Cell Rep. Phys. Sci.* 2: 100669.
30 Huang, F.J., Chen, Y., Wang, Y.Q., and Xia, F. (2019). Tunable superamphiphobic surfaces: a platform for naked-eye Atp detection. *Anal. Bioanal.Chem.* 411: 4721–4727.

Index

UV-Visible Photocatalysis for Clean Energy Production and Pollution Remediation: Materials, Reaction Mechanisms, and Applications, First Edition. Edited by Xinchen Wang, Masakazu Anpo, and Xianzhi Fu.
© 2023 WILEY-VCH GmbH. Published 2023 by WILEY-VCH GmbH.

f

g

h

q

r

s

t

u